# Section 11

# Heat-Transfer Equipment*

**Richard L. Shilling, P.E., B.S.M., B.E.M.E.** *Vice President of Engineering, Koch Heat Transfer Company LP; American Society of Mechanical Engineers (Section Editor, Shell-and-Tube Heat Exchangers, Hairpin/Double-Pipe Heat Exchangers, Air-Cooled Heat Exchangers, Heating and Cooling of Tanks, Fouling and Scaling, Heat Exchangers for Solids, Thermal Insulation, Thermal Design of Evaporators, Evaporators)*

**Patrick M. Bernhagen, P.E., B.S.M.E.** *Sales Manager—Fired Heater, Foster Wheeler North America Corp.; American Society of Mechanical Engineers (Compact and Nontubular Heat Exchangers)*

**Victor M. Goldschmidt, Ph.D., P.E.** *Professor Emeritus, Mechanical Engineering, Purdue University (Air Conditioning)*

**Predrag S. Hrnjak, Ph.D., V.Res.** *Assistant Professor, University of Illinois at Urbana-Champaign; Principal Investigator—U of I Air Conditioning and Refrigeration Center; Assistant Professor, University of Belgrade; International Institute of Chemical Engineers; American Society of Heat, Refrigerating, and Air Conditioning Engineers (Refrigeration)*

**David Johnson, P.E., M.S.C.E.** *Heat Exchanger Specialist, A&A Technology, B.P. p.l.c.; American Institute of Chemical Engineers; American Society of Mechanical Engineers; API Subcommittee on Heat Transfer Equipment; API 660/ISO 16812, API 661/ISO 13706, API 662/ISO 15547 (Thermal Design of Heat Exchangers, Condensers, Reboilers)*

**Klaus D. Timmerhaus, Ph.D., P.E.** *Professor and President's Teaching Scholar, University of Colorado; Fellow, American Institute of Chemical Engineers, American Society for Engineering Education, American Association for the Advancement of Science; Member, American Astronautical Society, National Academy of Engineering, Austrian Academy of Science, International Institute of Refrigeration, American Society of Heat, Refrigerating, and Air Conditioning Engineers, American Society of Environmental Engineers, Engineering Society for Advancing Mobility on Land, Sea, Air, and Space, Sigma Xi, The Research Society (Cryogenic Processes)*



### THERMAL DESIGN OF HEAT-TRANSFER EQUIPMENT

Introduction to Thermal Design . . . . 11-4
- Approach to Heat-Exchanger Design . . . . 11-4
- Overall Heat-Transfer Coefficient . . . . 11-4
- Mean Temperature Difference . . . . 11-4
- Countercurrent or Cocurrent Flow . . . . 11-4
- Reversed, Mixed, or Cross-Flow . . . . 11-5

Thermal Design for Single-Phase Heat Transfer . . . . 11-5
- Double-Pipe Heat Exchangers . . . . 11-5
- Baffled Shell-and-Tube Exchangers . . . . 11-7

Thermal Design of Condensers . . . . 11-11
- Single-Component Condenser . . . . 11-11
- Multicomponent Condensers . . . . 11-12

Thermal Design of Reboilers . . . . 11-13

*The prior and substantial contributions of Frank L. Rubin (Section Editor, Sixth Edition) and Dr. Kenneth J. Bell (Thermal Design of Heat Exchangers, Condensers, Reboilers), Dr. Thomas M. Flynn (Cryogenic Processes), and F. C. Standiford (Thermal Design of Evaporators, Evaporators), who were authors for the Seventh Edition, are gratefully acknowledged.

Kettle Reboilers . . . 11-13
Vertical Thermosiphon Reboilers . . . 11-13
Forced-Recirculation Reboilers . . . 11-13
Thermal Design of Evaporators . . . 11-13
Forced-Circulation Evaporators . . . 11-14
Long-Tube Vertical Evaporators . . . 11-14
Short-Tube Vertical Evaporators . . . 11-15
Miscellaneous Evaporator Types . . . 11-16
Heat Transfer from Various Metal Surfaces . . . 11-16
Effect of Fluid Properties on Heat Transfer . . . 11-17
Effect of Noncondensables on Heat Transfer . . . 11-18
Batch Operations: Heating and Cooling of Vessels . . . 11-18
Nomenclature . . . 11-18
Applications . . . 11-18
Effect of External Heat Loss or Gain . . . 11-19
Internal Coil or Jacket Plus External Heat Exchange . . . 11-19
Equivalent-Area Concept . . . 11-19
Nonagitated Batches . . . 11-20
Storage Tanks . . . 11-20
Thermal Design of Tank Coils . . . 11-20
Nomenclature . . . 11-20
Maintenance of Temperature . . . 11-20
Heating . . . 11-20
Heating and Cooling of Tanks . . . 11-21
Tank Coils . . . 11-21
Teflon Immersion Coils . . . 11-22
Bayonet Heaters . . . 11-22
External Coils and Tracers . . . 11-22
Jacketed Vessels . . . 11-22
Extended or Finned Surfaces . . . 11-22
Finned-Surface Application . . . 11-22
High Fins . . . 11-23
Low Fins . . . 11-23
Fouling and Scaling . . . 11-23
Control of Fouling . . . 11-23
Fouling Transients and Operating Periods . . . 11-24
Removal of Fouling Deposits . . . 11-24
Fouling Resistances . . . 11-24
Typical Heat-Transfer Coefficients . . . 11-24
Thermal Design for Solids Processing . . . 11-24
Conductive Heat Transfer . . . 11-24
Contactive (Direct) Heat Transfer . . . 11-29
Convective Heat Transfer . . . 11-30
Radiative Heat Transfer . . . 11-30
Scraped-Surface Exchangers . . . 11-31

## TEMA-STYLE SHELL-AND-TUBE HEAT EXCHANGERS

Types and Definitions . . . 11-33
TEMA Numbering and Type Designations . . . 11-33
Functional Definitions . . . 11-35
General Design Considerations . . . 11-35
Selection of Flow Path . . . 11-35
Construction Codes . . . 11-35
Tube Bundle Vibration . . . 11-36
Testing . . . 11-36
Principal Types of Construction . . . 11-36
Fixed-Tube-Sheet Heat Exchangers . . . 11-36
U-Tube Heat Exchanger . . . 11-37
Packed-Lantern-Ring Exchanger . . . 11-39
Outside-Packed Floating-Head Exchanger . . . 11-39
Internal Floating-Head Exchanger . . . 11-40
Pull-Through Floating-Head Exchanger . . . 11-40
Falling-Film Exchangers . . . 11-40
Tube-Side Construction . . . 11-41
Tube-Side Header . . . 11-41
Special High-Pressure Closures . . . 11-41
Tube-Side Passes . . . 11-41
Tubes . . . 11-41
Rolled Tube Joints . . . 11-41
Welded Tube Joints . . . 11-41
Double-Tube-Sheet Joints . . . 11-43
Shell-Side Construction . . . 11-43
Shell Sizes . . . 11-43
Shell-Side Arrangements . . . 11-43
Baffles and Tube Bundles . . . 11-43
Segmental Baffles . . . 11-43
Rod Baffles . . . 11-43
Tie Rods and Spacers . . . 11-44
Impingement Baffle . . . 11-44
Vapor Distribution . . . 11-44
Tube-Bundle Bypassing . . . 11-44
Helical Baffless . . . 11-45
Longitudinal Flow Baffles . . . 11-45
Corrosion in Heat Exchangers . . . 11-45
Materials of Construction . . . 11-45
Bimetallic Tubes . . . 11-45
Clad Tube Sheets . . . 11-46
Nonmetallic Construction . . . 11-46
Fabrication . . . 11-46
Shell-and-Tube Exchanger Costs . . . 11-46

## HAIRPIN/DOUBLE-PIPE HEAT EXCHANGERS

Principles of Construction . . . 11-48
Finned Double Pipes . . . 11-48
Multitube Hairpins . . . 11-48
Design Applications . . . 11-49

## AIR-COOLED HEAT EXCHANGERS

Air Cooled Heat Exchangers . . . 11-49
Forced and Induced Draft . . . 11-49
Tube Bundle . . . 11-50
Tubing . . . 11-51
Finned-Tube Construction . . . 11-51
Fans . . . 11-51
Fan Drivers . . . 11-51
Fan Ring and Plenum Chambers . . . 11-52
Air-Flow Control . . . 11-52
Air Recirculation . . . 11-52
Trim Coolers . . . 11-52
Humidification Chambers . . . 11-52
Evaporative Cooling . . . 11-53
Steam Condensers . . . 11-53
Air-Cooled Overhead Condensers . . . 11-53
Air-Cooled Heat-Exchanger Costs . . . 11-53
Design Considerations . . . 11-53

## COMPACT AND NONTUBULAR HEAT EXCHANGERS

Compact Heat Exchangers . . . 11-54
Plate-and-Frame Exchangers . . . 11-54
Gasketed-Plate Exchangers . . . 11-54
Description . . . 11-54
Applications . . . 11-54
Design . . . 11-55
Welded- and Brazed-Plate Exchangers . . . 11-57
Combination Welded-Plate Exchangers . . . 11-57
Spiral-Plate Exchangers . . . 11-57
Description . . . 11-57
Applications . . . 11-57
Design . . . 11-57
Brazed-Plate-Fin Heat Exchangers . . . 11-58
Design and Application . . . 11-58
Plate-Fin Tubular Exchangers (PFE) . . . 11-58
Description . . . 11-58
Applications . . . 11-58
Design . . . 11-58
Printed-Circuit Heat Exchangers . . . 11-58
Spiral-Tube Exchangers (STE) . . . 11-59
Description . . . 11-59
Applications . . . 11-59
Design . . . 11-59
Graphite Heat Exchangers . . . 11-59
Description . . . 11-59
Applications and Design . . . 11-59
Cascade Coolers . . . 11-59
Bayonet-Tube Exchangers . . . 11-59
Atmospheric Sections . . . 11-60
Nonmetallic Heat Exchangers . . . 11-60
PVDF Heat Exchangers . . . 11-60
Ceramic Heat Exchangers . . . 11-60
Teflon Heat Exchangers . . . 11-60

## HEAT EXCHANGERS FOR SOLIDS

Equipment for Solidification . . . 11-60
Table Type . . . 11-61
Agitated-Pan Type . . . 11-61
Vibratory Type . . . 11-61
Belt Types . . . 11-61
Rotating-Drum Type . . . 11-62
Rotating-Shelf Type . . . 11-62

Equipment for Fusion of Solids . . . 11-63
Horizontal-Tank Type . . . 11-63
Vertical Agitated-Kettle Type . . . 11-63
Mill Type . . . 11-63
Heat-Transfer Equipment for Sheeted Solids . . . 11-63
Cylinder Heat-Transfer Units . . . 11-63
Heat-Transfer Equipment for Divided Solids . . . 11-64
Fluidized-Bed Type . . . 11-65
Moving-Bed Type . . . 11-65
Agitated-Pan Type . . . 11-65
Kneading Devices . . . 11-65
Shelf Devices . . . 11-66
Rotating-Shell Devices . . . 11-66
Conveyor-Belt Devices . . . 11-67
Spiral-Conveyor Devices . . . 11-67
Double-Cone Blending Devices . . . 11-68
Vibratory-Conveyor Devices . . . 11-68
Elevator Devices . . . 11-69
Pneumatic-Conveying Devices . . . 11-69
Vacuum-Shelf Types . . . 11-70


## THERMAL INSULATION


Insulation Materials . . . 11-70
Materials . . . 11-70
Thermal Conductivity (*K* Factor) . . . 11-70
Finishes . . . 11-70
System Selection . . . 11-71
Cryogenic High Vacuum . . . 11-71
Low Temperature . . . 11-71
Moderate and High Temperature . . . 11-72
Economic Thickness of Insulation . . . 11-72
Recommended Thickness of Insulation . . . 11-73
Example 1 . . . 11-76
Example 2 . . . 11-76
Example 3 . . . 11-76
Installation Practice . . . 11-76
Pipe . . . 11-76
Method of Securing . . . 11-76
Double Layer . . . 11-76
Finish . . . 11-76
Tanks, Vessels, and Equipment . . . 11-76
Method of Securing . . . 11-76
Finish . . . 11-76


## AIR CONDITIONING


Introduction . . . 11-76
Comfort Air Conditioning . . . 11-76
Industrial Air Conditioning . . . 11-76
Ventilation . . . 11-76
Air-Conditioning Equipment . . . 11-77
Central Systems . . . 11-77
Unitary Refrigerant-Based Air-Conditioning Systems . . . 11-77
Load Calculation . . . 11-77


## REFRIGERATION


Introduction . . . 11-78
Basic Principles . . . 11-78
Basic Refrigeration Methods . . . 11-79
Mechanical Refrigeration (Vapor-Compression Systems) . . . 11-79
Vapor-Compression Cycles . . . 11-79
Multistage Systems . . . 11-79
Cascade System . . . 11-82
Equipment . . . 11-82
Compressors . . . 11-82
Positive-Displacement Compressors . . . 11-83
Centrifugal Compressors . . . 11-85
Condensers . . . 11-85
Evaporators . . . 11-87
System Analysis . . . 11-87
System, Equipment, and Refrigerant Selection . . . 11-90
Other Refrigerant Systems Applied in the Industry . . . 11-90
Absorption Refrigeration Systems . . . 11-90
Steam-Jet (Ejector) Systems . . . 11-94
Multistage Systems . . . 11-96
Capacity Control . . . 11-96
Refrigerants . . . 11-96
Secondary Refrigerants (Antifreezes or Brines) . . . 11-97
Organic Compounds (Inhibited Glycols) . . . 11-98
Safety in Refrigeration Systems . . . 11-98


## CRYOGENIC PROCESSES


Introduction . . . 11-99
Properties of Cryogenic Fluids . . . 11-99
Properties of Solids . . . 11-99
Structural Properties at Low Temperatures . . . 11-99
Thermal Properties at Low Temperatures . . . 11-100
Electrical Properties at Low Temperatures . . . 11-100
Superconductivity . . . 11-100
Refrigeration and Liquifaction . . . 11-100
Principles . . . 11-100
Expansion Types of Refrigerators . . . 11-100
Miniature Refrigerators . . . 11-103
Thermodynamic Analyses of Cycles . . . 11-103
Process Equipment . . . 11-103
Heat Exchangers . . . 11-103
Expanders . . . 11-104
Separation and Purification Systems . . . 11-104
Air-Separation Systems . . . 11-104
Helium and Natural-Gas Systems Separation . . . 11-106
Gas Purification . . . 11-106
Storage and Transfer Systems . . . 11-107
Insulation Principles . . . 11-107
Types of Insulation . . . 11-107
Storage and Transfer Systems . . . 11-108
Cryogenic Instrumentation . . . 11-108
Pressure . . . 11-109
Liquid Level . . . 11-109
Flow . . . 11-109
Temperature . . . 11-109
Safety . . . 11-109
Physiological Hazards . . . 11-109
Materials and Construction Hazards . . . 11-109
Flammability and Explosion Hazards . . . 11-110
High-Pressure Gas Hazards . . . 11-110
Summary . . . 11-110


## EVAPORATORS


Primary Design Problems . . . 11-110
Heat Transfer . . . 11-110
Vapor-Liquid Separation . . . 11-110
Selection Problems . . . 11-110
Product Quality . . . 11-110
Evaporator Types and Applications . . . 11-111
Forced-Circulation Evaporators . . . 11-111
Swirl Flow Evaporators . . . 11-111
Short-Tube Vertical Evaporators . . . 11-112
Long-Tube Vertical Evaporators . . . 11-112
Horizontal-Tube Evaporators . . . 11-113
Miscellaneous Forms of Heating Surface . . . 11-114
Evaporators without Heating Surfaces . . . 11-114
Utilization of Temperature Difference . . . 11-114
Vapor-Liquid Separation . . . 11-114
Evaporator Arrangement . . . 11-116
Single-Effect Evaporators . . . 11-116
Thermocompression . . . 11-116
Multiple-Effect Evaporation . . . 11-116
Seawater Evaporators . . . 11-117
Evaporator Calculations . . . 11-118
Single-Effect Evaporators . . . 11-118
Thermocompression Evaporators . . . 11-118
Flash Evaporators . . . 11-118
Multiple-Effect Evaporators . . . 11-119
Optimization . . . 11-119
Evaporator Accessories . . . 11-119
Condensers . . . 11-119
Vent Systems . . . 11-120
Salt Removal . . . 11-120
Evaporator Operation . . . 11-121

# THERMAL DESIGN OF HEAT-TRANSFER EQUIPMENT

## INTRODUCTION TO THERMAL DESIGN

Designers commonly use computer software to design heat exchangers. The best sources of such software are Heat Transfer Research, Inc. (HTRI), and Heat Transfer and Fluid Flow Services (HTFS), a division of ASPENTECH. These are companies that develop proprietary correlations based on their research and provide software that utilizes these correlations. However, it is important that engineers understand the fundamental principles that lie beneath the framework of the software. Therefore, design methods for several important classes of process heat-transfer equipment are presented in the following portions of Sec. 11. Mechanical descriptions and specifications of equipment are given in this section and should be read in conjunction with the use of this material. It is impossible to present here a comprehensive treatment of heat-exchanger selection, design, and application. The best general references in this field are Hewitt, Shires, and Bott, *Process Heat Transfer,* CRC Press, Boca Raton, FL, 1994; and Schlünder (ed.), *Heat Exchanger Design Handbook,* Begell House, New York, 2002.

**Approach to Heat-Exchanger Design** The proper use of basic heat-transfer knowledge in the design of practical heat-transfer equipment is an art. Designers must be constantly aware of the differences between the idealized conditions for and under which the basic knowledge was obtained and the real conditions of the mechanical expression of their design and its environment. The result must satisfy process and operational requirements (such as availability, flexibility, and maintainability) and do so economically. An important part of any design process is to consider and offset the consequences of error in the basic knowledge, in its subsequent incorporation into a design method, in the translation of design into equipment, or in the operation of the equipment and the process. Heat-exchanger design is not a highly accurate art under the best of conditions.

The design of a process heat exchanger usually proceeds through the following steps:

1. Process conditions (stream compositions, flow rates, temperatures, pressures) must be specified.
2. Required physical properties over the temperature and pressure ranges of interest must be obtained.
3. The type of heat exchanger to be employed is chosen.
4. A preliminary estimate of the size of the exchanger is made, using a heat-transfer coefficient appropriate to the fluids, the process, and the equipment.
5. A first design is chosen, complete in all details necessary to carry out the design calculations.
6. The design chosen in step 5 is evaluated, or *rated,* as to its ability to meet the process specifications with respect to both heat transfer and pressure drop.
7. On the basis of the result of step 6, a new configuration is chosen if necessary and step 6 is repeated. If the first design was inadequate to meet the required heat load, it is usually necessary to increase the size of the exchanger while still remaining within specified or feasible limits of pressure drop, tube length, shell diameter, etc. This will sometimes mean going to multiple-exchanger configurations. If the first design more than meets heat-load requirements or does not use all the allowable pressure drop, a less expensive exchanger can usually be designed to fulfill process requirements.
8. The final design should meet process requirements (within reasonable expectations of error) at lowest cost. The lowest cost should include operation and maintenance costs and credit for ability to meet long-term process changes, as well as installed (capital) cost.

Exchangers should not be selected entirely on a lowest-first-cost basis, which frequently results in future penalties.

**Overall Heat-Transfer Coefficient** The basic design equation for a heat exchanger is

$$dA = dQ/U\,\Delta T \tag{11-1}$$

where $dA$ is the element of surface area required to transfer an amount of heat $dQ$ at a point in the exchanger where the overall heat-transfer coefficient is $U$ and where the overall bulk temperature difference between the two streams is $\Delta T$. The overall heat-transfer coefficient is related to the individual film heat-transfer coefficients and fouling and wall resistances by Eq. (11-2). Basing $U_o$ on the outside surface area $A_o$ results in

$$U_o = \frac{1}{1/h_o + R_{do} + xA_o/k_w A_{wm} + (1/h_i + R_{di})A_o/A_i} \tag{11-2}$$

Equation (11-1) can be formally integrated to give the outside area required to transfer the total heat load $Q_T$:

$$A_o = \int_0^{Q_T} \frac{dQ}{U_o \Delta T} \tag{11-3}$$

To integrate Eq. (11-3), $U_o$ and $\Delta T$ must be known as functions of $Q$. For some problems, $U_o$ varies strongly and nonlinearly throughout the exchanger. In these cases, it is necessary to evaluate $U_o$ and $\Delta T$ at several intermediate values and numerically or graphically integrate. For many practical cases, it is possible to calculate a constant mean overall coefficient $U_{om}$ from Eq. (11-2) and define a corresponding mean value of $\Delta T_m$, such that

$$A_o = Q_T/U_{om}\,\Delta T_m \tag{11-4}$$

Care must be taken that $U_o$ does not vary too strongly, that the proper equations and conditions are chosen for calculating the individual coefficients, and that the mean temperature difference is the correct one for the specified exchanger configuration.

**Mean Temperature Difference** The temperature difference between the two fluids in the heat exchanger will, in general, vary from point to point. The mean temperature difference ($\Delta T_m$ or MTD) can be calculated from the terminal temperatures of the two streams if the following assumptions are valid:

1. All elements of a given fluid stream have the same thermal history in passing through the exchanger.*
2. The exchanger operates at steady state.
3. The specific heat is constant for each stream (or if either stream undergoes an isothermal phase transition).
4. The overall heat-transfer coefficient is constant.
5. Heat losses are negligible.

**Countercurrent or Cocurrent Flow** If the flow of the streams is either *completely* countercurrent or completely cocurrent or if one or both streams are isothermal (condensing or vaporizing a pure component with negligible pressure change), the correct MTD is the logarithmic-mean temperature difference (LMTD), defined as

$$\text{LMTD} = \Delta T_{lm} = \frac{(t_1' - t_2'') - (t_2' - t_1'')}{\ln\left(\dfrac{t_1' - t_2''}{t_2' - t_1''}\right)} \tag{11-5a}$$

for *countercurrent flow* (Fig. 11-1a) and

$$\text{LMTD} = \Delta T_{lm} = \frac{(t_1' - t_1'') - (t_2' - t_2'')}{\ln\left(\dfrac{t_1' - t_1''}{t_2' - t_2''}\right)} \tag{11-5b}$$

for *cocurrent flow* (Fig. 11-1*b*)

*This assumption is vital but is usually omitted or less satisfactorily stated as "each stream is well mixed at each point." In a heat exchanger with substantial bypassing of the heat-transfer surface, e.g., a typical baffled shell-and-tube exchanger, this condition is not satisfied. However, the error is in some degree offset if the same MTD formulation used in reducing experimental heat-transfer data to obtain the basic correlation is used in applying the correlation to design a heat exchanger. The compensation is not in general exact, and insight and judgment are required in the use of the MTD formulations. Particularly, in the design of an exchanger with a very close temperature approach, bypassing may result in an exchanger that is inefficient and even thermodynamically incapable of meeting specified outlet temperatures.

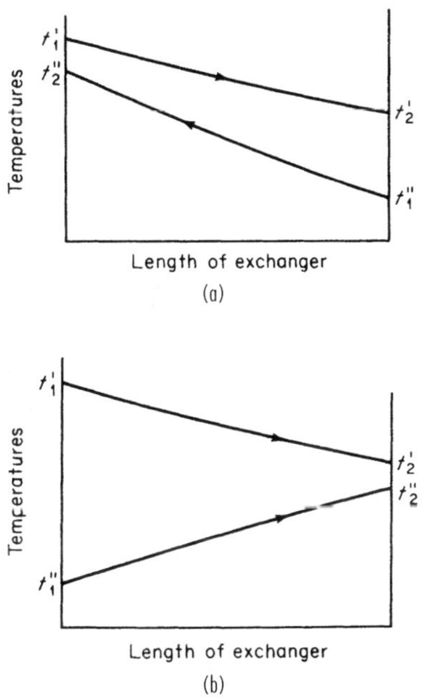


**FIG. 11-1** Temperature profiles in heat exchangers. (*a*) Countercurrent. (*b*) Cocurrent.

If $U$ is not constant but a linear function of $\Delta T$, the correct value of $U_{om}\Delta T_m$ to use in Eq. (11-4) is [Colburn, *Ind. Eng. Chem.*, **25,** 873 (1933)]

$$U_{om}\Delta T_m = \frac{U_o''(t_1' - t_2'') - U_o'(t_2' - t_1'')}{\ln\left(\dfrac{U_o''(t_1' - t_2'')}{U_o'(t_2' - t_1'')}\right)} \qquad (11\text{-}6a)$$

for *countercurrent flow,* where $U_o''$ is the overall coefficient evaluated when the stream temperatures are $t_1'$ and $t_2''$ and $U_o'$ is evaluated at $t_2'$ and $t_1''$. The corresponding equation for *cocurrent flow* is

$$U_{om}\Delta T_m = \frac{U_o''(t_1' - t_1'') - U_o'(t_2' - t_2'')}{\ln\left(\dfrac{U_o''(t_1' - t_1'')}{U_o'(t_2' - t_2'')}\right)} \qquad (11\text{-}6b)$$

where $U_o'$ is evaluated at $t_2'$ and $t_2''$ and $U_o''$ is evaluated at $t_1'$ and $t_1''$. To use these equations, it is necessary to calculate two values of $U_o$.*

The use of Eq. (11-6) will frequently give satisfactory results even if $U_o$ is not strictly linear with temperature difference.

**Reversed, Mixed, or Cross-Flow** If the flow pattern in the exchanger is not completely countercurrent or cocurrent, it is necessary to apply a **correction factor** $F_T$ by which the LMTD is multiplied to obtain the appropriate MTD. These corrections have been mathematically derived for flow patterns of interest, still by making assumptions 1 to 5 [see Bowman, Mueller, and Nagle, *Trans. Am. Soc. Mech. Eng.*, **62,** 283 (1940) or Hewitt, et al. op. cit.]. For a common flow pattern, the 1-2 exchanger (Fig. 11-2), the correction factor $F_T$ is given in Fig. 11-4*a,* which is also valid for finding $F_T$ for a 1-2 exchanger in which the shell-side flow direction is reversed from that shown in Fig. 11-2. Figure 11-4*a* is also applicable with negligible error to exchangers with one shell pass and any number of tube passes. Values of $F_T$ less than 0.8 (0.75 at the very lowest) are generally unacceptable because the exchanger configuration chosen is inefficient; the chart is difficult to read accurately; and even a small violation of the first assumption

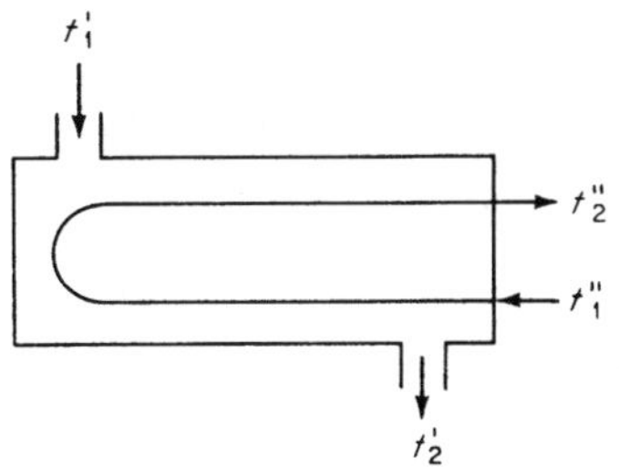


**FIG. 11-2** Diagram of a 1-2 exchanger (one well-baffled shell pass and two tube passes with an equal number of tubes in each pass).

underlying the MTD will invalidate the mathematical derivation and lead to a thermodynamically inoperable exchanger.

Correction-factor charts are also available for exchangers with more than one shell pass provided by a longitudinal shell-side baffle. However, these exchangers are seldom used in practice because of mechanical complications in their construction. Also thermal and physical leakages across the longitudinal baffle further reduce the mean temperature difference and are not properly incorporated into the correction-factor charts. Such charts are useful, however, when it is necessary to construct a multiple-shell exchanger train such as that shown in Fig. 11-3 and are included here for two, three, four, and six *separate identical* shells and two or more tube passes per shell in Fig. 11-4*b, c, d,* and *e.* If only one tube pass per shell is required, the piping can and should be arranged to provide pure countercurrent flow, in which case the LMTD is used with no correction.

Cross-flow exchangers of various kinds are also important and require correction to be applied to the LMTD calculated by assuming countercurrent flow. Several cases are given in Fig. 11-4*f, g, h, i,* and *j.*

Many other MTD correction-factor charts have been prepared for various configurations. The $F_T$ charts are often employed to make approximate corrections for configurations even in cases for which they are not completely valid.

## THERMAL DESIGN FOR SINGLE-PHASE HEAT TRANSFER

**Double-Pipe Heat Exchangers** The design of double-pipe heat exchangers is straightforward. It is generally conservative to

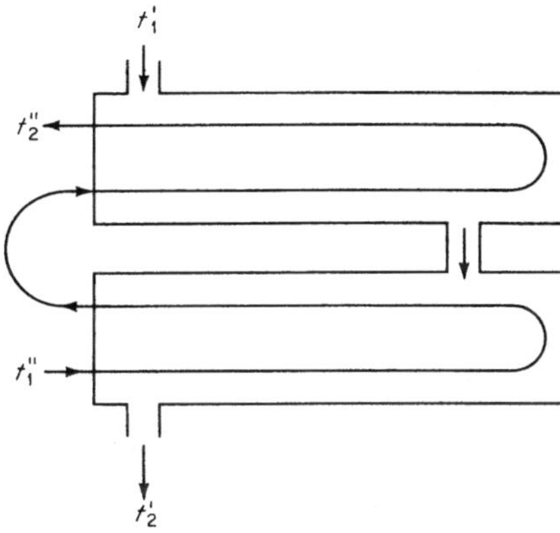


**FIG. 11-3** Diagram of a 2-4 exchanger (two separate identical well-baffled shells and four or more tube passes).

*This task can be avoided if a hydrocarbon stream is the limiting resistance by the use of the caloric temperature charts developed by Colburn [*Ind. Eng. Chem.*, **25,** 873 (1933)].

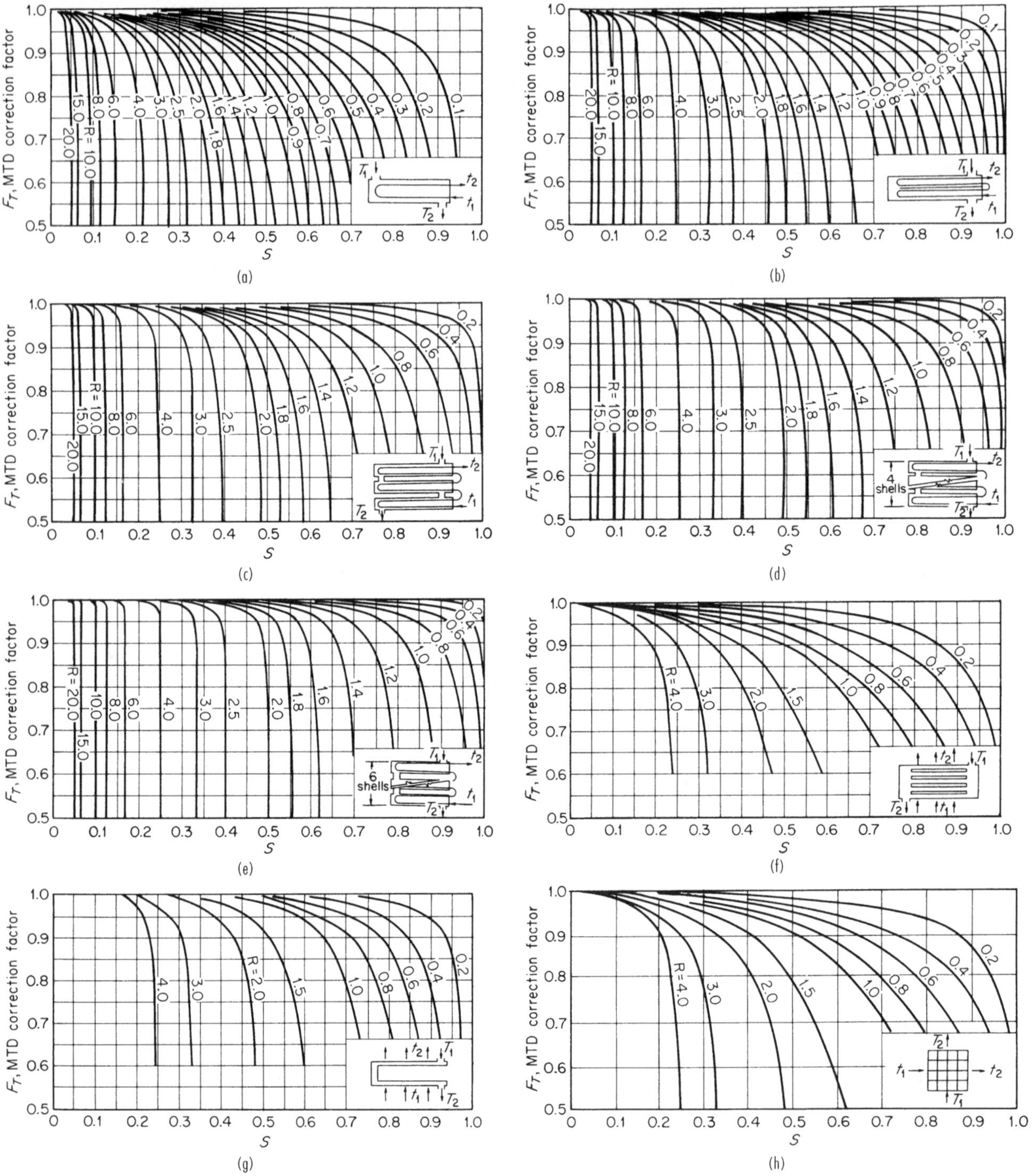


**FIG. 11-4** LMTD correction factors for heat exchangers. In all charts, $R = (T_1 - T_2)/(t_2 - t_1)$ and $S = (t_2 - t_1)/(T_1 - t_1)$. (*a*) One shell pass, two or more tube passes. (*b*) Two shell passes, four or more tube passes. (*c*) Three shell passes, six or more tube passes. (*d*) Four shell passes, eight or more tube passes. (*e*) Six shell passes, twelve or more tube passes. (*f*) Cross-flow, one shell pass, one or more parallel rows of tubes. (*g*) Cross-flow, two passes, two rows of tubes; for more than two passes, use $F_T = 1.0$. (*h*) Cross-flow, one shell pass, one tube pass, both fluids unmixed. (*i*) Cross-flow (drip type), two horizontal passes with U-bend connections (trombone type). (*j*) Cross-flow (drip type), helical coils with two turns.

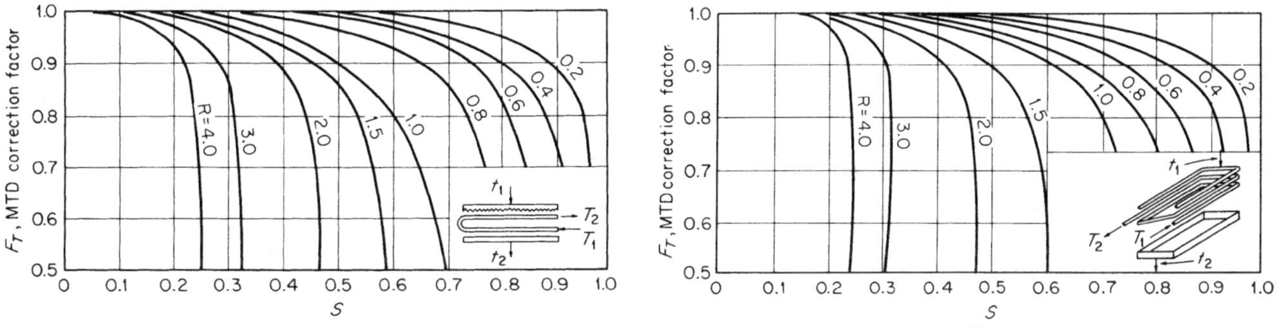


(i) (j)

**FIG. 11-4** *(Continued)*

neglect natural-convection and entrance effects in turbulent flow. In laminar flow, natural convection effects can increase the theoretical Graetz prediction by a factor of 3 or 4 for fully developed flows. Pressure drop is calculated by using the correlations given in Sec. 6.

If the inner tube is longitudinally finned on the outside surface, the equivalent diameter is used as the characteristic length in both the Reynolds-number and the heat-transfer correlations. The fin efficiency must also be known to calculate an *effective* outside area to use in Eq. (11-2).

Fittings contribute strongly to the pressure drop on the annulus side. General methods for predicting this are not reliable, and manufacturer's data should be used when available.

Double-pipe exchangers are often piped in complex series-parallel arrangements on both sides. The MTD to be used has been derived for some of these arrangements and is reported in Kern (*Process Heat Transfer*, McGraw-Hill, New York, 1950). More complex cases may require trial-and-error balancing of the heat loads and rate equations for subsections or even for individual exchangers in the bank.

**Baffled Shell-and-Tube Exchangers** The method given here is based on the research summarized in Final Report, Cooperative Research Program on Shell and Tube Heat Exchangers, Univ. Del. Eng. Exp. Sta. Bull. 5 (June 1963). The method assumes that the shell-side heat transfer and pressure-drop characteristics are equal to those of the ideal tube bank corresponding to the cross-flow sections of the exchanger, modified for the distortion of flow pattern introduced by the baffles and the presence of leakage and bypass flow through the various clearances required by mechanical construction.

It is assumed that process conditions and physical properties are known and the following are known or specified: tube outside diameter $D_o$, tube geometrical arrangement (unit cell), shell inside diameter $D_s$, shell outer tube limit $D_{otl}$, baffle cut $l_c$, baffle spacing $l_s$, and number of sealing strips $N_{ss}$. The effective tube length between tube sheets $L$ may be either specified or calculated after the heat-transfer coefficient has been determined. If additional specific information (e.g., tube-baffle clearance) is available, the exact values (instead of estimates) of certain parameters may be used in the calculation with some improvement in accuracy. To complete the rating, it is necessary to know also the tube material and wall thickness or inside diameter.

This rating method, though apparently generally the best in the open literature, is not extremely accurate. An exhaustive study by Palen and Taborek [*Chem. Eng. Prog. Symp. Ser.* 92, **65,** 53 (1969)] showed that this method predicted shell-side coefficients from about 50 percent low to 100 percent high, while the pressure-drop range was from about 50 percent low to 200 percent high. The mean error for heat transfer was about 15 percent low (safe) for all Reynolds numbers, while the mean error for pressure drop was from about 5 percent low (unsafe) at Reynolds numbers above 1000 to about 100 percent high at Reynolds numbers below 10.

***Calculation of Shell-Side Geometrical Parameters***

1. *Total number of tubes in exchanger* $N_t$. If not known by direct count, estimate using Eq. (11-74) or (11-75).

2. *Tube pitch parallel to flow* $p_p$ *and normal to flow* $p_n$. These quantities are needed only for estimating other parameters. If a detailed drawing of the exchanger is available, it is better to obtain these other parameters by direct count or calculation. The pitches are described by Fig. 11-5 and read therefrom for common tube layouts.

3. *Number of tube rows crossed in one cross-flow section* $N_c$. Count from exchanger drawing or estimate from

$$N_c = \frac{D_s[1 - 2(l_c/D_s)]}{p_p} \qquad (11\text{-}7)$$

4. *Fraction of total tubes in cross-flow* $F_c$

$$F_c = \frac{1}{\pi}\left[\pi + 2\,\frac{D_s - 2l_c}{D_{otl}}\,\sin\left(\cos^{-1}\frac{D_s - 2l_c}{D_{otl}}\right) - 2\cos^{-1}\frac{D_s - 2l_c}{D_{otl}}\right] \qquad (11\text{-}8)$$

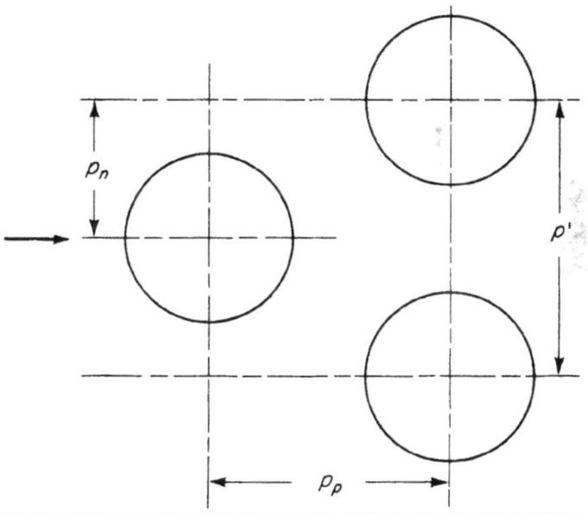


| Tube O.D. $D_o$, in. | Tube pitch $p'$, in. | Layout | $p_p$, in. | $p_n$, in. |
|---|---|---|---|---|
| 0.625 | 0.812 | 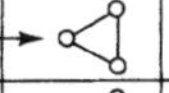 | 0.704 | 0.406 |
| 0.750 | 0.938 | 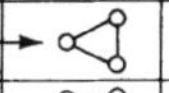 | 0.814 | 0.469 |
| 0.750 | 1 | 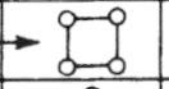 | 1.000 | 1.000 |
| 0.750 | 1 | 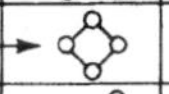 | 0.707 | 0.707 |
| 0.750 | 1 | 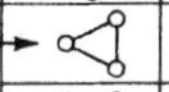 | 0.866 | 0.500 |
| 1.000 | 1.250 | 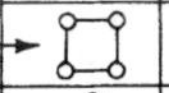 | 1.250 | 1.250 |
| 1.000 | 1.250 | 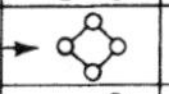 | 0.884 | 0.884 |
| 1.000 | 1.250 | 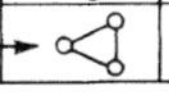 | 1.082 | 0.625 |

**FIG. 11-5** Values of tube pitch for common tube layouts. To convert inches to meters, multiply by 0.0254. Not that $D_o$, $p'$, $p_p$, and $p_n$ have units of inches.

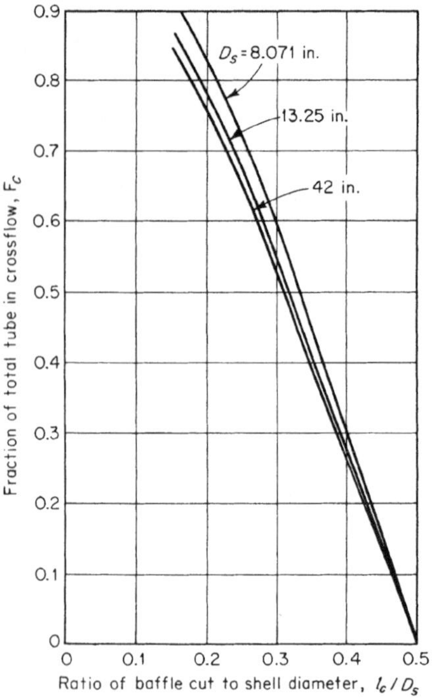


**FIG. 11-6** Estimation of fraction of tubes in cross-flow $F_c$ [Eq. (11-8)]. To convert inches to meters, multiply by 0.0254. Note that $l_c$ and $D_s$ have units of inches.

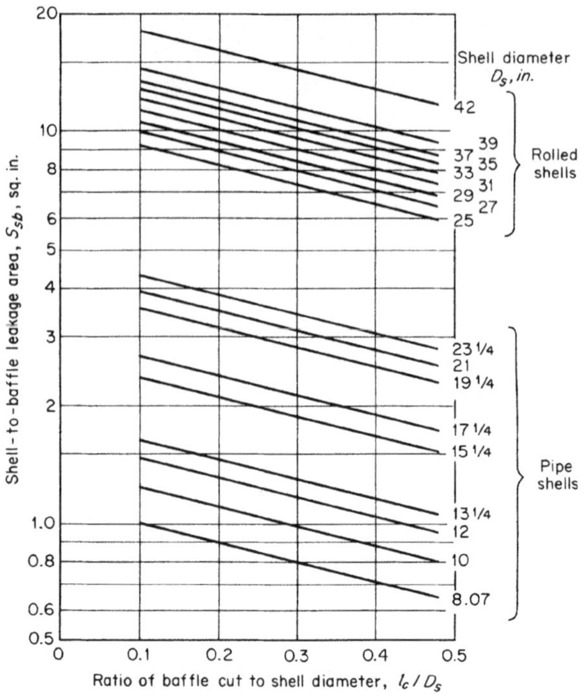


**FIG. 11-7** Estimation of shell-to-baffle leakage area [Eq. (11-13)]. To convert inches to meters, multiply by 0.0254; to convert square inches to square meters, multiply by (6.45)($10^{-4}$). Note that $l_c$ and $D_s$ have units of inches.

$F_c$ is plotted in Fig. 11-6. This figure is strictly applicable only to split-ring, floating-head construction but may be used for other situations with minor error.

5. *Number of effective cross-flow rows in each window* $N_{cw}$

$$N_{cw} = \frac{0.8l_c}{p_p} \tag{11-9}$$

6. *Cross-flow area at or near centerline for one cross-flow section* $S_m$

*a.* For rotated and in-line square layouts:

$$S_m = l_s\left[D_s - D_{otl} + \frac{D_{otl} - D_o}{p_n}(p' - D_o)\right] \text{ m}^2 \text{ (ft}^2\text{)} \tag{11-10a}$$

*b.* For triangular layouts:

$$S_m = l_s\left[D_s - D_{otl} + \frac{D_{otl} - D_o}{p'}(p' - D_o)\right] \text{ m}^2 \text{ (ft}^2\text{)} \tag{11-10b}$$

7. *Fraction of cross-flow area available for bypass flow* $F_{bp}$

$$F_{bp} = \frac{(D_s - D_{otl})l_s}{S_m} \tag{11-11}$$

8. *Tube-to-baffle leakage area for one baffle* $S_{tb}$. Estimate from

$$S_{tb} = bD_oN_T(1 + F_c) \text{ m}^2 \text{ (ft}^2\text{)} \tag{11-12}$$

where $b = (6.223)(10^{-4})$ (SI) or $(1.701)(10^{-4})$ (U.S. customary). These values are based on Tubular Exchanger Manufacturers Association (TEMA) Class R construction which specifies ⅓₂-in diametral clearance between tube and baffle. Values should be modified if extra tight or loose construction is specified or if clogging by dirt is anticipated.

9. *Shell-to-baffle leakage area for one baffle* $S_{sb}$. If diametral shell-baffle clearance $\delta_{sb}$ is known, $S_{sb}$ can be calculated from

$$S_{sb} = \frac{D_s\,\delta_{sb}}{2}\left[\pi - \cos^{-1}\left(1 - \frac{2l_c}{D_s}\right)\right] \quad \text{m}^2 \text{ (ft}^2\text{)} \tag{11-13}$$

where the value of the term $\cos^{-1}(1 - 2l_c/D_s)$ is in radians and is between 0 and $\pi/2$. $S_{sb}$ is plotted in Fig. 11-7, based on TEMA Class R standards. Since pipe shells are generally limited to diameters below 24 in, the larger sizes are shown by using the rolled-shell specification. Allowance should be made for especially tight or loose construction.

10. *Area for flow through window* $S_w$. This area is obtained as the difference between the gross window area $S_{wg}$ and the window area occupied by tubes $S_{wt}$:

$$S_w = S_{wg} - S_{wt} \tag{11-14}$$

$$S_{wg} = \frac{D_s^2}{4}\left[\cos^{-1}\left(1 - 2\frac{l_c}{D_s}\right) - \left(1 - 2\frac{l_c}{D_s}\right)\sqrt{1 - \left(1 - 2\frac{l_c}{D_s}\right)^2}\,\right] \quad \text{m}^2 \text{ (ft}^2\text{)} \tag{11-15}$$

$S_{wg}$ is plotted in Fig. 11-8. $S_{wt}$ can be calculated from

$$S_{wt} = (N_T/8)(1 - F_c)\pi D_o^2 \quad \text{m}^2 \text{ (ft}^2\text{)} \tag{11-16}$$

11. *Equivalent diameter of window* $D_w$ [*required only if laminar flow, defined as* $(N_{Re})_s \le 100$, *exists*]

$$D_w = \frac{4S_w}{(\pi/2)N_T(1 - F_c)D_o + D_s\theta_b} \quad \text{m (ft)} \tag{11-17}$$

where $\theta_b$ is the baffle-cut angle given by

$$\theta_b = 2\cos^{-1}\left(1 - \frac{2l_c}{D_s}\right) \quad \text{rad} \tag{11-18}$$

12. *Number of baffles* $N_b$

$$N_b = \frac{L - 2le}{l_s} + 1 \tag{11-19}$$

where *le* is the entrance/exit baffle spacing, often different from the central baffle spacing. The effective tube length $L$ must be known to calculate $N_b$, which is needed to calculate shell-side pressure drop. In designing an exchanger, the shell-side coefficient may be calculated and the required exchanger length for heat transfer obtained before $N_b$ is calculated.

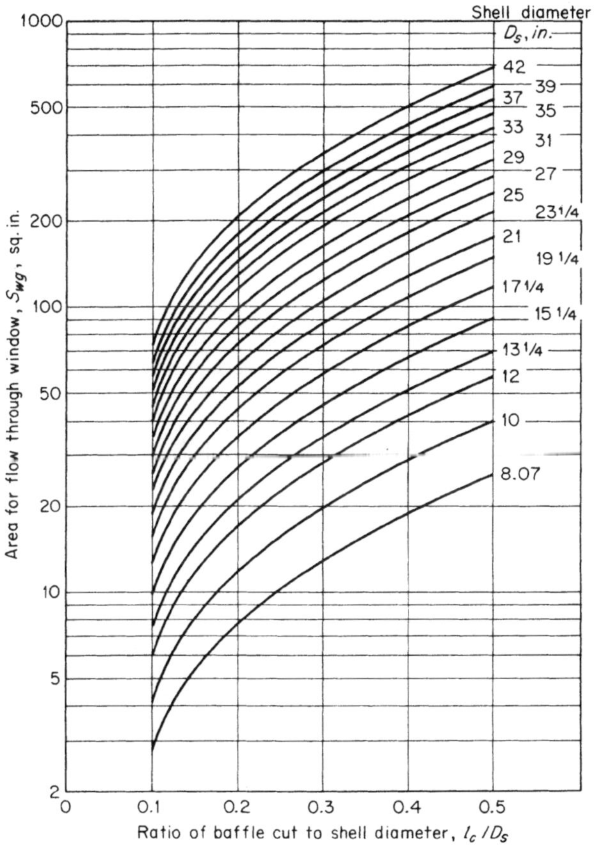


**FIG. 11-8** Estimation of window cross-flow area [Eq. (11-15)]. To convert inches to meters, multiply by 0.0254. Note that $l_c$ and $D_s$ have units of inches.

### *Shell-Side Heat-Transfer Coefficient Calculation*

1. Calculate the *shell-side Reynolds number* $(N_{Re})_s$.

$$(N_{Re})_s = D_o W / \mu_b S_m \tag{11-20}$$

where $W$ = mass flow rate and $\mu_b$ = viscosity at bulk temperature. The arithmetic mean bulk shell-side fluid temperature is usually adequate to evaluate all bulk properties of the shell-side fluid. For large temperature ranges or for viscosity that is very sensitive to temperature change, special care must be taken, such as using Eq. (11-6).

2. Find $j_k$ from the ideal-tube bank curve for a given tube layout at the calculated value of $(N_{Re})_s$, using Fig. 11-9, which is adapted from ideal-tube-bank data obtained at Delaware by Bergelin et al. [*Trans. Am. Soc. Mech. Eng.*, **74,** 953 (1952) and the Grimison correlation [*Trans. Am. Soc. Mech. Eng.*, **59,** 583 (1937)].

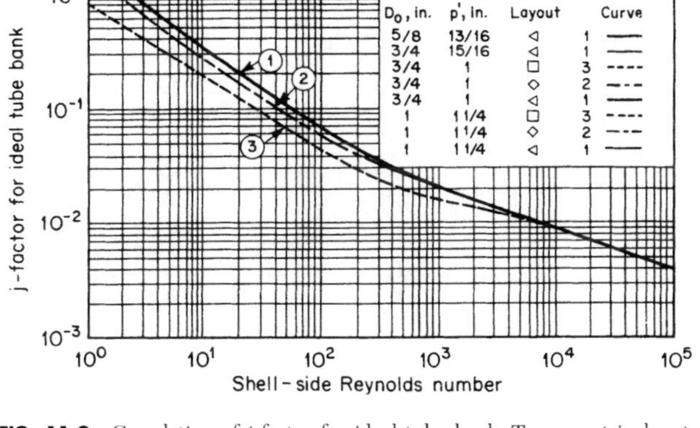


**FIG. 11-9** Correlation of $j$ factor for ideal tube bank. To convert inches to meters, multiply by 0.0254. Note that $p'$ and $D_o$ have units of inches.

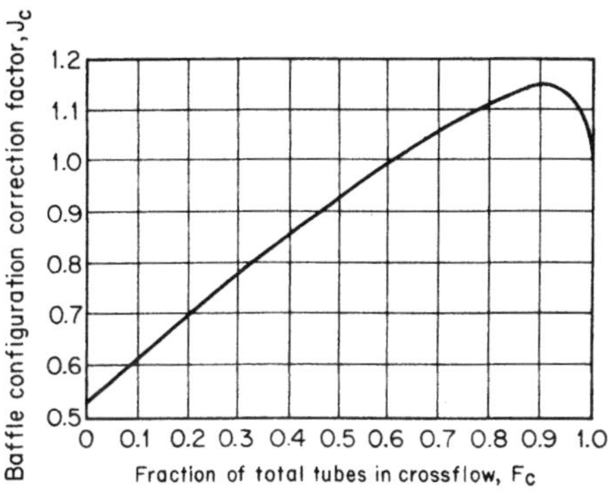


**FIG. 11-10** Correction factor for baffle-configuration effects.

3. Calculate the shell-side *heat-transfer coefficient for an ideal tube bank* $h_k$.

$$h_k = j_k c \frac{W}{S_m} \left(\frac{k}{c\mu}\right)^{2/3} \left(\frac{\mu_b}{\mu_w}\right)^{0.14} \tag{11-21}$$

where $c$ is the specific heat, $k$ is the thermal conductivity, and $\mu_w$ is the viscosity evaluated at the mean surface temperature.

4. Find the correction factor for baffle-configuration effects $J_c$ from Fig. 11-10.

5. Find the correction factor for baffle-leakage effects $J_l$ from Fig. 11-11.

6. Find the correction factor for bundle-bypassing effects $J_b$ from Fig. 11-12

7. Find the correction factor for adverse temperature-gradient buildup at low Reynolds number $J_r$:

*a.* If $(N_{Re})_s < 100$, find $J_r^\circ$ from Fig. 11-13, knowing $N_b$ and $(N_c + N_{cw})$.

*b.* If $(N_{Re})_s \leq 20$, $J_r = J_r^\circ$.

*c.* If $20 < (N_{Re})_s < 100$, find $J_r$ from Fig. 11-14, knowing $J_r^\circ$ and $(N_{Re})_s$.

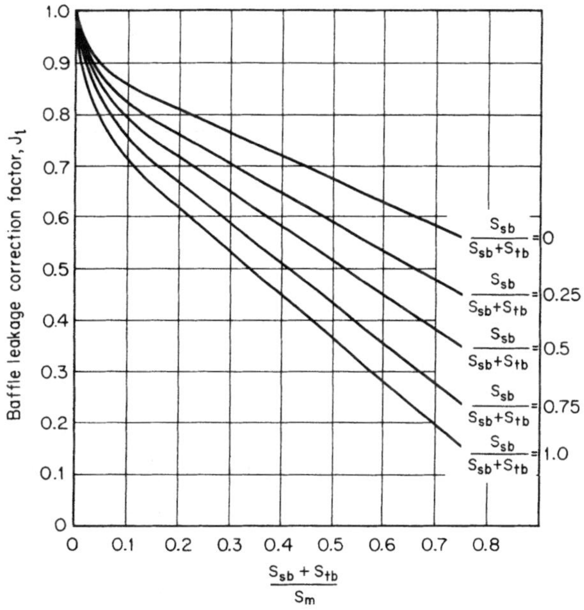


**FIG. 11-11** Correction factor for baffle-leakage effects.

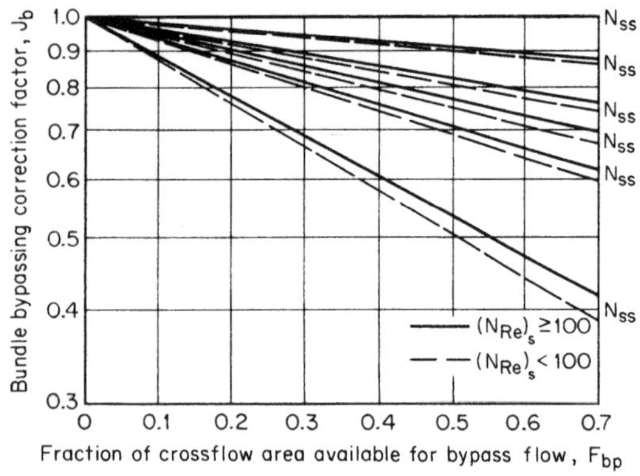


**FIG. 11-12** Correction factor for bypass flow.

8. Calculate the *shell-side heat-transfer coefficient for the exchanger* $h_s$ from

$$h_s = h_k J_c J_l J_b J_r \tag{11-22}$$

***Shell-Side Pressure-Drop Calculation***

1. Find $f_k$ from the ideal-tube-bank friction-factor curve for the given tube layout at the calculated value of $(N_{Re})_s$, using Fig. 11-15a for triangular and rotated square arrays and Fig. 11-15b for in-line square arrays. These curves are adapted from Bergelin et al. and Grimison (loc. cit.).

2. Calculate the *pressure drop for an ideal cross-flow section.*

$$\Delta P_{bk} = b \frac{f_k W^2 N_c}{\rho S_m^2} \left(\frac{\mu_w}{\mu_b}\right)^{0.14} \tag{11-23}$$

where $b = (2.0)(10^{-3})$ (SI) or $(9.9)(10^{-5})$ (U.S. customary).

3. Calculate the *pressure drop for an ideal window section.* If $(N_{Re})_s \geq 100$,

$$\Delta P_{wk} = b \frac{W^2(2 + 0.6N_{cw})}{S_m S_w \rho} \tag{11-24a}$$

where $b = (5)(10^{-4})$ (SI) or $(2.49)(10^{-5})$ (U.S. customary).

If $(N_{Re})_s < 100$,

$$\Delta P_{wk} = b_1 \frac{\mu_b W}{S_m S_w \rho}\left(\frac{N_{cw}}{p' - D_o} + \frac{l_s}{D_w^2}\right) + b_2 \frac{W^2}{S_m S_w \rho} \tag{11-24b}$$

where $b_1 = (1.681)(10^{-5})$ (SI) or $(1.08)(10^{-4})$ (U.S. customary), and $b_2 = (9.99)(10^{-4})$ (SI) or $(4.97)(10^{-5})$ (U.S. customary).

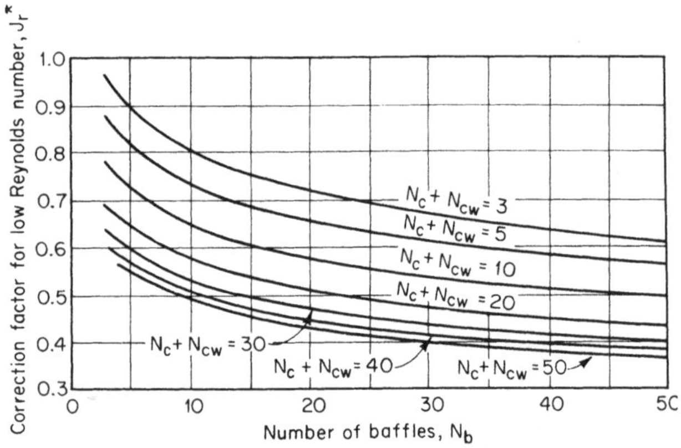


**FIG. 11-13** Basic correction factor for adverse temperature gradient at low Reynolds numbers.

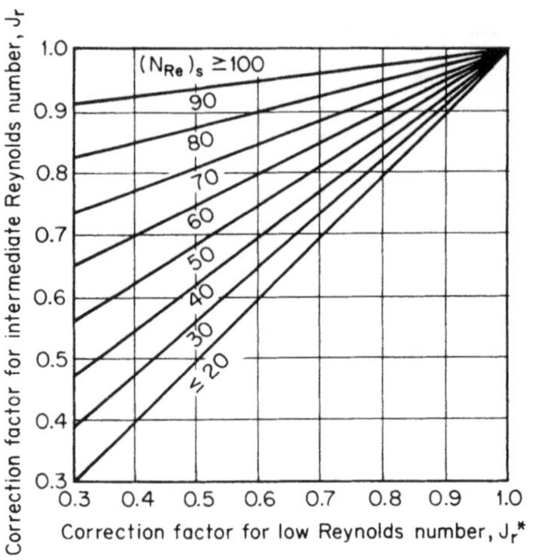


**FIG. 11-14** Correction factor for adverse temperature gradient at intermediate Reynolds numbers.

4. Find the correction factor for the effect of baffle leakage on pressure drop $R_l$ from Fig. 11-16. Curves shown are not to be extrapolated beyond the points shown.

5. Find the correction factor for bundle bypass $R_b$ from Fig. 11-17.

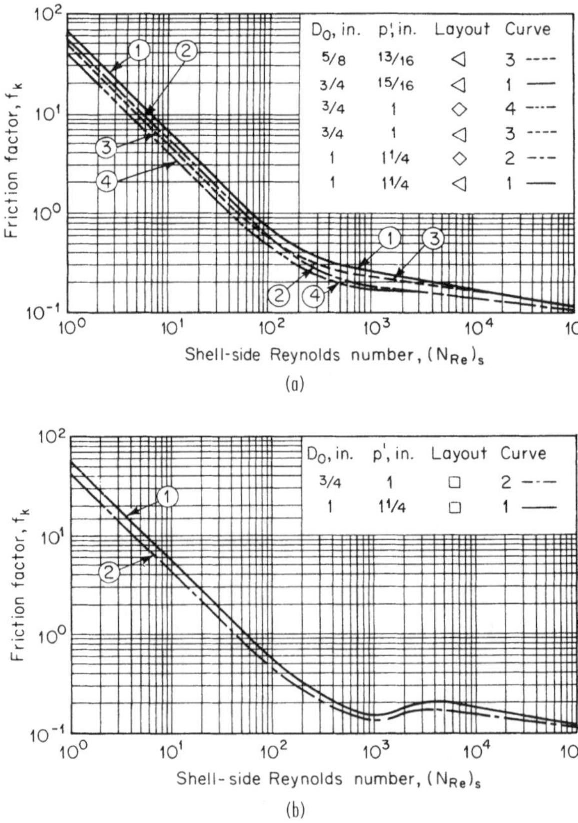


**FIG. 11-15** Correction of friction factors for ideal tube banks. (*a*) Triangular and rotated square arrays. (*b*) In-line square arrays.

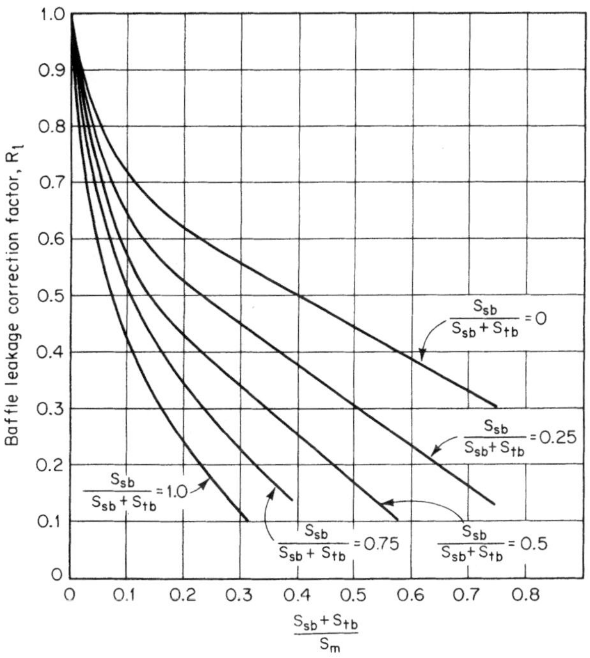


**FIG. 11-16** Correction factor for baffle-leakage effect on pressure drop.

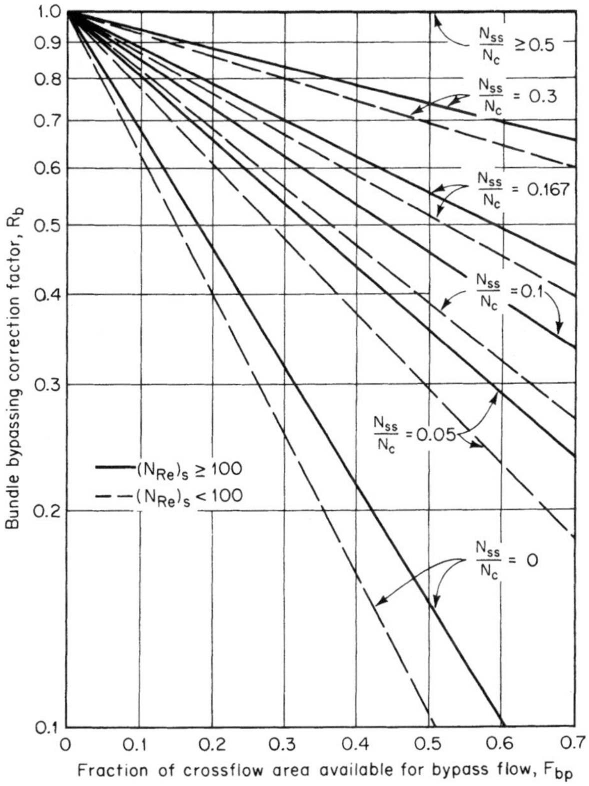


**FIG. 11-17** Correction factor on pressure drop for bypass flow.

6. Calculate the *pressure drop across the shell side* (excluding nozzles). Units for pressure drop are lbf/ft².

$$\Delta P_s = [(N_b - 1)(\Delta P_{bk})R_b + N_b\, \Delta P_{wk}]R_l + 2\, \Delta P_{bk} R_b \left(1 + \frac{N_{cw}}{N_c}\right) \tag{11-25}$$

The values of $h_s$ and $\Delta P_s$ calculated by this procedure are for clean exchangers and are intended to be as accurate as possible, not conservative. A fouled exchanger will generally give lower heat-transfer rates, as reflected by the dirt resistances incorporated into Eq. (11-2), and higher pressure drops. Some estimate of **fouling effects** on pressure drop may be made by using the methods just given by assuming that the fouling deposit blocks the leakage and possibly the bypass areas. The fouling may also decrease the clearance between tubes and significantly increase the pressure drop in cross-flow.

## THERMAL DESIGN OF CONDENSERS

### Single-Component Condensers

***Mean Temperature Difference*** In condensing a single component at its saturation temperature, the entire resistance to heat transfer on the condensing side is generally assumed to be in the layer of condensate. A mean condensing coefficient is calculated from the appropriate correlation and combined with the other resistances in Eq. (11-2). The overall coefficient is then used with the LMTD (no $F_T$ correction is necessary for isothermal condensation) to give the required area, even though the condensing coefficient and hence $U$ are not constant throughout the condenser.

If the vapor is **superheated** at the inlet, the vapor may first be desuperheated by sensible heat transfer from the vapor. This occurs if the surface temperature is above the saturation temperature, and a single-phase heat-transfer correlation is used. If the surface is below the saturation temperature, condensation will occur directly from the superheated vapor, and the effective coefficient is determined from the appropriate condensation correlation, using the saturation temperature in the LMTD. To determine whether or not condensation will occur directly from the superheated vapor, calculate the surface temperature by assuming single-phase heat transfer.

$$T_{\text{surface}} = T_{\text{vapor}} - \frac{U}{h}(T_{\text{vapor}} - T_{\text{coolant}}) \tag{11-26}$$

where $h$ is the sensible heat-transfer coefficient for the vapor, $U$ is calculated by using $h$, and both are on the same area basis. If $T_{\text{surface}} > T_{\text{saturation}}$, no condensation occurs at that point and the heat flux is actually higher than if $T_{\text{surface}} \leq T_{\text{saturation}}$ and condensation did occur. It is generally conservative to design a pure-component desuperheater-condenser as if the entire heat load were transferred by condensation, using the saturation temperature in the LMTD.

The design of an integral **condensate subcooling section** is more difficult, especially if close temperature approach is required. The condensate layer on the surface is on the average subcooled by one-third to one-half of the temperature drop across the film, and this is often sufficient if the condensate is not reheated by raining through the vapor. If the condensing-subcooling process is carried out inside tubes or in the shell of a vertical condenser, the single-phase subcooling section can be treated separately, giving an area that is added onto that needed for condensation. If the subcooling is achieved on the shell side of a horizontal condenser by flooding some of the bottom tubes with a weir or level controller, the rate and heat-balance equations must be solved for each section to obtain the area required.

**Pressure drop** on the condensing side reduces the final condensing temperature and the MTD and should always be checked. In designs requiring close approach between inlet coolant and exit condensate (subcooled or not), underestimation of pressure drop on the condensing side can lead to an exchanger that cannot meet specified terminal temperatures. Since pressure-drop calculations in two-phase flows such as condensation are relatively inaccurate, designers must consider carefully the consequences of a larger-than-calculated pressure drop.

***Horizontal In-Shell Condensers*** The mean **condensing coefficient** for the outside of a bank of horizontal tubes is calculated from Eq. (5-93) for a single tube, corrected for the number of tubes in a vertical row. For undisturbed laminar flow over all the tubes, Eq. (5-97) is, for realistic condenser sizes, overly conservative because of rippling, splashing, and turbulent flow (*Process Heat Transfer,* McGraw-Hill, New York, 1950). Kern proposed an exponent of −1⁄6 on the basis of experience, while Freon-11 data of Short and Brown (*General Discussion on Heat Transfer,* Institute of Mechanical Engineers, London, 1951) indicate independence of the number of tube rows. It seems reasonable to use no correction for inviscid liquids and Kern's correction for viscous condensates. For a cylindrical tube bundle, where $N$ varies, it is customary to take $N$ equal to two-thirds of the maximum or centerline value.

Baffles in a horizontal in-shell condenser are oriented with the cuts vertical to facilitate drainage and eliminate the possibility of flooding in the upward cross-flow sections. **Pressure drop** on the vapor side can be estimated by the data and method of Diehl and Unruh [*Pet. Refiner,* **36**(10), 147 (1957); **37**(10), 124 (1958)].

High vapor velocities across the tubes enhance the condensing coefficient. There is no correlation in the open literature to permit designers to take advantage of this. Since the vapor flow rate varies along the length, an incremental calculation procedure would be required in any case. In general, the pressure drops required to gain significant benefit are above those allowed in most process applications.

***Vertical In-Shell Condensers*** Condensers are often designed so that condensation occurs on the outside of vertical tubes. Equation (5-88) is valid as long as the condensate film is laminar. When it becomes turbulent, Fig. 5-10 or Colburn's equation [*Trans. Am. Inst. Chem. Eng.,* **30,** 187 (1933–1934)] may be used.

Some judgment is required in the use of these correlations because of construction features of the condenser. The tubes must be supported by baffles, usually with maximum cut (45 percent of the shell diameter) and maximum spacing to minimize pressure drop. The flow of the condensate is interrupted by the baffles, which may draw off or redistribute the liquid and which will also cause some splashing of free-falling drops onto the tubes.

For **subcooling,** a liquid inventory may be maintained in the bottom end of the shell by means of a weir or a liquid-level-controller. The subcooling heat-transfer coefficient is given by the correlations for natural convection on a vertical surface [Eqs. (5-33*a*), (5-33*b*)], with the pool assumed to be well mixed (isothermal) at the subcooled condensate exit temperature. Pressure drop may be estimated by the shell-side procedure.

***Horizontal In-Tube Condensers*** Condensation of a vapor inside horizontal tubes occurs in kettle and horizontal thermosiphon reboilers and in air-cooled condensers. In-tube condensation also offers certain advantages for condensation of multicomponent mixtures, discussed in the subsection "Multicomponent Condensers." The various in-tube correlations are closely connected to the **two-phase flow pattern** in the tube [*Chem. Eng. Prog. Symp. Ser.,* **66**(102), 150 (1970)]. At low flow rates, when gravity dominates the flow pattern, Eq. (5-101) may be used. At high flow rates, the flow and heat transfer are governed by vapor shear on the condensate film, and Eq. (5-100*a*) is valid. A simple and generally conservative procedure is to calculate the coefficient for a given case by both correlations and use the *larger* one.

Pressure drop during condensation inside horizontal tubes can be computed by using the correlations for two-phase flow given in Sec. 6 and neglecting the pressure recovery due to deceleration of the flow.

***Vertical In-Tube Condensation*** Vertical-tube condensers are generally designed so that vapor and liquid flow cocurrently downward; if pressure drop is not a limiting consideration, this configuration can result in higher heat-transfer coefficients than shell-side condensation and has particular advantages for multicomponent condensation. If gravity controls, the mean heat-transfer coefficient for condensation is given by Figs. 5-9 and 5-10. If vapor shear controls, Eq. (5-99*a*) is applicable. It is generally conservative to calculate the coefficients by both methods and choose the *higher* value. The pressure drop can be calculated by using the Lockhart-Martinelli method [*Chem. Eng. Prog.,* **45,** 39 (1945)] for friction loss, neglecting momentum and hydrostatic effects.

Vertical in-tube condensers are often designed for **reflux or knock-back application** in reactors or distillation columns. In this case, vapor flow is upward, countercurrent to the liquid flow on the tube wall; the vapor shear acts to thicken and retard the drainage of the condensate film, reducing the coefficient. Neither the fluid dynamics nor the heat transfer is well understood in this case, but Soliman, Schuster, and Berenson [*J. Heat Transfer,* **90,** 267–276 (1968)] discuss the problem and suggest a computational method. The Diehl-Koppany correlation [*Chem. Eng. Prog. Symp. Ser.* 92, **65** (1969)] may be used to estimate the maximum allowable vapor velocity at the tube inlet. If the vapor velocity is great enough, the liquid film will be carried upward; this design has been employed in a few cases in which only part of the stream is to be condensed. This velocity cannot be accurately computed, and a very conservative (high) outlet velocity must be used if unstable flow and flooding are to be avoided; 3 times the vapor velocity given by the Diehl-Koppany correlation for incipient flooding has been suggested as the design value for completely stable operation.

## Multicomponent Condensers

***Thermodynamic and Mass-Transfer Considerations*** *Multicomponent vapor mixture* includes several different cases: all the components may be liquids at the lowest temperature reached in the condensing side, or there may be components which dissolve substantially in the condensate even though their boiling points are below the exit temperature, or one or more components may be both noncondensable and nearly insoluble.

Multicomponent condensation always involves sensible-heat changes in the vapor and liquid along with the latent-heat load. Compositions of both phases in general change through the condenser, and **concentration gradients** exist in both phases. Temperature and concentration profiles and transport rates at a point in the condenser usually cannot be calculated, but the binary cases have been treated: condensation of one component in the presence of a completely insoluble gas [Colburn and Hougen, *Ind. Eng. Chem.,* **26,** 1178–1182 (1934); and Colburn and Edison, *Ind. Eng. Chem.,* **33,** 457–458 (1941)] and condensation of a binary vapor [Colburn and Drew, *Trans. Am. Inst. Chem. Eng.,* **33,** 196–215 (1937)]. It is necessary to know or calculate diffusion coefficients for the system, and a reasonable approximate method to avoid this difficulty and the reiterative calculations is desirable. To integrate the point conditions over the total condensation requires the temperature, composition enthalpy, and flow-rate profiles as functions of the heat removed. These are calculated from component thermodynamic data if the vapor and liquid are assumed to be in equilibrium at the local vapor temperature. This assumption is not exactly true, since the condensate and the liquid-vapor interface (where equilibrium does exist) are intermediate in temperature between the coolant and the vapor.

In calculating the condensing curve, it is generally assumed that the vapor and liquid flow collinearly and in intimate contact so that composition equilibrium is maintained between the total streams at all points. If, however, the condensate drops out of the vapor (as can happen in horizontal shell-side condensation) and flows to the exit without further interaction, the remaining vapor becomes excessively enriched in light components with a decrease in condensing temperature and in the temperature difference between vapor and coolant. The result may be not only a small reduction in the amount of heat transferred in the condenser but also an inability to condense totally the light ends even at reduced throughput or with the addition of more surface. To prevent the liquid from segregating, in-tube condensation is preferred in critical cases.

***Thermal Design*** If the controlling resistance for heat and mass transfer in the vapor is sensible-heat removal from the cooling vapor, the following design equation is obtained:

$$A = \int_0^{Q_T} \frac{1 + U'Z_H/h_{sv}}{U'(T_v - T_c)}\, dQ \qquad (11\text{-}27)$$

$U'$ is the overall heat-transfer coefficient between the vapor-liquid interface and the coolant, including condensate film, dirt and wall resistances, and coolant. The condensate film coefficient is calculated from the appropriate equation or correlation for pure vapor condensation for the geometry and flow regime involved, using mean liquid properties. $Z_H$ is the ratio of the sensible heat removed from the vapor-gas stream to the total heat transferred; this quantity is obtained from thermodynamic calculations and may vary substantially from one end of the condenser to the other, especially when removing vapor from a noncondensable gas. The sensible-heat-transfer coefficient for the vapor-gas stream $h_{sv}$ is calculated by using the appropriate correlation or design method for the geometry involved, neglecting the presence of the liquid. As the vapor condenses, this coefficient decreases and must be calculated at several points in the process. $T_v$ and $T_c$ are temperatures of the vapor and of the coolant respectively. This procedure is similar in principle to that of Ward [*Petro/Chem. Eng.*, **32**(11), 42–48 (1960)]. It may be nonconservative for condensing steam and other high-latent-heat substances, in which case it may be necessary to increase the calculated area by 25 to 50 percent.

Pressure drop on the condensing side may be estimated by judicious application of the methods suggested for pure-component condensation, taking into account the generally nonlinear decrease of vapor-gas flow rate with heat removal.

## THERMAL DESIGN OF REBOILERS

For a **single-component reboiler design,** attention is focused upon the mechanism of heat and momentum transfer at the hot surface. In *multicomponent systems,* the light components are preferentially vaporized at the surface, and the process becomes limited by their rate of diffusion. The net effect is to decrease the effective temperature difference between the hot surface and the bulk of the boiling liquid. If one attempts to vaporize too high a fraction of the feed liquid to the reboiler, the temperature difference between surface and liquid is reduced to the point that nucleation and vapor generation on the surface are suppressed and heat transfer to the liquid proceeds at the lower rate associated with single-phase natural convection. The only safe procedure in design for wide-boiling-range mixtures is to vaporize such a limited fraction of the feed that the boiling point of the remaining liquid mixture is still at least 5.5°C (10°F) below the surface temperature. Positive flow of the unvaporized liquid through and out of the reboiler should be provided.

**Kettle Reboilers** It has been generally assumed that kettle reboilers operate in the pool boiling mode, but with a lower peak heat flux because of vapor binding and blanketing of the upper tubes in the bundle. There is some evidence that vapor generation in the bundle causes a high circulation rate through the bundle. The result is that, at the lower heat fluxes, the kettle reboiler actually gives higher heat-transfer coefficients than a single tube. Present understanding of the recirculation phenomenon is insufficient to take advantage of this in design. Available nucleate pool boiling correlations are only very approximate, failing to account for differences in the nucleation characteristics of different surfaces. The Mostinski correlation [Eq. (5-102)] and the McNelly correlation [Eq. (5-103)] are generally the best for single components or narrow-boiling-range mixtures at low fluxes, though they may give errors of 40 to 50 percent. Experimental heat-transfer coefficients for pool boiling of a given liquid on a given surface should be used if available. The bundle **peak heat flux** is a function of tube-bundle geometry, especially of tube-packing density; in the absence of better information, the Palen-Small modification [Eq. (5-108)] of the Zuber maximum-heat-flux correlation is recommended.

A general method for analyzing kettle reboiler performance is by Fair and Klip, *Chem. Eng. Prog.* **79**(3), 86 (1983). It is effectively limited to computer application.

Kettle reboilers are generally assumed to require negligible pressure drop. It is important to provide good longitudinal liquid flow paths within the shell so that the liquid is uniformly distributed along the entire length of the tubes and excessive local vaporization and vapor binding are avoided.

This method may also be used for the thermal design of **horizontal thermosiphon reboilers.** The recirculation rate and pressure profile of the thermosiphon loop can be calculated by the methods of Fair [*Pet. Refiner,* **39**(2), 105–123 (1960)].

**Vertical Thermosiphon Reboilers** Vertical thermosiphon reboilers operate by natural circulation of the liquid from the still through the downcomer to the reboiler and of the two-phase mixture from the reboiler through the return piping. The flow is induced by the hydrostatic pressure imbalance between the liquid in the downcomer and the two-phase mixture in the reboiler tubes. Thermosiphons do not require any pump for recirculation and are generally regarded as less likely to foul in service because of the relatively high two-phase velocities obtained in the tubes. Heavy components are not likely to accumulate in the thermosiphon, but they are more difficult to design satisfactorily than kettle reboilers, especially in vacuum operation. Several shortcut methods have been suggested for thermosiphon design, but they must generally be used with caution. The method due to Fair (loc. cit.), based upon two-phase flow correlations, is the most complete in the open literature but requires a computer for practical use. Fair also suggests a shortcut method that is satisfactory for preliminary design and can be reasonably done by hand.

**Forced-Recirculation Reboilers** In forced-recirculation reboilers, a pump is used to ensure circulation of the liquid past the heattransfer surface. Force-recirculation reboilers may be designed so that boiling occurs inside vertical tubes, inside horizontal tubes, or on the shell side. For forced boiling inside vertical tubes, Fair's method (loc. cit.) may be employed, making only the minor modification that the recirculation rate is fixed and does not need to be balanced against the pressure available in the downcomer. Excess pressure required to circulate the two-phase fluid through the tubes and back into the column is supplied by the pump, which must develop a positive pressure increase in the liquid.

Fair's method may also be modified to design forced-recirculation reboilers with horizontal tubes. In this case the hydrostatic-head-pressure effect through the tubes is zero but must be considered in the two-phase return lines to the column.

The same procedure may be applied in principle to design of forced-recirculation reboilers with shell-side vapor generation. Little is known about two-phase flow on the shell side, but a reasonable estimate of the friction pressure drop can be made from the data of Diehl and Unruh [*Pet. Refiner,* **36**(10), 147 (1957); **37**(10), 124 (1958)]. No void-fraction data are available to permit accurate estimation of the hydrostatic or acceleration terms. These may be roughly estimated by assuming homogeneous flow.

## THERMAL DESIGN OF EVAPORATORS

Heat duties of evaporator heating surfaces are usually determined by conventional heat and material balance calculations. Heating surface areas are normally, but not always taken as those in contact with the material being evaporated. It is the heat transfer $\Delta T$ that presents the most difficulty in deriving or applying heat-transfer coefficients. The total $\Delta T$ between heat source and heat sink is never all available for heat transfer. Since energy usually is carried to and from an evaporator body or effect by condensible vapors, loss in pressure represents a loss in $\Delta T$. Such losses include pressure drop through entrainment separators, friction in vapor piping, and acceleration losses into and out of the piping. The latter loss has often been overlooked, even though it can be many times greater than the friction loss. Similarly, friction and acceleration losses past the heating surface, such as in a falling film evaporator, cause a loss of $\Delta T$ that may or may not have been included in the heat transfer $\Delta T$ when reporting experimental results. Boiling-point rise, the difference between the boiling point of the solution and the condensing point of the solvent at the same pressure, is another loss. Experimental data are almost always corrected for boiling-point rise, but plant data are suspect when based on temperature measurements because vapor at the point of measurement may still contain some superheat, which represents but a very small fraction of the heat given up when the vapor condenses but may represent a substantial fraction of the actual net $\Delta T$ available for heat transfer. A $\Delta T$ loss that must be considered in forced-circulation evaporators is that due to temperature rise through the heater, a consequence of the heat being absorbed there as sensible heat. A further loss may occur when the heater effluent flashes as it enters the vapor-liquid separator. Some of the liquid may not reach the surface and flash to equilibrium with the vapor pressure in the separator, instead of recirculating to the heater, raising the average temperature at which heat is absorbed and further reducing the net $\Delta T$. Whether or not these $\Delta T$ losses are allowed for in the heat-transfer coefficients reported depends on the method of measurement. Simply basing the liquid temperature on the measured vapor head pressure may ignore both—or only the latter if temperature rise through the heater is estimated separately from known heat input and circulation rate. In general, when calculating overall heat-transfer coefficients from individual-film coefficients, all of these losses must be allowed for, while when using reported overall coefficients care must be exercised to determine which losses may already have been included in the heat transfer $\Delta T$.

**Forced-Circulation Evaporators** In evaporators of this type in which hydrostatic head prevents boiling at the heating surface, **heat-transfer coefficients** can be predicted from the usual correlations for condensing steam (Fig. 5-10) and forced-convection sensible heating [Eq. (5-50)]. The liquid film coefficient is improved if boiling is not completely suppressed. When only the film next to the wall is above the boiling point, Boarts, Badger, and Meisenberg [*Ind. Eng. Chem.*, **29,** 912 (1937)] found that results could be correlated by Eq. (5-50) by using a constant of 0.0278 instead of 0.023. In such cases, the course of the liquid temperature can still be calculated from known circulation rate and heat input.

When the bulk of the liquid is boiling in part of the tube length, the film coefficient is even higher. However, the liquid temperature starts dropping as soon as full boiling develops, and it is difficult to estimate the course of the temperature curve. It is certainly safe to estimate heat transfer on the basis that no bulk boiling occurs. Fragen and Badger [*Ind. Eng. Chem.*, **28,** 534 (1936)] obtained an **empirical correlation** of overall heat-transfer coefficients in this type of evaporator, based on the $\Delta T$ at the heater inlet:

In U.S. customary units

$$U = 2020D^{0.57}(V_s)^{3.6/L}/\mu^{0.25}\,\Delta T^{0.1} \qquad (11\text{-}28)$$

where $D$ = mean tube diameter, $V_s$ = inlet velocity, $L$ = tube length, and $\mu$ = liquid viscosity. This equation is based primarily on experiments with copper tubes of 0.022 m (8/8 in) outside diameter, 0.00165 m (16 gauge), 2.44 m (8 ft) long, but it includes some work with 0.0127-m (½-in) tubes 2.44 m (8 ft) long and 0.0254-m (1-in) tubes 3.66 m (12 ft) long.

**Long-Tube Vertical Evaporators** In the rising-film version of this type of evaporator, there is usually a nonboiling zone in the bottom section and a boiling zone in the top section. The length of the nonboiling zone depends on heat-transfer characteristics in the two zones and on pressure drop during two-phase flow in the boiling zone. The work of Martinelli and coworkers [Lockhart and Martinelli, *Chem. Eng. Prog.*, **45,** 39–48 (January 1949); and Martinelli and Nelson, *Trans. Am. Soc. Mech. Eng.*, **70,** 695–702 (August 1948)] permits a prediction of pressure drop, and a number of correlations are available for estimating film coefficients of heat transfer in the two zones. In estimating pressure drop, integrated curves similar to those presented by Martinelli and Nelson are the easiest to use. The curves for pure water are shown in Figs. 11-18 and 11-19, based on the assumption that the flow of both vapor and liquid would be turbulent if each were flowing alone in the tube. Similar curves can be prepared if one or both flows are laminar or if the properties of the liquid differ appreciably from the properties of pure water. The **acceleration pressure drop** $\Delta P_a$ is calculated from the equation

$$\Delta P_a = br_2G^2/32.2 \qquad (11\text{-}29)$$

where $b = (2.6)(10^7)$(SI) and 1.0 (U.S. customary) and using $r_2$ from Fig. 11-18. The frictional pressure drop is derived from Fig. 11-19, which shows the ratio of two-phase pressure drop to that of the entering liquid flowing alone.

Pressure drop due to hydrostatic head can be calculated from liquid holdup $R_1$. For nonfoaming dilute aqueous solutions, $R_1$ can be estimated from $R_1 = 1/[1 + 2.5(V/L)(\rho_1/\rho_v)^{1/2}]$. Liquid holdup, which represents the ratio of liquid-only velocity to actual liquid velocity, also appears to be the principal determinant of the convective coefficient in the boiling zone (Dengler, Sc.D. thesis, MIT, 1952). In other words, the convective coefficient is that calculated from Eq. (5-50) by using the liquid-only velocity divided by $R_1$ in the Reynolds number. Nucleate boiling augments convective heat transfer, primarily when $\Delta T$'s are high and the convective coefficient is low [Chen, *Ind. Eng. Chem. Process Des. Dev.*, **5,** 322 (1966)].

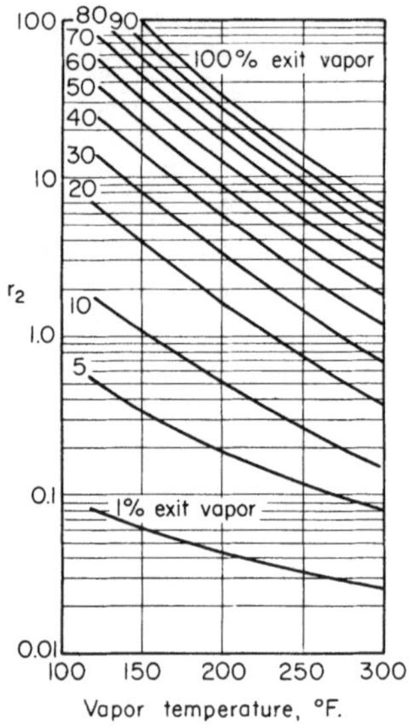


**FIG. 11-18** Acceleration losses in boiling flow. °C = (°F − 32)/1.8.

Film coefficients for the **boiling of liquids other than water** have been investigated. Coulson and McNelly [*Trans. Inst. Chem. Eng.*, **34,** 247 (1956)] derived the following relation, which also correlated the data of Badger and coworkers [*Chem. Metall. Eng.*, **46,** 640 (1939); *Chem. Eng.*, **61**(2), 183 (1954); and *Trans. Am. Inst. Chem. Eng.*, **33,** 392 (1937); **35,** 17 (1939); **36,** 759 (1940)] on water:

$$N_{\text{Nu}} = (1.3 + b\,D)(N_{\text{Pr}})_l^{0.9}(N_{\text{Re}})_l^{0.23}(N_{\text{Re}})_g^{0.34}\left(\frac{\rho_l}{\rho_g}\right)^{0.25}\left(\frac{\mu_g}{\mu_l}\right) \qquad (11\text{-}30)$$

where $b$ = 128 (SI) or 39 (U.S. customary), $N_{\text{Nu}}$ = Nusselt number based on liquid thermal conductivity, $D$ = tube diameter, and the remaining terms are dimensionless groupings of liquid Prandtl number, liquid Reynolds number, vapor Reynolds number, and ratios of densities and viscosities. The Reynolds numbers are calculated on the basis of each fluid flowing by itself in the tube.

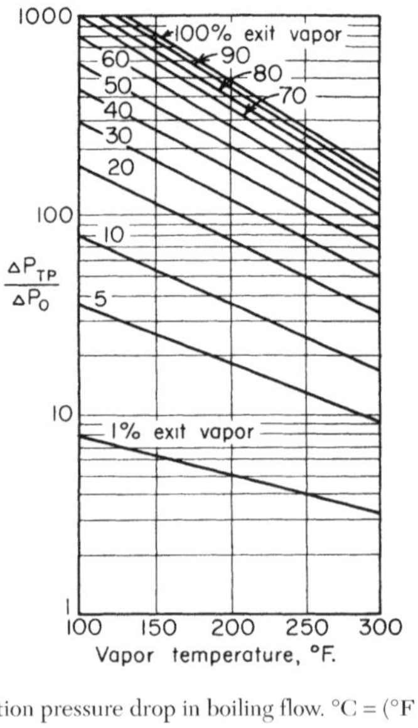


**FIG. 11-19** Friction pressure drop in boiling flow. °C = (°F − 32)/1.8.

Additional corrections must be applied when the fraction of vapor is so high that the remaining liquid does not wet the tube wall or when the velocity of the mixture at the tube exits approaches sonic velocity. McAdams, Woods, and Bryan (*Trans. Am. Soc. Mech. Eng.*, 1940), Dengler and Addoms (loc. cit.), and Stroebe, Baker, and Badger [*Ind. Eng. Chem.*, **31,** 200 (1939)] encountered dry-wall conditions and reduced coefficients when the weight fraction of vapor exceeded about 80 percent. Schweppe and Foust [*Chem. Eng. Prog.*, **49,** *Symp. Ser.* 5, 77 (1953)] and Harvey and Foust (ibid., p. 91) found that "sonic choking" occurred at surprisingly low flow rates.

The simplified method of calculation outlined includes no allowance for the **effect of surface tension.** Stroebe, Baker, and Badger (loc. cit.) found that by adding a small amount of surface-active agent the boiling-film coefficient varied inversely as the square of the surface tension. Coulson and Mehta [*Trans. Inst. Chem. Eng.*, **31,** 208 (1953)] found the exponent to be −1.4. The higher coefficients at low surface tension are offset to some extent by a higher pressure drop, probably because the more intimate mixture existing at low surface tension causes the liquid fraction to be accelerated to a velocity closer to that of the vapor. The pressure drop due to acceleration $\Delta P_a$ derived from Fig. 11-18 allows for some slippage. In the limiting case, such as might be approached at low surface tension, the acceleration pressure drop in which "fog" flow is assumed (no slippage) can be determined from the equation

$$\Delta P'_a = \frac{y(V_g - V_l)G^2}{g_c} \tag{11-31}$$

where $y$ = fraction vapor by weight
$V_g$, $V_l$ = specific volume gas, liquid
$G$ = mass velocity

While the foregoing methods are valuable for detailed evaporator design or for evaluating the effect of changes in conditions on performance, they are cumbersome to use when making preliminary designs or cost estimates. Figure 11-20 gives the general range of **overall long-tube vertical- (LTV) evaporator heat-transfer coefficients** usually encountered in commercial practice. The higher coefficients are encountered when evaporating dilute solutions and the lower range when evaporating viscous liquids. The dashed curve represents the approximate lower limit, for liquids with viscosities of about 0.1 Pa·s (100 cP). The LTV evaporator does not work well at low temperature differences, as indicated by the results shown in Fig. 11-21 for seawater in 0.051-m (2-in), 0.0028-m (12-gauge) brass tubes 7.32 m (24 ft) long (W. L. Badger Associates, Inc., U.S. Department of the Interior, Office of Saline Water Rep. 26, December 1959, OTS Publ. PB 161290). The feed was at its boiling point at the vapor-head pressure, and feed rates varied from 0.025 to 0.050 kg/(s·tube) [200 to 400 lb/(h·tube)] at the higher temperature to 0.038 to 0.125 kg/(s·tube) [300 to 1000 lb/(h·tube)] at the lowest temperature.

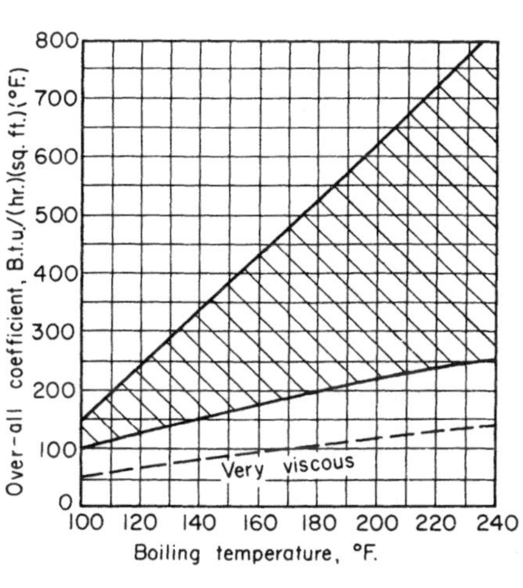


**FIG. 11-20** General range of long-tube vertical- (LTV) evaporator coefficients. °C = (°F − 32)/1.8; to convert British thermal units per hour-square foot-degrees Fahrenheit to joules per square meter-second-kelvins, multiply by 5.6783.

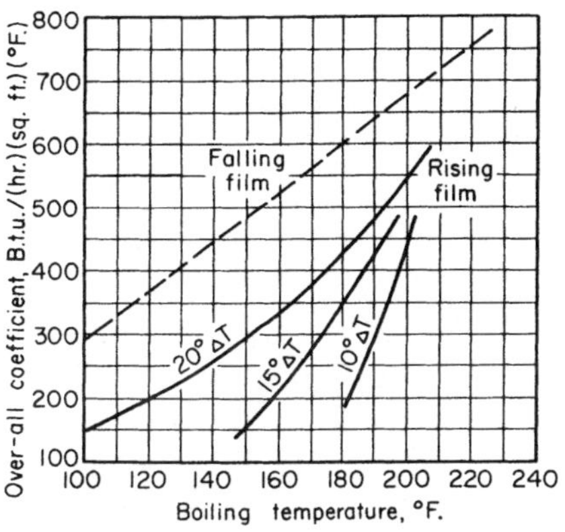


**FIG. 11-21** Heat-transfer coefficients in LTV seawater evaporators. °C = (°F − 32)/1.8; to convert British thermal units per hour square foot-degrees Fahrenheit to joules per square meter-second-kelvins, multiply by 5.6783.

Falling film evaporators find their widest use at low temperature differences—also at low temperatures. Under most operating conditions encountered, heat transfer is almost all by pure convection, with a negligible contribution from nucleate boiling. Film coefficients on the condensing side may be estimated from Dukler's correlation, [*Chem. Eng. Prog.* **55,** 62 1950]. The same Dukler correlation presents curves covering falling film heat transfer to nonboiling liquids that are equally applicable to the falling film evaporator [Sinek and Young, *Chem. Eng. Prog.* **58,** No. 12, 74 (1962)]. Kunz and Yerazunis [*J. Heat Transfer* **8,** 413 (1969)] have since extended the range of physical properties covered, as shown in Fig. 11-22. The boiling point in the tubes of such an evaporator is higher than in the vapor head because of both frictional-pressure drop and the head needed to accelerate the vapor to the tube-exit velocity. These factors, which can easily be predicted, make the overall apparent coefficients somewhat lower than those for nonboiling conditions. Figure 11-21 shows overall apparent heat-transfer coefficients determined in a falling-film seawater evaporator using the same tubes and flow rates as for the rising-film tests (W. L. Badger Associates, Inc., loc. cit.).

**Short-Tube Vertical Evaporators** Coefficients can be estimated by the same detailed method described for recirculating LTV evaporators. Performance is primarily a function of temperature level, temperature difference, and viscosity. While liquid level can also have an important influence, this is usually encountered only at levels lower than considered safe in commercial operation. **Overall heat-transfer coefficients** are shown in Fig. 11-23 for a basket-type evaporator (one with an annular downtake) when boiling water with 0.051-m (2-in) outside-diameter 0.0028-m-wall (12-gauge), 1.22-m-(4-ft-) long steel tubes [Badger and Shepard, *Chem. Metall. Eng.*, **23,** 281 (1920)]. Liquid level was maintained at the top tube sheet. Foust, Baker, and Badger [*Ind. Eng. Chem.*, **31,** 206 (1939)] measured recirculating velocities and heat-transfer coefficients in the same evaporator except with 0.064-m (2.5-in) 0.0034-m-wall (10-gauge), 1.22-m- (4-ft-) long tubes and temperature differences from 7 to 26°C (12 to 46°F). In the normal range of liquid levels, their results can be expressed as

$$U_c = \frac{b(\Delta T_c)^{0.22} N_{Pr}^{0.4}}{(V_g - V_l)^{0.37}} \tag{11-32}$$

where $b$ = 153 (SI) or 375 (U.S. customary) and the subscript $c$ refers to true liquid temperature, which under these conditions was about 0.56°C (1°F) above the vapor-head temperature. This work was done with water.

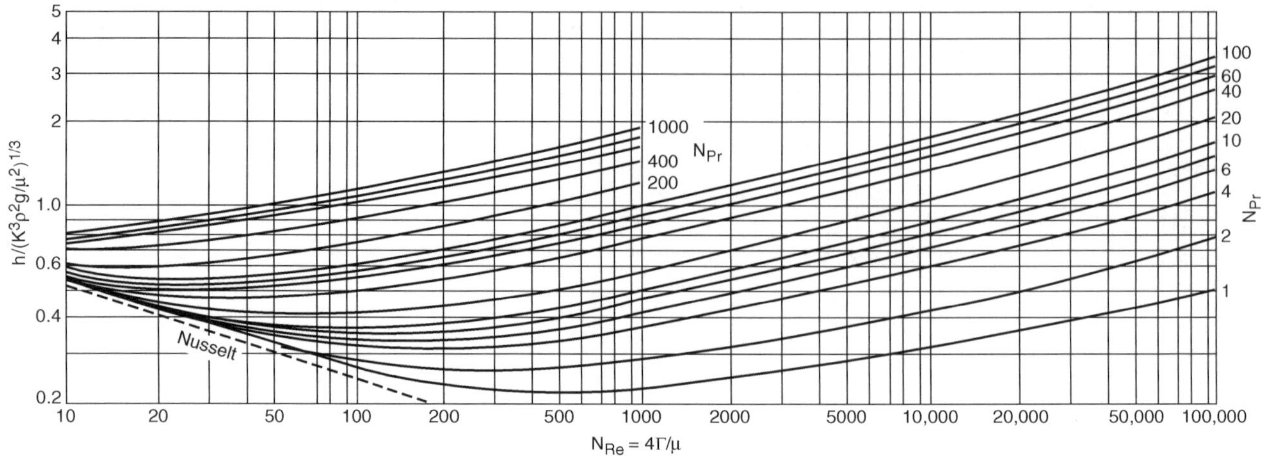


**FIG. 11-22** Kunz and Yerazunis Correlation for falling-film heat transfer.

No detailed tests have been reported for the performance of propeller calandrias. Not enough is known regarding the performance of the propellers themselves under the cavitating conditions usually encountered to permit predicting circulation rates. In many cases, it appears that the propeller does no good in accelerating heat transfer over the transfer for natural circulation (Fig. 11-23).

**Miscellaneous Evaporator Types** **Horizontal-tube evaporators** operating with partially or fully submerged heating surfaces behave in much the same way as short-tube verticals, and heat-transfer coefficients are of the same order of magnitude. Some test results for water were published by Badger [*Trans. Am. Inst. Chem. Eng.*, **13,** 139 (1921)]. When operating unsubmerged, their heat transfer performance is roughly comparable to the falling-film vertical tube evaporator. Condensing coefficients inside the tubes can be derived from Nusselt's theory which, based on a constant-heat flux rather than a constant film $\Delta T$, gives:

$$\frac{h}{(k^3\rho^2g/\mu^2)^{1/3}} = 1.59(4\Gamma/\mu)^{-1/3} \qquad (11\text{-}33a)$$

For the boiling side, a correlation based on seawater tests gave:

$$\frac{h}{(k^3\rho^2g/\mu^2)^{1/3}} = 0.0147(4\Gamma/\mu)^{1/3}(D)^{-1/3} \qquad (11\text{-}33b)$$

where $\Gamma$ is based on feed-rate per unit length of the top tube in each vertical row of tubes and $D$ is in meters.

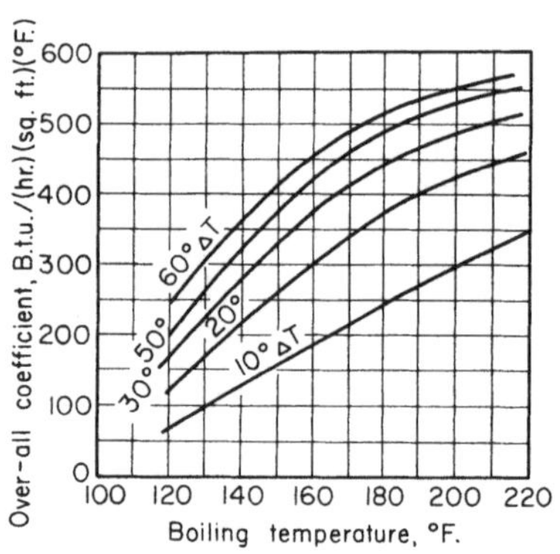


**FIG. 11-23** Heat-transfer coefficients for water in short-tube evaporators. °C = (°F − 32)/1.8; to convert British thermal units per hour-square foot-degrees Fahrenheit to joules per square meter-second-kelvins, multiply by 5.6783.

Heat-transfer coefficients in clean coiled-tube evaporators for seawater are shown in Fig. 11-24 [Hillier, *Proc. Inst. Mech. Eng. (London)*, **1B**(7), 295 (1953)]. The tubes were of copper.

Heat-transfer coefficients in **agitated-film evaporators** depend primarily on liquid viscosity. This type is usually justifiable only for very viscous materials. Figure 11-25 shows general ranges of overall coefficients [Hauschild, *Chem. Ing. Tech.*, **25,** 573 (1953); Lindsey, *Chem. Eng.*, **60**(4), 227 (1953); and Leniger and Veldstra, *Chem. Ing. Tech.*, **31,** 493 (1959)]. When used with nonviscous fluids, a wiped-film evaporator having fluted external surfaces can exhibit very high coefficients (Lustenader et al., *Trans. Am. Soc. Mech. Eng.*, Paper 59-SA-30, 1959), although at a probably unwarranted first cost.

**Heat Transfer from Various Metal Surfaces** In an early work, Pridgeon and Badger [*Ind. Eng. Chem.*, **16,** 474 (1924)] published test results on copper and iron tubes in a horizontal-tube evaporator that indicated an extreme **effect of surface cleanliness** on heat-transfer coefficients. However, the high degree of cleanliness needed for high coefficients was difficult to achieve, and the tube layout and liquid level were changed during the course of the tests so as to make direct comparison of results difficult. Other workers have found little or no effect of conditions of surface or tube material on boiling-film

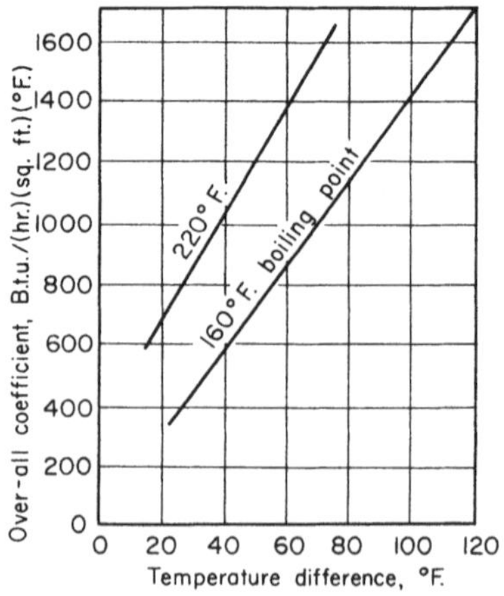


**FIG. 11-24** Heat-transfer coefficients for seawater in coil-tube evaporators. °C = (°F − 32)/1.8; to convert British thermal units per hour-square foot-degrees Fahrenheit to joules per square meter-second-kelvins, multiply by 5.6783.

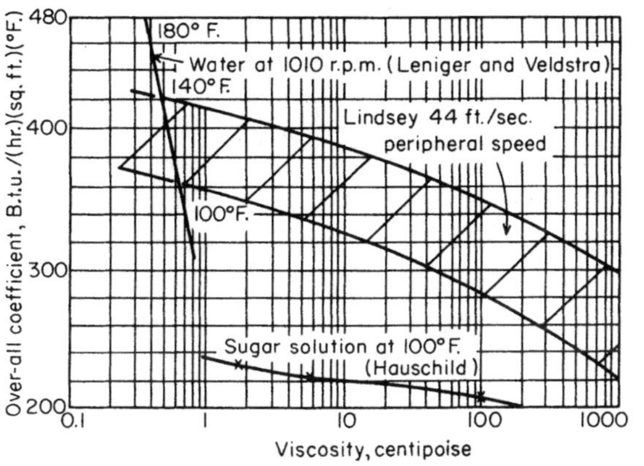


**FIG. 11-25** Overall heat-transfer coefficients in agitated-film evaporators. °C = (°F − 32)/1.8; to convert British thermal units per hour square foot degrees Fahrenheit to joules per square meter-second-kelvins, multiply by 5.6783; to convert centipoises to pascal-seconds, multiply by $10^{-3}$.

coefficients in the range of commercial operating conditions [Averin, *Izv. Akad. Nauk SSSR Otd. Tekh. Nauk*, no. 3, p. 116, 1954; and Coulson and McNelly, *Trans. Inst. Chem. Eng.*, **34**, 247 (1956)].

Work in connection with desalination of seawater has shown that **specially modified surfaces** can have a profound effect on heat-transfer coefficients in evaporators. Figure 11-26 (Alexander and Hoffman, Oak Ridge National Laboratory TM-2203) compares overall coefficients for some of these surfaces when boiling fresh water in 0.051-m (2-in) tubes 2.44-m (8-ft) long at atmospheric pressure in both upflow and downflow. The area basis used was the nominal outside area. Tube 20 was a smooth 0.0016-m- (0.062-in-) wall aluminum brass tube that had accumulated about 6 years of fouling in seawater

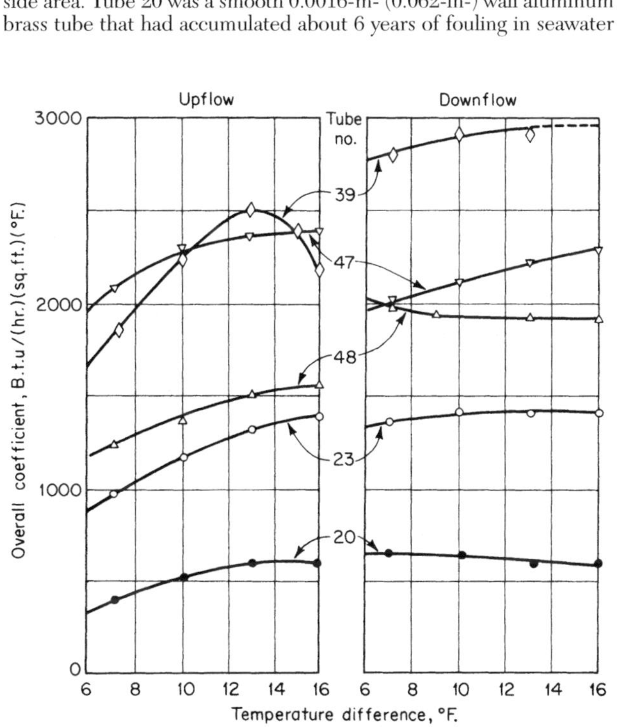


**FIG. 11-26** Heat-transfer coefficients for enhanced surfaces. °C = (°F − 32)/1.8; to convert British thermal units per hour-square foot-degrees Fahrenheit to joules per square meter-second-kelvins, multiply by 5.6783. *(By permission from Oak Ridge National Laboratory TM-2203.)*

service and exhibited a fouling resistance of about $(2.6)(10^{-5})$ $(m^2 \cdot s \cdot K)/J$ [0.00015 $(ft^2 \cdot h \cdot °F)/Btu$]. Tube 23 was a clean aluminum tube with 20 spiral corrugations of 0.0032-m (⅛-in) radius on a 0.254-m (10-in) pitch indented into the tube. Tube 48 was a clean copper tube that had 50 longitudinal flutes pressed into the wall (General Electric double-flute profile, Diedrich, U.S. Patent 3,244,601, Apr. 5, 1966). Tubes 47 and 39 had a specially patterned porous sintered-metal deposit on the boiling side to promote nucleate boiling (Minton, U.S. Patent 3,384,154, May 21, 1968). Both of these tubes also had steam-side coatings to promote dropwise condensation—parylene for tube 47 and gold plating for tube 39.

Of these special surfaces, only the **double-fluted tube** has seen extended services. Most of the gain in heat-transfer coefficient is due to the condensing side; the flutes tend to collect the condensate and leave the lands bare [Carnavos, *Proc. First Int. Symp. Water Desalination*, **2**, 205 (1965)]. The condensing-film coefficient (based on the actual outside area, which is 28 percent greater than the nominal area) may be approximated from the equation

$$h = b\left(\frac{k^3\rho^2 g}{\mu^2}\right)^{1/3}\left(\frac{\mu\lambda}{L}\right)^{1/3}\left(\frac{q}{A}\right)^{-0.833} \tag{11-34a}$$

where $b$ = 2100 (SI) or 1180 (U.S. customary). The boiling-side coefficient (based on actual inside area) for salt water in downflow may be approximated from the equation

$$h = 0.035(k^3\rho^2 g/\mu^2)^{1/3}(4\Gamma/\mu)^{1/3} \tag{11-34b}$$

The boiling-film coefficient is about 30 percent lower for pure water than it is for salt water or seawater. There is as yet no accepted explanation for the superior performance in salt water. This phenomenon is also seen in evaporation from smooth tubes.

**Effect of Fluid Properties on Heat Transfer** Most of the heat-transfer data reported in the preceding paragraphs were obtained with water or with dilute solutions having properties close to those of water. Heat transfer with other materials will depend on the type of evaporator used. For forced-circulation evaporators, methods have been presented to calculate the effect of changes in fluid properties. For natural-circulation evaporators, **viscosity** is the most important variable as far as aqueous solutions are concerned. Badger (*Heat Transfer and Evaporation*, Chemical Catalog, New York, 1926, pp. 133–134) found that, as a rough rule, overall heat-transfer coefficients varied in inverse proportion to viscosity if the boiling film was the main resistance to heat transfer. When handling molasses solutions in a forced-circulation evaporator in which boiling was allowed to occur in the tubes, Coates and Badger [*Trans. Am. Inst. Chem. Eng.*, **32**, 49 (1936)] found that from 0.005 to 0.03 Pa·s (5 to 30 cP) the overall heat-transfer coefficient could be represented by $U = b/\mu_f^{1.24}$, where $b$ = 2.55 (SI) or 7043 (U.S. customary). Fragen and Badger [*Ind. Eng. Chem.*, **28**, 534 (1936)] correlated overall coefficients on sugar and sulfite liquor in the same evaporator for viscosities to 0.242 Pa·s (242 cP) and found a relationship that included the viscosity raised only to the 0.25 power.

Little work has been published on the effect of viscosity on heat transfer in the long-tube vertical evaporator. Cessna, Leintz, and Badger [*Trans. Am. Inst. Chem. Eng.*, **36**, 759 (1940)] found that the overall coefficient in the nonboiling zone varied inversely as the 0.7 power of viscosity (with sugar solutions). Coulson and Mehta [*Trans. Inst. Chem. Eng.*, **31**, 208 (1953)] found the exponent to be −0.44, and Stroebe, Baker, and Badger (loc. cit.) arrived at an exponent of −0.3 for the effect of viscosity on the film coefficient in the boiling zone.

Kerr (Louisiana Agr. Exp. Sta. Bull. 149) obtained plant data shown in Fig. 11-27 on various types of full-sized evaporators for cane sugar. These are invariably forward-feed evaporators concentrating to about 50° Brix, corresponding to a viscosity on the order of 0.005 Pa·s (5 cP) in the last effect. In Fig. 11-27 curve *A* is for short-tube verticals with central downtake, *B* is for standard horizontal tube evaporators, *C* is for Lillie evaporators (which were horizontal-tube machines with no liquor level but having recirculating liquor showered over the tubes), and *D* is for long-tube vertical evaporators. These curves show apparent coefficients, but sugar solutions have boiling-point rises low enough not to affect the results noticeably. Kerr also obtained the data

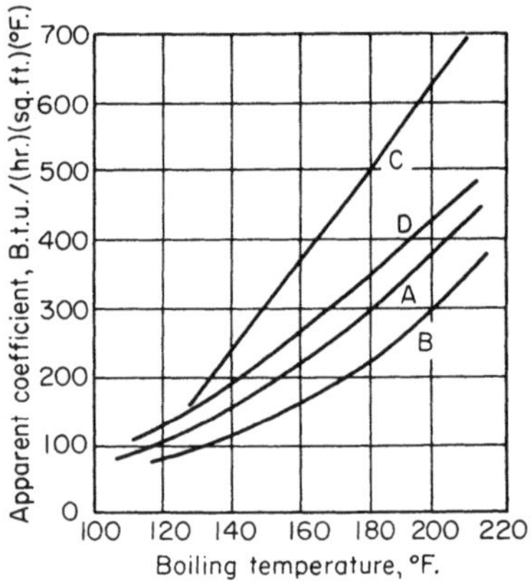


**FIG. 11-27** Kerr's tests with full-sized sugar evaporators. °C = (°F − 32)/1.8; to convert British thermal units per hour-square foot-degrees Fahrenheit to joules per square meter-second-kelvins, multiply by 5.6783.

shown in Fig. 11-28 on a laboratory short-tube vertical evaporator with 0.44- by 0.61-m (1¾- by 24-in) tubes. This work was done with sugar juices boiling at 57°C (135°F) and an 11°C (20°F) temperature difference.

**Effect of Noncondensables on Heat Transfer** Most of the heat transfer in evaporators does not occur from pure steam but from vapor evolved in a preceding effect. This vapor usually contains inert gases—from air leakage if the preceding effect was under vacuum, from air entrained or dissolved in the feed, or from gases liberated by decomposition reactions. To prevent these inerts from seriously impeding heat transfer, the gases must be channeled past the heating surface and vented from the system while the gas concentration is still quite low. The influence of inert gases on heat transfer is due partially to the effect on $\Delta T$ of lowering the partial pressure and hence condensing temperature of the steam. The primary effect, however, results from the formation at the heating surface of an insulating blanket of gas through which the steam must diffuse before it can condense. The latter effect can be treated as an added resistance or fouling factor equal to $6.5 \times 10^{-5}$ times the local mole percent inert gas (in $J^{-1} \cdot s \cdot m^2 \cdot K$) [Standiford, *Chem. Eng. Prog.*, **75,** 59–62 (July 1979)]. The effect on $\Delta T$ is readily calculated from Dalton's law. Inert-gas concentrations may vary by a factor of 100 or more between vapor inlet and vent outlet, so these relationships should be integrated through the tube bundle.

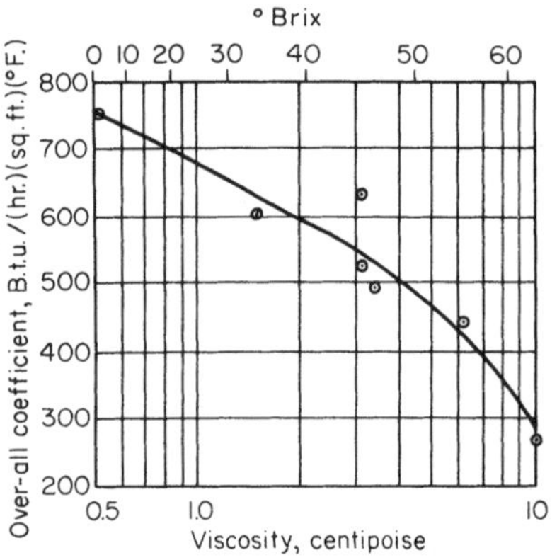


**FIG. 11-28** Effect of viscosity on heat transfer in short-tube vertical evaporator. To convert centipoises to pascal-seconds, multiply by $10^{-3}$; to convert British thermal units per hour-square foot-degrees Fahrenheit to joules per square meter-second-kelvins, multiply by 5.6783.

## BATCH OPERATIONS: HEATING AND COOLING OF VESSELS

**Nomenclature** (Use consistent units.) $A$ = heat-transfer surface; $C$, $c$ = specific heats of hot and cold fluids respectively; $L_0$ = flow rate of liquid added to tank; $M$ = mass of fluid in tank; $T$, $t$ = temperature of hot and cold fluids respectively; $T_1$, $t_1$ = temperatures at beginning of heating or cooling period or at inlet; $T_2$, $t_2$ = temperature at end of period or at outlet; $T_0$, $t_0$ = temperature of liquid added to tank; $U$ = coefficient of heat transfer; and $W$, $w$ = flow rate through external exchanger of hot and cold fluids respectively.

**Applications** One typical application in heat transfer with batch operations is the heating of a reactor mix, maintaining temperature during a reaction period, and then cooling the products after the reaction is complete. This subsection is concerned with the heating and cooling of such systems in either unknown or specified periods.

The technique for deriving expressions relating time for heating or cooling agitated batches to coil or jacket area, heat-transfer coefficients, and the heat capacity of the vessel contents was developed by Bowman, Mueller, and Nagle [*Trans. Am. Soc. Mech. Eng.*, **62,** 283–294 (1940)] and extended by Fisher [*Ind. Eng. Chem.*, **36,** 939–942 (1944)] and Chaddock and Sanders [*Trans. Am. Inst. Chem. Eng.*, **40,** 203–210 (1944)] to external heat exchangers. Kern (*Process Heat Transfer*, McGraw-Hill, New York, 1950, Chap. 18) collected and published the results of these investigators.

The assumptions made were that (1) $U$ is constant for the process and over the entire surface, (2) liquid flow rates are constant, (3) specific heats are constant for the process, (4) the heating or cooling medium has a constant inlet temperature, (5) agitation produces a uniform batch fluid temperature, (6) no partial phase changes occur, and (7) heat losses are negligible. The developed equations are as follows. If any of the assumptions do not apply to a system being designed, new equations should be developed or appropriate corrections made. Heat exchangers are counterflow except for the 1-2 exchangers, which are one-shell-pass, two-tube-pass, parallel-flow counterflow.

***Coil-in-Tank or Jacketed Vessel: Isothermal Heating Medium***

$$\ln (T_1 - t_1)/(T_1 - t_2) = UA\theta/Mc \qquad (11\text{-}35)$$

***Cooling-in-Tank or Jacketed Vessel: Isothermal Cooling Medium***

$$\ln (T_1 - t_1)/(T_2 - t_1) = UA\theta/MC \qquad (11\text{-}35a)$$

***Coil-in-Tank or Jacketed Vessel: Nonisothermal Heating Medium***

$$\ln \frac{T_1 - t_1}{T_1 - t_2} = \frac{WC}{Mc}\left(\frac{K_1 - 1}{K_1}\right)\theta \qquad (11\text{-}35b)$$

where $K_1 = e^{UA/WC}$

***Coil-in-Tank: Nonisothermal Cooling Medium***

$$\ln \frac{T_1 - t_1}{T_2 - t_1} = \frac{wc}{MC}\left(\frac{K_2 - 1}{K_2}\right)\theta \qquad (11\text{-}35c)$$

where $K_2 = e^{UA/wc}$

***External Heat Exchanger: Isothermal Heating Medium***

$$\ln \frac{T_1 - t_1}{T_1 - t_2} = \frac{wc}{Mc}\left(\frac{K_2 - 1}{K_2}\right)\theta \qquad (11\text{-}35d)$$

***External Exchanger: Isothermal Cooling Medium***

$$\ln \frac{T_1 - t_1}{T_2 - t_1} = \frac{WC}{MC}\left(\frac{K_1 - 1}{K_1}\right)\theta \qquad (11\text{-}35e)$$

***External Exchanger: Nonisothermal Heating Medium***

$$\ln \frac{T_1 - t_1}{T_1 - t_2} = \left(\frac{K_3 - 1}{M}\right)\left(\frac{wWC}{K_3 wc - WC}\right)\theta \qquad (11\text{-}35f)$$

where $K_3 = e^{UA(1/WC - 1/wc)}$

***External Exchanger: Nonisothermal Cooling Medium***

$$\ln \frac{T_1 - t_1}{T_2 - t_1} = \left(\frac{K_4 - 1}{M}\right)\left(\frac{Wwc}{K_4 wc - WC}\right)\theta \qquad (11\text{-}35g)$$

where $K_4 = e^{UA(1/WC - 1/wc)}$

***External Exchanger with Liquid Continuously Added to Tank: Isothermal Heating Medium***

$$\ln \frac{t_1 - t_0 - \dfrac{w}{L_0}\left(\dfrac{K_2 - 1}{K_2}\right)(T_1 - t_1)}{t_2 - t_0 - \dfrac{w}{L_0}\left(\dfrac{K_2 - 1}{K_2}\right)(T_1 - t_2)} = \left[\frac{w}{L_0}\left(\frac{K_2 - 1}{K_2}\right) + 1\right] \ln \frac{M + L_0\theta}{M} \qquad (11\text{-}35h)$$

If the addition of liquid to the tank causes an average endothermic or exothermic heat of solution, $\pm q_s$J/kg (Btu/lb) of makeup, it may be included by adding $\pm q_s/c_0$ to both the numerator and the denominator of the left side. The subscript 0 refers to the makeup.

***External Exchanger with Liquid Continuously Added to Tank: Isothermal Cooling Medium***

$$\ln \frac{T_0 - T_1 - \dfrac{W}{L_0}\left(\dfrac{K_1 - 1}{K_1}\right)(T_1 - t_1)}{T_0 - T_2 - \dfrac{W}{L_0}\left(\dfrac{K_1 - 1}{K_1}\right)(T_2 - t_1)} = \left[1 - \frac{W}{L_0}\left(\frac{K_1 - 1}{K_1}\right)\right] \ln \frac{M + L_0\theta}{M} \qquad (11\text{-}35i)$$

The heat-of-solution effects can be included by adding $\pm q_s/C_0$ to both the numerator and the denominator of the left side.

***External Exchanger with Liquid Continuously Added to Tank: Nonisothermal Heating Medium***

$$\ln \frac{t_0 - t_1 + \dfrac{wWC(K_5 - 1)(T_1 - t_1)}{L_0(K_5WC - wc)}}{t_0 - t_2 + \dfrac{wWC(K_5 - 1)(T_1 - t_2)}{L_0(K_5WC - wc\ )}} = \left[\frac{wWC(K_5 - 1)}{L_0(K_5Wc - wc)} + 1\right] \ln \frac{M + L_0\theta}{M} \qquad (11\text{-}35j)$$

where $K_5 = e^{(UA/wc)(1 - wc/WC)}$

The heat-of-solution effects can be included by adding $\pm q_s/c_0$ to both the numerator and the denominator of the left side.

***External Exchanger with Liquid Continuously Added to Tank: Nonisothermal Cooling Medium***

$$\ln \frac{T_0 - T_1 - \dfrac{Wwc(K_6 - 1)(T_1 - t_1)}{L_0(K_6wc - WC)}}{T_0 - T_2 - \dfrac{Wwc(K_6 - 1)(T_2 - t_1)}{L_0(K_6wc - \ WC)}} = \left[\frac{Wwc(K_6 - 1)}{L_0(K_6wc - WC)} + 1\right] \ln \frac{M + L_0\theta}{M} \qquad (11\text{-}35k)$$

where $K_6 = e^{(UA/WC)(1 - WC/wc)}$

The heat-of-solution effects can be included by adding $\pm q_s/C_0$ to both the numerator and the denominator of the left side.

***Heating and Cooling Agitated Batches: 1-2 Parallel Flow-Counterflow***

$$\frac{UA}{wc} = \frac{1}{\sqrt{R^2 + 1}} \ln \frac{2 - S(R + 1 - \sqrt{R^2 + 1})}{2 - S(R + 1 + \sqrt{R^2 + 1})} \qquad (11\text{-}35l)$$

$$R = \frac{T_1 - T_2}{t' - t} = \frac{wc}{WC} \qquad \text{and} \qquad S = \frac{t' - t}{T_1 - t}$$

$$\frac{2 - S(R + 1 - \sqrt{R^2 + 1})}{2 - S(R + 1 + \sqrt{R^2 + 1})} = e^{(UA/wc)\sqrt{R^2 + 1}} = K_7 \qquad (11\text{-}35m)$$

$$S = \frac{2(K_7 - 1)}{K_7(R + 1 + R^2 + 1) - (R + 1 - R^2 + 1)}$$

***External 1-2 Exchanger: Heating***

$$\ln (T_1 - t_1)/(T_1 - t_2) = (Sw/M)\theta \qquad (11\text{-}35n)$$

***External 1-2 Exchanger: Cooling***

$$\ln (T_1 - t_1)/(T_2 - t_1) = S(wc/MC)\theta \qquad (11\text{-}35o)$$

The cases of multipass exchangers with liquid continuously added to the tank are covered by Kern, as cited earlier. An alternative method for all multipass-exchanger gases, including those presented as well as cases with two or more shells in series, is as follows:

1. Determine $UA$ for using the applicable equations for counterflow heat exchangers.
2. Use the initial batch temperature $T_1$ or $t_1$.
3. Calculate the outlet temperature from the exchanger of each fluid. (This will require trial-and-error methods.)
4. Note the $F_T$ correction factor for the corrected mean temperature difference. (See Fig. 11-4.)
5. Repeat steps 2, 3, and 4 by using the final batch temperature $T_2$ and $t_2$.
6. Use the average of the two values for $F$, then increase the required multipass $UA$ as follows:

$$UA(\text{multipass}) = UA(\text{counterflow})/F_T$$

In general, values of $F_T$ below 0.8 are uneconomical and should be avoided. $F_T$ can be raised by increasing the flow rate of either or both of the flow streams. Increasing flow rates to give values well above 0.8 is a matter of economic justification.

If $F_T$ varies widely from one end of the range to the other, $F_T$ should be determined for one or more intermediate points. The average should then be determined for each step which has been established and the average of these taken for use in step 6.

**Effect of External Heat Loss or Gain** If heat loss or gain through the vessel walls cannot be neglected, equations which include this heat transfer can be developed by using energy balances similar to those used for the derivations of equations given previously. Basically, these equations must be modified by adding a heat-loss or heat-gain term.

A simpler procedure, which is probably acceptable for most practical cases, is to ratio $UA$ or $\theta$ either up or down in accordance with the required modification in total heat load over time $\theta$.

Another procedure, which is more accurate for the external-heat-exchanger cases, is to use an equivalent value for $MC$ (for a vessel being heated) derived from the following energy balance:

$$Q = (Mc)_e(t_2 - t_1) = Mc(t_2 - t_1) + U'A'(MTD')\theta \qquad (11\text{-}35p)$$

where $Q$ is the total heat transferred over time $\theta$, $U'A'$ is the heat-transfer coefficient for heat loss times the area for heat loss, and $MTD'$ is the mean temperature difference for the heat loss.

A similar energy balance would apply to a vessel being cooled.

**Internal Coil or Jacket Plus External Heat Exchanger** This case can be most simply handled by treating it as two separate problems. $M$ is divided into two separate masses $M_1$ and $(M - M_1)$, and the appropriate equations given earlier are applied to each part of the system. Time $\theta$, of course, must be the same for both parts.

**Equivalent-Area Concept** The preceding equations for batch operations, particularly Eq. 11-35 can be applied for the calculation of heat loss from tanks which are allowed to cool over an extended period of time. However, different surfaces of a tank, such as the top (which would not be in contact with the tank contents) and the bottom, may have coefficients of heat transfer which are different from those of the vertical tank walls. The simplest way to resolve this difficulty is to use an equivalent area $A_e$ in the appropriate equations where

$$A_e = A_bU_b/U_s + A_tU_t/U_s + A_s \qquad (11\text{-}35q)$$

**TABLE 11-1 Typical Values for Use with Eqs. (11-36) to (11-44)***

| Application | Fluid | $U_s$ | $A_s$ |
|---|---|---|---|
| Tanks on legs, outdoors, not insulated | Oil | 3.7 | $0.22\,A_t + A_b + A_s$ |
| | Water at 150°F. | 5.1 | $0.16\,A_t + A_b + A_s$ |
| Tanks on legs, outdoors, insulated 1 in. | Oil | 0.45 | $0.7\,A_t + A_b + A_s$ |
| | Water | 0.43 | $0.67\,A_t + A_b + A_s$ |
| Tanks on legs, indoors, not insulated | Oil | 1.5 | $0.53\,A_t + A_b + A_s$ |
| | Water | 1.8 | $0.35\,A_t + A_b + A_s$ |
| Tanks on legs, indoors, insulated 1 in. | Oil | 0.36 | $0.8\,A_t + A_b + A_s$ |
| | Water | 0.37 | $0.73\,A_t + A_b + A_s$ |
| Flat-bottom tanks,† outdoors, not insulated | Oil | 3.7 | $0.22\,A_t + A_s + 0.43\,D_t$ |
| | Water | 5.1 | $0.16\,A_t + A_s + 0.31\,D_t$ |
| Flat-bottom tanks,† outdoors, insulated 1 in. | Oil | 0.36 | $0.7\,A_t + A_s + 3.9\,D_t$ |
| | Water | 0.37 | $0.16\,A_t + A_s + 3.7\,D_t$ |
| Flat-bottom tanks, indoors, not insulated | Oil | 1.5 | $0.53\,A_t + A_s + 1.1\,D_t$ |
| | Water | 1.8 | $0.35\,A_t + A_s + 0.9\,D_t$ |
| Flat-bottom tanks, indoors, insulated 1 in. | Oil | 0.36 | $0.8\,A_t + A_s + 4.4\,D_t$ |
| | Water | 0.37 | $0.73\,A_t + A_s + 4.5\,D_t$ |

*Based on typical coefficients.
†The ratio $(t - t_g)(t - t')$ assumed at 0.85 for outdoor tanks. °C = (°F − 32)/1.8; to convert British thermal units per hour-square foot-degrees Fahrenheit to joules per square meter-second-kelvins, multiply by 5.6783.

and the subscripts *b, s,* and *t* refer to the bottom, sides, and top respectively. $U$ is usually taken as $U_s$. Table 11-1 lists typical values for $U_s$ and expressions for $A_e$ for various tank configurations.

**Nonagitated Batches** Cases in which vessel contents are vertically stratified, rather than uniform in temperature, have been treated by Kern (op. cit.). These are of little practical importance except for tall, slender vessels heated or cooled with external exchangers. The result is that a smaller exchanger is required than for an equivalent agitated batch system that is uniform.

**Storage Tanks** The equations for batch operations with agitation may be applied to storage tanks even though the tanks are not agitated. This approach gives conservative results. The important cases (nonsteady state) are:

1. *Tanks cool; contents remain liquid.* This case is relatively simple and can easily be handled by the equations given earlier.

2. *Tanks cool, contents partially freeze, and solids drop to bottom or rise to top.* This case requires a two-step calculation. The first step is handled as in case 1. The second step is calculated by assuming an isothermal system at the freezing point. It is possible, given time and a sufficiently low ambient temperature, for tank contents to freeze solid.

3. *Tanks cool and partially freeze; solids form a layer of self-insulation.* This complex case, which has been known to occur with heavy hydrocarbons and mixtures of hydrocarbons, has been discussed by Stuhlbarg [*Pet. Refiner,* **38,** 143 (Apr. 1, 1959)]. The contents in the center of such tanks have been known to remain warm and liquid even after several years of cooling.

It is very important that a melt-out riser be installed whenever tank contents are expected to freeze on prolonged shutdown. The purpose is to provide a molten chimney through the crust for relief of thermal expansion or cavitation if fluids are to be pumped out or recirculated through an external exchanger. An external heat tracer, properly located, will serve the same purpose but may require more remelt time before pumping can be started.

## THERMAL DESIGN OF TANK COILS

The thermal design of tank coils involves the determination of the area of heat-transfer surface required to maintain the contents of the tank at a constant temperature or to raise or lower the temperature of the contents by a specified magnitude over a fixed time.

**Nomenclature** $A$ = area; $A_b$ = area of tank bottom; $A_c$ = area of coil; $A_e$ = equivalent area; $A_s$ = area of sides; $A_t$ = area of top; $A_1$ = equivalent area receiving heat from external coils; $A_2$ = equivalent area not covered with external coils; $D_t$ = diameter of tank; $F$ = design (safety) factor; $h$ = film coefficient; $h_a$ = coefficient of ambient air; $h_c$ = coefficient of coil; $h_h$ = coefficient of heating medium; $h_i$ = coefficient of liquid phase of tank contents or tube-side coefficient referred to outside of coil; $h_z$ = coefficient of insulation; $k$ = thermal conductivity; $k_g$ = thermal conductivity of ground below tank; $M$ = mass of tank contents when full; $t$ = temperature; $t_a$ = temperature of ambient air; $t_d$ = temperature of dead-air space; $t_f$ = temperature of contents at end of heating; $t_g$ = temperature of ground below tank; $t_h$ = temperature of heating medium; $t_0$ = temperature of contents at beginning of heating; $U$ = overall coefficient; $U_b$ = coefficient at tank bottom; $U_c$ = coefficient of coil; $U_d$ = coefficient of dead air to the tank contents; $U_i$ = coefficient through insulation; $U_s$ = coefficient at sides; $U_t$ = coefficient at top; and $U_2$ = coefficient at area $A_2$.

Typical coil coefficients are listed in Table 11-2. More exact values can be calculated by using the methods for natural convection or forced convection given elsewhere in this section.

**Maintenance of Temperature** Tanks are often maintained at temperature with internal coils if the following equations are assumed to be applicable:

$$q = U_s A_e (T - t') \tag{11-36}$$

and

$$A_c = q/U_c(MTD) \tag{11-36a}$$

These make no allowance for unexpected shutdowns. One method of allowing for shutdown is to add a safety factor to Eq. 11-36*a*.

In the case of a tank maintained at temperature with internal coils, the coils are usually designed to cover only a portion of the tank. The temperature $t_d$ of the dead-air space between the coils and the tank is obtained from

$$U_d A_1 (t_d - t) = U_2 A_2 (t - t') \tag{11-37}$$

The heat load is

$$q = U_d A_1 (t_d - t) + A_1 U_i (t_d - t') \tag{11-38}$$

The coil area is

$$A_c = \frac{qF}{U_c (t_h - t_d)_m} \tag{11-39}$$

where $F$ is a safety factor.

### Heating

***Heating with Internal Coil from Initial Temperature for Specified Time***

$$Q = Wc(t_f - t_o) \tag{11-40}$$

$$A_c = \left[\frac{Q}{\theta_h} + U_s A_e \left(\frac{t_f + t_o}{2} - t'\right)\right] \left[\frac{1}{U_c[t_h - (t_f + t_o)/2]}\right] (F) \tag{11-41}$$

where $\theta_h$ is the length of heating period. This equation may also be used when the tank contents have cooled from $t_f$ to $t_o$ and must be

**TABLE 11-2 Overall Heat-Transfer Coefficients for Coils Immersed in Liquids**

$U$ Expressed as Btu/(h · ft² · °F)

| Substance inside coil | Substance outside coil | Coil material | Agitation | $U$ |
|---|---|---|---|---|
| Steam | Water | Lead | Agitated | 70 |
| Steam | Sugar and molasses solutions | Copper | None | 50–240 |
| Steam | Boiling aqueous solution | | | 600 |
| Cold water | Dilute organic dye intermediate | Lead | Turboagitator at 95 r.p.m. | 300 |
| Cold water | Warm water | Wrought iron | Air bubbled into water surrounding coil | 150–300 |
| Cold water | Hot water | Lead | 0.40 r.p.m. paddle stirrer | 90–360 |
| Brine | Amino acids | | 30 r.p.m. | 100 |
| Cold water | 25% oleum at 60°C. | Wrought iron | Agitated | 20 |
| Water | Aqueous solution | Lead | 500 r.p.m. sleeve propeller | 250 |
| Water | 8% NaOH | | 22 r.p.m. | 155 |
| Steam | Fatty acid | Copper (pancake) | None | 96–100 |
| Milk | Water | | Agitation | 300 |
| Cold water | Hot water | Copper | None | 105–180 |
| 60°F. water | 50% aqueous sugar solution | Lead | Mild | 50–60 |
| Steam and hydrogen at 1500 lb./sq. in. | 60°F. water | Steel | | 100–165 |
| Steam 110–146 lb./sq. in. gage | Vegetable oil | Steel | None | 23–29 |
| Steam | Vegetable oil | Steel | Various | 39–72 |
| Cold water | Vegetable oil | Steel | Various | 29–72 |

NOTES: Chilton, Drew, and Jebens [*Ind. Eng. Chem.*, **36**, 510 (1944)] give film coefficients for heating and cooling agitated fluids using a coil in a jacketed vessel.
Because of the many factors affecting heat transfer, such as viscosity, temperature difference, and coil size, the values in this table should be used primarily for preliminary design estimates and checking calculated coefficients.
°C = (°F − 32)/1.8; to convert British thermal units per hour-square foot-degrees Fahrenheit to joules per square meter-second-kelvins, multiply by 5.6783.

reheated to $t_f$. If the contents cool during a time $\theta_c$, the temperature at the end of this cooling period is obtained from

$$\ln\left(\frac{t_f - t'}{t_o - t'}\right) = \frac{U_s A_e \theta_c}{Wc} \qquad (11\text{-}42)$$

***Heating with External Coil from Initial Temperature for Specified Time*** The temperature of the dead-air space is obtained from

$$U_d A_1[t_d - 0.5(t_f - t_o)] = U_2 A_2[0.5(t_f - t_o) - t'] + Q/\theta_h \qquad (11\text{-}43)$$

The heat load is

$$q = U_i A_1(t_d - t') + U_2 A_2[0.5(t_f - t_o) - t'] + Q/\theta_h \qquad (11\text{-}44)$$

The coil area is obtained from Eq. 11-39.

The safety factor used in the calculations is a matter of judgment based on confidence in the design. A value of 1.10 is normally not considered excessive. Typical design parameters are shown in Tables 11-1 and 11-2.

## HEATING AND COOLING OF TANKS

**Tank Coils** **Pipe tank coils** are made in a wide variety of configurations, depending upon the application and shape of the vessel. **Helical** and **spiral** coils are most commonly shop-fabricated, while the **hairpin** pattern is generally field-fabricated. The helical coils are used principally in process tanks and pressure vessels when large areas for rapid heating or cooling are required. In general, heating coils are placed low in the tank, and cooling coils are placed high or distributed uniformly through the vertical height.

Stocks which tend to solidify on cooling require uniform coverage of the bottom or agitation. A maximum spacing of 0.6 m (2 ft) between turns of 50.8-mm (2-in) and larger pipe and a close approach to the tank wall are recommended. For smaller pipe or for low-temperature heating media, closer spacing should be used. In the case of the common hairpin coils in vertical cylindrical tanks, this means adding an encircling ring within 152 mm (6 in) of the tank wall (see Fig. 11-29*a* for this and other typical coil layouts). The coils should be set directly on the bottom or raised not more than 50.8 to 152 mm (2 to 6 in), depending upon the difficulty of remelting the solids, in order to permit free movement of product within the vessel. The coil inlet should be above the liquid level (or an internal melt-out riser installed) to provide a molten path for liquid expansion or venting of vapors.

Coils may be sloped to facilitate drainage. When it is impossible to do so and remain close enough to the bottom to get proper remelting, the coils should be blown out after usage in cold weather to avoid damage by freezing.

Most coils are firmly clamped (but not welded) to supports. **Supports** should allow expansion but be rigid enough to prevent uncontrolled motion (see Fig. 11-29*b*). Nuts and bolts should be securely fastened. Reinforcement of the inlet and outlet connections through the tank wall is recommended, since bending stresses due to thermal expansion are usually high at such points.

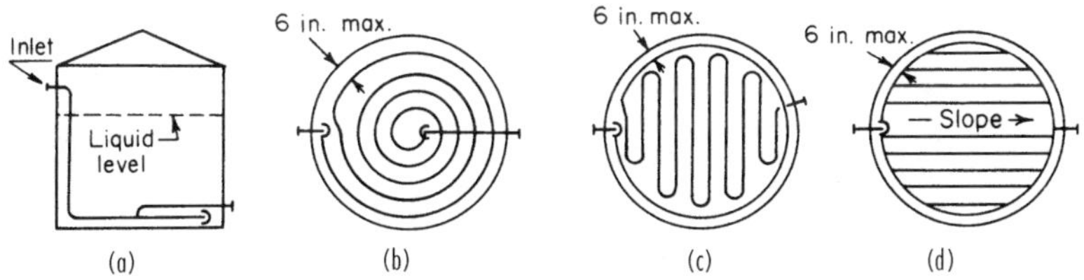


**FIG. 11-29a** Typical coil designs for good bottom coverage. (*a*) Elevated inlet on spiral coil. (*b*) Spiral with recircling ring. (*c*) Hairpin with encircling ring. (*d*) Ring header type.

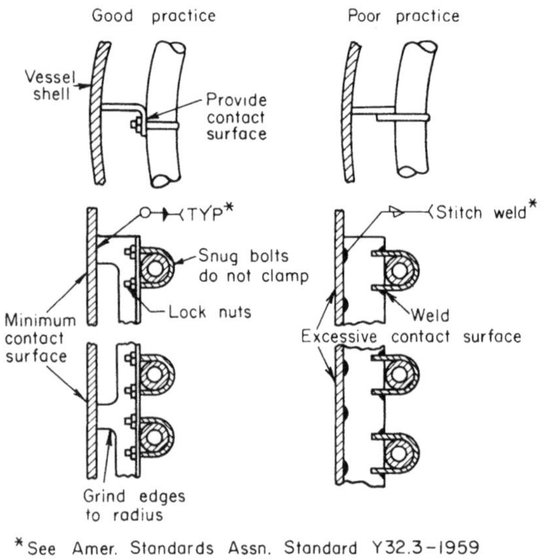


**FIG. 11-29b** Right and wrong ways to support coils. [*Chem. Eng.*, 172 (May 16, 1960).]

In general, 50.8- and 63.4-mm (2- and 2½-in) coils are the most economical for shop fabrication and 38.1- and 50.8-mm (1½- and 2-in) for field fabrication. The tube-side heat-transfer coefficient, high-pressure, or layout problems may lead to the use of smaller-size pipe.

The wall thickness selected varies with the service and material. Carbon steel coils are often made from schedule 80 or heavier pipe to allow for corrosion. When stainless steel or other high-alloy coils are not subject to corrosion or excessive pressure, they may be of schedule 5 or 10 pipe to keep costs at a minimum, although high-quality welding is required for these thin walls to assure trouble-free service.

Methods for calculating heat loss from tanks and the sizing of tank coils have been published by Stuhlbarg [*Pet. Refiner,* **38,** 143 (April 1959)].

**Fin-tube coils** are used for fluids which have poor heat-transfer characteristics to provide more surface for the same configuration at reduced cost or when temperature-driven fouling is to be minimized. Fin tubing is not generally used when bottom coverage is important. **Fin-tube tank heaters** are compact prefabricated bundles which can be brought into tanks through manholes. These are normally installed vertically with longitudinal fins to produce good convection currents. To keep the heaters low in the tank, they can be installed horizontally with helical fins or with perforated longitudinal fins to prevent entrapment. Fin tubing is often used for heat-sensitive material because of the lower surface temperature for the same heating medium, resulting in a lesser tendency to foul.

**Plate or panel** coils made from two metal sheets with one or both embossed to form passages for a heating or cooling medium can be used in lieu of pipe coils. Panel coils are relatively light in weight, easy to install, and easily removed for cleaning. They are available in a range of standard sizes and in both flat and curved patterns. Process tanks have been built by using panel coils for the sides or bottom. A serpentine construction is generally utilized when liquid flows through the unit. Header-type construction is used with steam or other condensing media.

Standard **glass coils** with 0.18 to 11.1 m$^2$ (2 to 120 ft$^2$) of heat-transfer surface are available. Also available are plate-type units made of **impervious graphite.**

**Teflon Immersion Coils** Immersion coils made of Teflon fluorocarbon resin are available with 2.5-mm (0.10-in) ID tubes to increase overall heat-transfer efficiency. The flexible bundles are available with 100, 160, 280, 500, and 650 tubes with standard lengths varying in 0.6-m (2-ft) increments between 1.2 and 4.8 m (4 and 16 ft). These coils are most commonly used in metal-finishing baths and are adaptable to service in reaction vessels, crystallizers, and tanks where corrosive fluids are used.

**Bayonet Heaters** A bayonet-tube element consists of an outer and an inner tube. These elements are inserted into tanks and process vessels for heating and cooling purposes. Often the outer tube is of expensive alloy or nonmetallic (e.g., glass, impervious graphite), while the inner tube is of carbon steel. In glass construction, elements with 50.8- or 76.2-mm (2- or 3-in) glass pipe [with lengths to 2.7 m (9 ft)] are in contact with the external fluid, with an inner tube of metal.

**External Coils and Tracers** Tanks, vessels, and pipe lines can be equipped for heating or cooling purposes with external coils. These are generally 9.8 to 19 mm (⅜ to ¾ in) so as to provide good distribution over the surface and are often of soft copper or aluminum, which can be bent by hand to the contour of the tank or line. When necessary to avoid "hot spots," the tracer is so mounted that it does not touch the tank.

External coils spaced away from the tank wall exhibit a coefficient of around 5.7 W/(m$^2$·°C) [1 Btu/(h·ft$^2$ of coil surface·°F)]. Direct contact with the tank wall produces higher coefficients, but these are difficult to predict since they are strongly dependent upon the degree of contact. The use of **heat-transfer cements** does improve performance. These puttylike materials of high thermal conductivity are troweled or caulked into the space between the coil and the tank or pipe surface.

Costs of the cements (in 1960) varied from 37 to 63 cents per pound, with requirements running from about 0.27 lb/ft of ⅜-in outside-diameter tubing to 1.48 lb/ft of 1-in pipe. Panel coils require ½ to 1 lb/ft$^2$. A rule of thumb for preliminary estimating is that the per-foot installed cost of tracer with cement is about double that of the tracer alone.

**Jacketed Vessels** Jacketing is often used for vessels needing frequent cleaning and for glass-lined vessels which are difficult to equip with internal coils. The jacket eliminates the need for the coil yet gives a better overall coefficient than external coils. However, only a limited heat-transfer area is available. The conventional jacket is of simple construction and is frequently used. It is most effective with a condensing vapor. A liquid heat-transfer fluid does not maintain uniform flow characteristics in such a jacket. Nozzles, which set up a swirling motion in the jacket, are effective in improving heat transfer. Wall thicknesses are often high unless reinforcement rings are installed.

**Spiral baffles,** which are sometimes installed for liquid services to improve heat transfer and prevent channeling, can be designed to serve as reinforcements. A spiral-wound channel welded to the vessel wall is an alternative to the spiral baffle which is more predictable in performance, since cross-baffle leakage is eliminated, and is reportedly lower in cost [Feichtinger, *Chem. Eng.,* **67,** 197 (Sept. 5, 1960)].

The half-pipe jacket is used when high jacket pressures are required. The flow pattern of a liquid heat-transfer fluid can be controlled and designed for effective heat transfer. The dimple jacket offers structural advantages and is the most economical for high jacket pressures. The low volumetric capacity produces a fast response to temperature changes.

## EXTENDED OR FINNED SURFACES

**Finned-Surface Application** Extended or finned surfaces are often used when one film coefficient is substantially lower than the other, the goal being to make $h_o A_{oe} \approx h_i A_i$. A few typical fin configurations are shown in Fig. 11-30*a*. Longitudinal fins are used in double-pipe exchangers. Transverse fins are used in cross-flow and shell-and-tube configurations. High transverse fins are used mainly with low-pressure gases; low fins are used for boiling and condensation of nonaqueous streams as well as for sensible-heat transfer. Finned surfaces have been proven to be a successful means of controlling temperature driven fouling such as coking and scaling. Fin spacing should be great enough to avoid entrapment of particulate matter in the fluid stream (5 mm minimum spacing).

The area added by the fin is not as efficient for heat transfer as bare tube surface owing to resistance to conduction through the fin. The

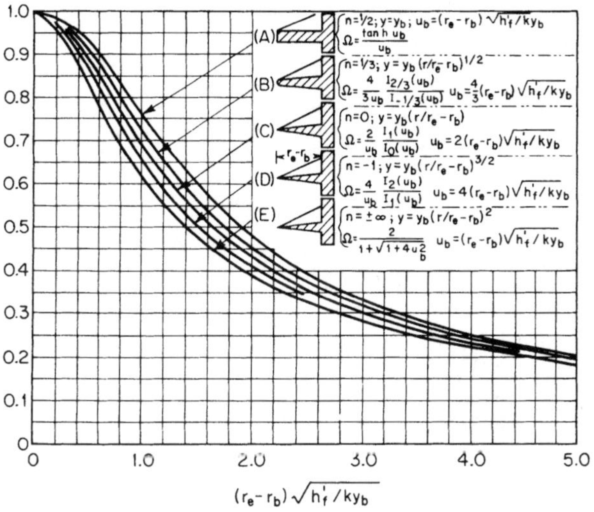


**FIG. 11-30*a*** Efficiencies for several longitudinal fin configurations.

effective heat-transfer area is

$$A_{oe} = A_{uf} + A_f\Omega \quad (11\text{-}45)$$

The fin efficiency is found from mathematically derived relations, in which the film heat-transfer coefficient is assumed to be constant over the entire fin and temperature gradients across the thickness of the fin have been neglected (see Kraus, *Extended Surfaces,* Spartan Books, Baltimore, 1963). The efficiency curves for some common fin configurations are given in Figs. 11-30*a* and 11-30*b*.

**High Fins** To calculate heat-transfer coefficients for cross-flow to a transversely finned surface, it is best to use a correlation based on experimental data for that surface. Such data are not often available, and a more general correlation must be used, making allowance for the possible error. Probably the best general correlation for bundles of finned tubes is given by Schmidt [*Kaltetechnik,* **15,** 98–102, 370–378 (1963)]:

$$hD_r/k = K(D_r\rho V'_{max}/\mu)^{0.625}R_f^{-0.375}N_{Pr}^{1/3} \quad (11\text{-}46)$$

where $K = 0.45$ for staggered tube arrays and 0.30 for in-line tube arrays: $D_r$ is the root or base diameter of the tube; $V'_{max}$ is the maximum velocity through the tube bank, i.e., the velocity through the minimum flow area between adjacent tubes; and $R_f$ is the ratio of the total outside surface area of the tube (including fins) to the surface of a tube having the same root diameter but without fins.

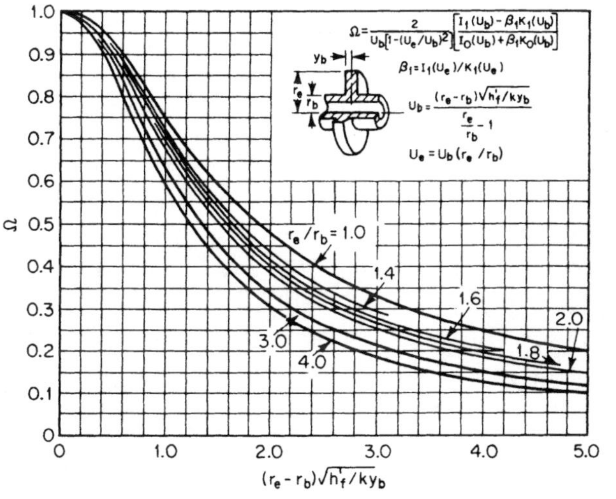


**FIG. 11-30*b*** Efficiencies for annular fins of constant thickness.

**Pressure drop** is particularly sensitive to geometrical parameters, and available correlations should be extrapolated to geometries different from those on which the correlation is based only with great caution and conservatism. The best correlation is that of Robinson and Briggs [*Chem. Eng. Prog.,* **62,** *Symp. Ser.* 64, 177–184 (1966)].

**Low Fins** Low-finned tubing is generally used in shell-and-tube configurations. For sensible-heat transfer, only minor modifications are needed to permit the shell-side method given earlier to be used for both heat transfer and pressure [see Briggs, Katz, and Young, *Chem. Eng. Prog.,* **59**(11), 49–59 (1963)]. For condensing on low-finned tubes in horizontal bundles, the Nusselt correlation is generally satisfactory for low-surface-tension [$\sigma < (3)(10^{-6})$N/m (30 dyn/cm)] condensates fins of finned surfaces should not be closely spaced for high-surface-tension condensates (notably water), which do not drain easily.

The modified Palen Small method can be employed for reboiler design using finned tubes, but the maximum flux is calculated from $A_o$, the total outside heat-transfer area including fins. The resulting value of $q_{max}$ refers to $A_o$.

## FOULING AND SCALING

Fouling refers to any change in the solid boundary separating two heat transfer fluids, whether by dirt accumulation or other means, which results in a decrease in the rate of heat transfer occurring across that boundary. Fouling may be classified by mechanism into six basic categories:

1. *Corrosion fouling.* The heat transfer surface reacts chemically with elements of the fluid stream producing a less conductive, corrosion layer on all or part of the surface.
2. *Biofouling.* Organisms present in the fluid stream are attracted to the warm heat-transfer surface where they attach, grow, and reproduce. The two subgroups are microbiofoulants such as slime and algae and macrobiofoulants such as snails and barnacles.
3. *Particulate fouling.* Particles held in suspension in the flow stream will deposit out on the heat-transfer surface in areas of sufficiently lower velocity.
4. *Chemical reaction fouling* (ex.—Coking). Chemical reaction of the fluid takes place on the heat-transfer surface producing an adhering solid product of reaction.
5. *Precipitation fouling* (ex.—Scaling). A fluid containing some dissolved material becomes supersaturated with respect to this material at the temperatures seen at the heat-transfer surface. This results in a crystallization of the material which "plates out" on the warmer surface.
6. *Freezing fouling.* Overcooling of a fluid below the fluid's freezing point at the heat-transfer surface causes solidification and coating of the heat-transfer surface.

**Control of Fouling** Once the combination of mechanisms contributing to a particular fouling problem are recognized, methods to substantially reduce the fouling rate may be implemented. For the case of **corrosion fouling,** the common solution is to choose a less corrosive material of construction balancing material cost with equipment life. In cases of **biofouling,** the use of copper alloys and/or chemical treatment of the fluid stream to control organism growth and reproduction are the most common solutions.

In the case of **particulate fouling,** one of the more common types, insuring a sufficient flow velocity and minimizing areas of lower velocities and stagnant flows to help keep particles in suspension is the most common means of dealing with the problem. For water, the recommended tubeside minimum velocity is about 0.9 to 1.0 m/s. This may not always be possible for moderate to high-viscosity fluids where the resulting pressure drop can be prohibitive.

Special care should be taken in the application of any velocity requirement to the shellside of segmental-baffled bundles due to the many different flow streams and velocities present during operation, the unavoidable existence of high-fouling areas of flow stagnation, and the danger of flow-induced tube vibration. In general, shellside-particulate fouling will be greatest for segmentally baffled bundles in the regions of low velocity and the TEMA-fouling factors (which are based upon the use of this bundle type) should be used. However, since the 1940's, there have been a host of successful, low-fouling exchangers developed, some tubular and some not, which have in common the elimination of the cross-flow plate baffle and provide practically no regions of flow stagnation at the heat-transfer surface. Some examples are the plate and frame exchanger, the spiral plate exchanger, and the twisted tube exchanger, all of which have dispensed with baffles altogether and use the heat-transfer surface itself for bundle support. The general rule for these designs is to provide between 25 and 30 percent excess surface to compensate for potential fouling, although this can vary in special applications.

For the remaining classifications—**polymerization, precipitation,** and **freezing**—fouling is the direct result of temperature extremes at the heat-transfer surface and is reduced by reducing the temperature difference between the heat-transfer surface and the bulk-fluid stream. Conventional wisdom says to increase velocity, thus increasing the local heat-transfer coefficient to bring the heat-transfer surface temperature closer to the bulk-fluid temperature. However, due to a practical limit on the amount of heat-transfer coefficient increase available by increasing velocity, this approach, although better than nothing, is often not satisfactory by itself.

A more effective means of reducing the temperature difference is by using, in concert with adequate velocities, some form of extended surface. As discussed by Shilling (*Proceedings of the 10th International Heat Transfer Conference,* Brighton, U.K., **4,** p. 423), this will tend to reduce the temperature extremes between fluid and heat transfer surface and not only reduce the rate of fouling but make the heat exchanger generally less sensitive to the effects of any fouling that does occur. In cases where unfinned tubing in a triangular tube layout would not be acceptable because fouling buildup and eventual mechanical cleaning are inevitable, extended surface should be used only when the exchanger construction allows access for cleaning.

**Fouling Transients and Operating Periods** Three common behaviors are noted in the development of a fouling film over a period of time. One is the so-called *asymptotic fouling* in which the speed of fouling resistance increase decreases over time as it approaches some asymptotic value beyond which no further fouling can occur. This is commonly found in temperature-driven fouling. A second is *linear fouling* in which the increase in fouling resistance follows a straight line over the time of operation. This could be experienced in a case of severe particulate fouling where the accumulation of dirt during the time of operation did not appreciably increase velocities to mitigate the problem. The third, *falling rate fouling,* is neither linear nor asymptotic but instead lies somewhere between these two extremes. The rate of fouling decreases with time but does not appear to approach an asymptotic maximum during the time of operation. This is the most common type of fouling in the process industry and is usually the result of a combination of different fouling mechanisms occurring together.

The optimum operating period between cleanings depends upon the rate and type of fouling, the heat exchanger used (i.e. baffle type, use of extended surface, and velocity and pressure drop design constraints), and the ease with which the heat exchanger may be removed from service for cleaning. As noted above, care must be taken in the use of fouling factors for exchanger design, especially if the exchanger configuration has been selected specifically to minimize fouling accumulation. An oversurfaced heat exchanger which will not foul enough to operate properly can be almost as much a problem as an undersized exchanger. This is especially true in steam-heated exchangers where the ratio of design MTD to minimum achievable MTD is less than U_clean divided by U_fouled.

**Removal of Fouling Deposits** Chemical removal of fouling can be achieved in some cases by weak acid, special solvents, and so on. Other deposits adhere weakly and can be washed off by periodic operation at very high velocities or by flushing with a high-velocity steam or water jet or using a sand-water slurry. These methods may be applied to both the shell side and tube side without pulling the bundle. Many fouling deposits, however, must be removed by positive mechanical action such as rodding, turbining, or scraping the surface. These techniques may be applied inside of tubes without pulling the bundle but can be applied on the shellside only after bundle removal. Even then there is limited access because of the tube pitch and rotated square or large triangular layouts are recommended. In many cases, it has been found that designs developed to minimize fouling often develop a fouling layer which is more easily removed.

**Fouling Resistances** There are no published methods for predicting fouling resistances a priori. The accumulated experience of exchanger designers and users was assembled more than 40 years ago based primarily upon segmental-baffled exchanger bundles and may be found in the *Standards of Tubular Exchanger Manufacturers Association* (TEMA). In the absence of other information, the fouling resistances contained therein may be used.

## TYPICAL HEAT-TRANSFER COEFFICIENTS

Typical overall heat-transfer coefficients are given in Tables 11-3 through 11-8. Values from these tables may be used for preliminary estimating purposes. They should not be used in place of the design methods described elsewhere in this section, although they may serve as a useful check on the results obtained by those design methods.

## THERMAL DESIGN FOR SOLIDS PROCESSING

Solids in divided form, such as powders, pellets, and lumps, are heated and/or cooled in chemical processing for a variety of objectives such as solidification or fusing (Sec. 11), drying and water removal (Sec. 20), solvent recovery (Secs. 13 and 20), sublimation (Sec. 17), chemical reactions (Sec. 20), and oxidation. For process and mechanical-design considerations, see the referenced sections.

**Thermal design** concerns itself with sizing the equipment to effect the heat transfer necessary to carry on the process. The design equation is the familiar one basic to all modes of heat transfer, namely,

$$A = Q/U\,\Delta t \tag{11-47}$$

where $A$ = effective heat-transfer surface, $Q$ = quantity of heat required to be transferred, $\Delta t$ = temperature difference of the process, and $U$ = overall heat-transfer coefficient. It is helpful to define the modes of heat transfer and the corresponding overall coefficient as $U_{co}$ = overall heat-transfer coefficient for (indirect through-a-wall) *conduction,* $U_{co}$ = overall heat-transfer coefficient for the little-used *convection* mechanism, $U_{ct}$ = heat-transfer coefficient for the *contactive* mechanism in which the gaseous-phase heat carrier passes directly through the solids bed, and $U_{ra}$ = heat-transfer coefficient for *radiation.*

There are two general methods for determining numerical values for $U_{co}$, $U_{cv}$, $U_{ct}$, and $U_{ra}$. One is by analysis of actual operating data. Values so obtained are used on geometrically similar systems of a size not too different from the equipment from which the data were obtained. The second method is predictive and is based on the material properties and certain operating parameters. Relative values of the coefficients for the various modes of heat transfer at temperatures up to 980°C (1800°F) are as follows (Holt, Paper 11, Fourth National Heat Transfer Conference, Buffalo, 1960):

| | |
|---|---|
| Convective | 1 |
| Radiant | 2 |
| Conductive | 20 |
| Contactive | 200 |

Because heat-transfer equipment for solids is generally an adaptation of a primarily material-handling device, the area of heat transfer is often small in relation to the overall size of the equipment. Also peculiar to solids heat transfer is that the $\Delta t$ varies for the different heat-transfer mechanisms. With a knowledge of these mechanisms, the $\Delta t$ term generally is readily estimated from temperature limitations imposed by the burden characteristics and/or the construction.

**Conductive Heat Transfer** Heat-transfer equipment in which heat is transferred by conduction is so constructed that the solids load (burden) is separated from the heating medium by a wall.

For a high proportion of applications, $\Delta t$ is the log-mean temperature difference. Values of $U_{co}$ are reported in Secs. 11, 15, 17, and 19.

**TABLE 11-3 Typical Overall Heat-Transfer Coefficients in Tubular Heat Exchangers**

$U$ = Btu/(°F · ft² · h)

| Shell side | Tube side | Design $U$ | Includes total dirt |
|---|---|---|---|
| **Liquid-liquid media** | | | |
| Aroclor 1248 | Jet fuels | 100–150 | 0.0015 |
| Cutback asphalt | Water | 10–20 | .01 |
| Demineralized water | Water | 300–500 | .001 |
| Ethanol amine (MEA or DEA) 10–25% solutions | Water or DEA, or MEA solutions | 140–200 | .003 |
| Fuel oil | Water | 15–25 | .007 |
| Fuel oil | Oil | 10–15 | .008 |
| Gasoline | Water | 60–100 | .003 |
| Heavy oils | Heavy oils | 10–40 | .004 |
| Heavy oils | Water | 15–50 | .005 |
| Hydrogen-rich reformer stream | Hydrogen-rich reformer stream | 90–120 | .002 |
| Kerosene or gas oil | Water | 25–50 | .005 |
| Kerosene or gas oil | Oil | 20–35 | .005 |
| Kerosene or jet fuels | Trichlorethylene | 40–50 | .0015 |
| Jacket water | Water | 230–300 | .002 |
| Lube oil (low viscosity) | Water | 25–50 | .002 |
| Lube oil (high viscosity) | Water | 40–80 | .003 |
| Lube oil | Oil | 11–20 | .006 |
| Naphtha | Water | 50–70 | .005 |
| Naphtha | Oil | 25–35 | .005 |
| Organic solvents | Water | 50–150 | .003 |
| Organic solvents | Brine | 35–90 | .003 |
| Organic solvents | Organic solvents | 20–60 | .002 |
| Tall oil derivatives, vegetable oil, etc. | Water | 20–50 | .004 |
| Water | Caustic soda solutions (10–30%) | 100–250 | .003 |
| Water | Water | 200–250 | .003 |
| Wax distillate | Water | 15–25 | .005 |
| Wax distillate | Oil | 13–23 | .005 |
| **Condensing vapor-liquid media** | | | |
| Alcohol vapor | Water | 100–200 | .002 |
| Asphalt (450°F.) | Dowtherm vapor | 40–60 | .006 |
| Dowtherm vapor | Tall oil and derivatives | 60–80 | .004 |

| Shell side | Tube side | Design $U$ | Includes total dirt |
|---|---|---|---|
| Dowtherm vapor | Dowtherm liquid | 80–120 | .0015 |
| Gas-plant tar | Steam | 40–50 | .0055 |
| High-boiling hydrocarbons V | Water | 20–50 | .003 |
| Low-boiling hydrocarbons A | Water | 80–200 | .003 |
| Hydrocarbon vapors (partial condenser) | Oil | 25–40 | .004 |
| Organic solvents A | Water | 100–200 | .003 |
| Organic solvents high NC, A | Water or brine | 20–60 | .003 |
| Organic solvents low NC, V | Water or brine | 50–120 | .003 |
| Kerosene | Water | 30–65 | .004 |
| Kerosene | Oil | 20–30 | .005 |
| Naphtha | Water | 50–75 | .005 |
| Naphtha | Oil | 20–30 | .005 |
| Stabilizer reflux vapors | Water | 80–120 | .003 |
| Steam | Feed water | 400–1000 | .0005 |
| Steam | No. 6 fuel oil | 15–25 | .0055 |
| Steam | No. 2 fuel oil | 60–90 | .0025 |
| Sulfur dioxide | Water | 150–200 | .003 |
| Tall-oil derivatives, vegetable oils (vapor) | Water | 20–50 | .004 |
| Water | Aromatic vapor-stream azeotrope | 40–80 | .005 |
| **Gas-liquid media** | | | |
| Air, $N_2$, etc. (compressed) | Water or brine | 40–80 | .005 |
| Air, $N_2$, etc., A | Water or brine | 10–50 | .005 |
| Water or brine | Air, $N_2$ (compressed) | 20–40 | .005 |
| Water or brine | Air, $N_2$, etc., A | 5–20 | .005 |
| Water | Hydrogen containing natural-gas mixtures | 80–125 | .003 |
| **Vaporizers** | | | |
| Anhydrous ammonia | Steam condensing | 150–300 | .0015 |
| Chlorine | Steam condensing | 150–300 | .0015 |
| Chlorine | Light heat-transfer oil | 40–60 | .0015 |
| Propane, butane, etc. | Steam condensing | 200–300 | .0015 |
| Water | Steam condensing | 250–400 | .0015 |

NC = noncondensable gas present.
V = vacuum.
A = atmospheric pressure.
Dirt (or fouling factor) units are (h · ft² · °F)/Btu.
To convert British thermal units per hour-square foot-degrees Fahrenheit to joules per square meter-second-kelvins, multiply by 5.6783; to convert hours per square foot-degree Fahrenheit-British thermal units to square meters per second-kelvin-joules, multiply by 0.1761.

**TABLE 11-4 Typical Overall Heat-Transfer Coefficients in Refinery Service**

Btu/(°F · ft² · h)

| Fluid | API gravity | Fouling factor (one stream) | Reboiler, steam-heated | Condenser, water-cooled* | Exchangers, liquid to liquid (tube-side fluid designation appears below) C | G | H | Reboiler (heating liquid designated below) C | G† | K | Condenser (cooling liquid designated below) D | F | G | J |
|---|---|---|---|---|---|---|---|---|---|---|---|---|---|---|
| A Propane | | 0.001 | 160 | 95 | 85 | 85 | 80 | 110 | 95 | 35 | | | | |
| B Butane | | .001 | 155 | 90 | 80 | 75 | 75 | 105 | 90 | 35 | 80 | 55 | 40 | 30 |
| C 400°F. end-point gasoline | 50 | .001 | 120 | 80 | 70 | 65 | 60 | 65 | 50 | 30 | | | | |
| D Virgin light naphtha | 70 | .001 | 140 | 85 | 70 | 55 | 55 | 75 | 60 | 35 | 75 | | | |
| E Virgin heavy naphtha | 45 | .001 | 95 | 75 | 65 | 55 | 50 | 55 | 45 | 30 | 70 | 50 | 35 | 30 |
| F Kerosene | 40 | .001 | 85 | 60 | 60 | 55 | 50 | | 45 | 25 | | 50 | 35 | 30 |
| G Light gas oil | 30 | .002 | 70 | 50 | 60 | 50 | 50 | | 40 | 25 | 70 | 45 | 30 | 30 |
| H Heavy gas oil | 22 | .003 | 60 | 45 | 55 | 50 | 45 | 50 | 40 | 20 | 70 | 40 | 30 | 20 |
| J Reduced crude | 17 | .005 | | | 55 | 45 | 40 | | | | | | | |
| K Heavy fuel oil (tar) | 10 | .005 | | | 50 | 40 | 35 | | | | | | | |

Fouling factor, water side 0.0002; heating or cooling streams are shown at top of columns as C, D, F, G, etc.; to convert British thermal units per hour-square foot-degrees Fahrenheit to joules per square meter-second-kelvins, multiply by 5.6783; to convert hours per square foot-degree Fahrenheit-British thermal units to square meters per second-kelvin-joules, multiply by 0.1761.

*Cooler, water-cooled, rates are about 5 percent lower.

†With heavy gas oil (H) as heating medium, rates are about 5 percent lower.

**TABLE 11-5 Overall Coefficients for Air-Cooled Exchangers on Bare-Tube Basis**

Btu/(°F · ft² · h)

| Condensing | Coefficient | Liquid cooling | Coefficient |
|---|---|---|---|
| Ammonia | 110 | Engine-jacket water | 125 |
| Freon-12 | 70 | Fuel oil | 25 |
| Gasoline | 80 | Light gas oil | 65 |
| Light hydrocarbons | 90 | Light hydrocarbons | 85 |
| Light naphtha | 75 | Light naphtha | 70 |
| Heavy naphtha | 65 | Reformer liquid streams | 70 |
| Reformer reactor effluent | 70 | Residuum | 15 |
| Low-pressure steam | 135 | Tar | 7 |
| Overhead vapors | 65 | | |
| **Gas cooling** | **Operating pressure, lb./sq. in. gage** | **Pressure drop, lb./sq. in.** | **Coefficient** |
| Air or flue gas | 50 | 0.1 to 0.5 | 10 |
| | 100 | 2 | 20 |
| | 100 | 5 | 30 |
| Hydrocarbon gas | 35 | 1 | 35 |
| | 125 | 3 | 55 |
| | 1000 | 5 | 80 |
| Ammonia reactor stream | | | 85 |

Bare-tube external surface is 0.262 ft²/ft.

Fin-tube surface/bare-tube surface ratio is 16.9.

To convert British thermal units per hour-square foot-degrees Fahrenheit to joules per square meter-second-kelvins, multiply by 5.6783; to convert pounds-force per square inch to kilopascals, multiply by 6.895.

A *predictive* equation for $U_{co}$ is

$$U_{co} = \left(\frac{h}{h - 2ca/d_m}\right)\left(\frac{2ca}{d_m}\right) \qquad (11\text{-}48)$$

where $h$ = wall film coefficient, $c$ = volumetric heat capacity, $d_m$ = depth of the burden, and $\alpha$ = thermal diffusivity. Relevant thermal properties of various materials are given in Table 11-9. For details of terminology, equation development, numerical values of terms in typical equipment and use, see Holt [*Chem. Eng.*, **69,** 107 (Jan. 8, 1962)].

Equation (11-48) is applicable to burdens in the solid, liquid, or gaseous phase, either static or in laminar motion; it is applicable to solidification equipment and to divided-solids equipment such as metal belts, moving trays, stationary vertical tubes, and stationary-shell fluidizers.

**Fixed (or packed) bed operation** occurs when the fluid velocity is low or the particle size is large so that fluidization does not occur. For such operation, Jakob (*Heat Transfer,* vol. 2, Wiley, New York, 1957) gives

$$hD_t/k = b_1 b D_t^{0.17}(D_pG/\mu)^{0.83}(c\mu/k) \qquad (11\text{-}49a)$$

where $b_1$ = 1.22 (SI) or 1.0 (U.S. customary), $h = U_{co}$ = overall coefficient between the inner container surface and the fluid stream,

$$b = .2366 + .0092\left(\frac{D_p}{D_t}\right) - 4.0672\left(\frac{D_p}{D_t}\right)^2 + 18.229\left(\frac{D_p}{D_t}\right)^3 - 11.837\left(\frac{D_p}{D_t}\right)^4 \qquad (11.49b)$$

$D_p$ = particle diameter, $D_t$ = vessel diameter, (note that $D_p/D_t$ has units of foot per foot in the equation), $G$ = superficial mass velocity, $k$ = fluid thermal conductivity, $\mu$ = fluid viscosity, and $c$ = fluid specific heat. Other correlations are those of Leva [*Ind. Eng. Chem.*, **42,** 2498 (1950)]:

$$h = 0.813\frac{k}{D_t}e^{-6D_p/D_t}\left(\frac{D_pG}{\mu}\right)^{0.90} \quad \text{for } \frac{D_p}{D_t} < 0.35 \qquad (11\text{-}50a)$$

$$h = 0.125\frac{k}{D_t}\left(\frac{D_pG}{\mu}\right)^{0.75} \quad \text{for } 0.35 < \frac{D_p}{D_t} < 0.60 \qquad (11\text{-}50b)$$

and Calderbank and Pogerski [*Trans. Inst. Chem. Eng.* (London), **35,** 195 (1957)]:

$$hD_p/k = 3.6(D_pG/\mu\epsilon_v)^{0.365} \qquad (11\text{-}51)$$

where $\epsilon_v$ = fraction voids in the bed.

**TABLE 11-6 Panel Coils Immersed in Liquid: Overall Average Heat-Transfer Coefficients***

$U$ expressed in Btu/(h · ft² · °F)

| | | Clean-surface coefficients | | Design coefficients, considering usual fouling in this service | |
|---|---|---|---|---|---|
| Hot side | Cold side | Natural convection | Forced convection | Natural convection | Forced convection |
| Heating applications: | | | | | |
| Steam | Watery solution | 250–500 | 300–550 | 100–200 | 150–275 |
| Steam | Light oils | 50–70 | 110–140 | 40–45 | 60–110 |
| Steam | Medium lube oil | 40–60 | 100–130 | 35–40 | 50–100 |
| Steam | Bunker C or No. 6 fuel oil | 20–40 | 70–90 | 15–30 | 60–80 |
| Steam | Tar or asphalt | 15–35 | 50–70 | 15–25 | 40–60 |
| Steam | Molten sulfur | 35–45 | 45–55 | 20–35 | 35–45 |
| Steam | Molten paraffin | 35–45 | 45–55 | 25–35 | 40–50 |
| Steam | Air or gases | 2–4 | 5–10 | 1–3 | 4–8 |
| Steam | Molasses or corn sirup | 20–40 | 70–90 | 15–30 | 60–80 |
| High temperature hot water | Watery solutions | 115–140 | 200–250 | 70–100 | 110–160 |
| High temperature heat-transfer oil | Tar or asphalt | 12–30 | 45–65 | 10–20 | 30–50 |
| Dowtherm or Aroclor | Tar or asphalt | 15–30 | 50–60 | 12–20 | 30–50 |
| Cooling applications: | | | | | |
| Water | Watery solution | 110–135 | 195–245 | 65–95 | 105–155 |
| Water | Quench oil | 10–15 | 25–45 | 7–10 | 15–25 |
| Water | Medium lube oil | 8–12 | 20–30 | 5–8 | 10–20 |
| Water | Molasses or corn sirup | 7–10 | 18–26 | 4–7 | 8–15 |
| Water | Air or gases | 2–4 | 5–10 | 1–3 | 4–8 |
| Freon or ammonia | Watery solution | 35–45 | 60–90 | 20–35 | 40–60 |
| Calcium or sodium brine | Watery solution | 100–120 | 175–200 | 50–75 | 80–125 |

*Tranter Manufacturing, Inc.

NOTE: To convert British thermal units per hour-square foot-degrees Fahrenheit to joules per square meter second-kelvins, multiply by 5.6783.

**TABLE 11-7 Jacketed Vessels: Overall Coefficients**

| Jacket fluid | Fluid in vessel | Wall material | Overall U* Btu/(h · ft² · °F) | Overall U* J/(m² · s · K) |
|---|---|---|---|---|
| Steam | Water | Stainless steel | 150–300 | 850–1700 |
| Steam | Aqueous solution | Stainless steel | 80–200 | 450–1140 |
| Steam | Organics | Stainless steel | 50–150 | 285–850 |
| Steam | Light oil | Stainless steel | 60–160 | 340–910 |
| Steam | Heavy oil | Stainless steel | 10–50 | 57–285 |
| Brine | Water | Stainless steel | 40–180 | 230–1625 |
| Brine | Aqueous solution | Stainless steel | 35–150 | 200–850 |
| Brine | Organics | Stainless steel | 30–120 | 170–680 |
| Brine | Light oil | Stainless steel | 35–130 | 200–740 |
| Brine | Heavy oil | Stainless steel | 10–30 | 57–170 |
| Heat-transfer oil | Water | Stainless steel | 50–200 | 285–1140 |
| Heat-transfer oil | Aqueous solution | Stainless steel | 40–170 | 230–965 |
| Heat-transfer oil | Organics | Stainless steel | 30–120 | 170–680 |
| Heat-transfer oil | Light oil | Stainless steel | 35–130 | 200–740 |
| Heat-transfer oil | Heavy oil | Stainless steel | 10–40 | 57–230 |
| Steam | Water | Glass-lined CS | 70–100 | 400–570 |
| Steam | Aqueous solution | Glass-lined CS | 50–85 | 285–480 |
| Steam | Organics | Glass-lined CS | 30–70 | 170–400 |
| Steam | Light oil | Glass-lined CS | 40–75 | 230–425 |
| Steam | Heavy oil | Glass-lined CS | 10–40 | 57–230 |
| Brine | Water | Glass-lined CS | 30–80 | 170–450 |
| Brine | Aqueous solution | Glass-lined CS | 25–70 | 140–400 |
| Brine | Organics | Glass-lined CS | 20–60 | 115–340 |
| Brine | Light oil | Glass-lined CS | 25–65 | 140–370 |
| Brine | Heavy oil | Glass-lined CS | 10–30 | 57–170 |
| Heat-transfer oil | Water | Glass-lined CS | 30–80 | 170–450 |
| Heat-transfer oil | Aqueous solution | Glass-lined CS | 25–70 | 140–400 |
| Heat-transfer oil | Organics | Glass-lined CS | 25–65 | 140–370 |
| Heat-transfer oil | Light oil | Glass-lined CS | 20–70 | 115–400 |
| Heat-transfer oil | Heavy oil | Glass-lined CS | 10–35 | 57–200 |

*Values listed are for moderate nonproximity agitation. CS = carbon steel.

A technique for calculating radial temperature gradients in a packed bed is given by Smith (*Chemical Engineering Kinetics*, McGraw-Hill, New York, 1956).

**Fluidization** occurs when the fluid flow rate is great enough so that the pressure drop across the bed equals the weight of the bed. As stated previously, the solids film thickness adjacent to the wall $d_m$ is difficult to measure and/or predict. Wen and Fau [*Chem. Eng.*, **64**(7), 254 (1957)] give for *external walls:*

$$h = bk(c_s\rho_s)^{0.4}(G\eta/\mu N_f)^{0.36} \qquad (11\text{-}51a)$$

where $b$ = 0.29 (SI) or 11.6 (U.S. customary), $c_s$ = heat capacity of solid, $\rho_s$ = particle density, $\eta$ = fluidization efficiency (Fig. 11-31) and $N_f$ = bed expansion ratio (Fig. 11-32). *For internal walls,* Wen

**TABLE 11-8 External Coils; Typical Overall Coefficients***

*U* expressed in Btu/(h · ft² · °F)

| Type of coil | Coil spacing, in.† | Fluid in coil | Fluid in vessel | Temp. range, °F. | U‡ without cement | U with heat-transfer cement |
|---|---|---|---|---|---|---|
| ⅜ in. o.d. copper tubing attached with bands at 24-in. spacing | 2 | 5 to 50 lb./sq. in. gage steam | Water under light agitation | 158–210 | 1–5 | 42–46 |
| | 3⅛ | | | 158–210 | 1–5 | 50–53 |
| | 6¼ | | | 158–210 | 1–5 | 60–64 |
| | 12½ or greater | | | 158–210 | 1–5 | 69–72 |
| ⅜ in. o.d. copper tubing attached with bands at 24-in. spacing | 2 | 50 lb./sq. in. gage steam | No. 6 fuel oil under light agitation | 158–258 | 1–5 | 20–30 |
| | 3⅛ | | | 158–258 | 1–5 | 25–38 |
| | 6¼ | | | 158–240 | 1–5 | 30–40 |
| | 12½ or greater | | | 158–238 | 1–5 | 35–46 |
| Panel coils | | 50 lb./sq. in. gage steam | Boiling water | 212 | 29 | 48–54 |
| | | Water | Water | 158–212 | 8–30 | 19–48 |
| | | Water | No. 6 fuel oil | 228–278 | 6–15 | 24–56 |
| | | | Water | 130–150 | 7 | 15 |
| | | | No. 6 fuel oil | 130–150 | 4 | 9–19 |

*Data courtesy of Thermon Manufacturing Co.

†External surface of tubing or side of panel coil facing tank.

‡For tubing, the coefficients are more dependent upon tightness of the coil against the tank than upon either fluid. The low end of the range is recommended.

NOTE: To convert British thermal units per hour-square foot-degrees Fahrenheit to joules per square meter-second-kelvins, multiply by 5.6783; to convert inches to meters, multiply by 0.0254; and to convert pounds-force per square inch to kilopascals, multiply by 6.895.

**TABLE 11-9 Thermal Properties of Various Materials as Affecting Conductive Heat Transfer**

| Material | Thermal conductivity, B.t.u./(hr.)(sq. ft.)(°F./ft.) | Volume specific heat, B.t.u./(cu. ft.)(°F.) | Thermal diffusivity, sq. ft./hr. |
|---|---|---|---|
| Air | 0.0183 | 0.016 | 1.143 |
| Water | 0.3766 | 62.5 | 0.0755 |
| Double steel plate, sand divider | 0.207 | 19.1 | 0.0108 |
| Sand | 0.207 | 19.1 | 0.0108 |
| Powdered iron | 0.0533 | 12.1 | 0.0044 |
| Magnetite iron ore | 0.212 | 63 | 0.0033 |
| Aerocat catalysts | 0.163 | 20 | 0.0062 |
| Table salt | 0.168 | 12.6 | 0.0133 |
| Bone char | 0.0877 | 16.9 | 0.0051 |
| Pitch coke | 0.333 | 16.2 | 0.0198 |
| Phenolformaldehyde resin granules | 0.0416 | 10.5 | 0.0042 |
| Phenolformaldehyde resin powder | 0.070 | 10 | 0.0070 |
| Powdered coal | 0.070 | 15 | 0.0047 |

To convert British thermal units per hour-square foot-degrees Fahrenheit to joules per meter-second-kelvins, multiply by 1.7307; to convert British thermal units per cubic foot-degrees Fahrenheit to joules per cubic meter-kelvins, multiply by (6.707)($10^4$); and to convert square feet per hour to square meters per second, multiply by (2.581)($10^{-5}$).

and Fau give

$$h_i = bhG^{-0.37} \tag{11-51b}$$

where $b$ = 0.78 (SI) or 9 (U.S. customary), $h_i$ is the coefficient for internal walls, and $h$ is calculated from Eq. (11-51*a*). $G_{mf}$, the minimum fluidizing velocity, is defined by

$$G_{mf} = \frac{b\rho_g^{1.1}(\rho_s - \rho_g)^{0.9}D_p^2}{\mu} \tag{11-51c}$$

where $b$ = (1.23)($10^{-2}$) (SI) or (5.23)($10^5$) (U.S. customary).

Wender and Cooper [*Am. Inst. Chem. Eng. J.*, **4**, 15 (1958)] developed an empirical correlation for *internal walls:*

$$\frac{hD_p/k}{1-\epsilon_v}\left(\frac{k}{c\rho}\right)^{0.43} = bC_R\left(\frac{D_pG}{\mu}\right)^{0.23}\left(\frac{c_s}{c_g}\right)^{0.80}\left(\frac{\rho_s}{\rho_g}\right)^{0.66} \tag{11-52a}$$

where $b$ = (3.51)($10^{-4}$) (SI) or 0.033 (U.S. customary) and $C_R$ = correction for displacement of the immersed tube from the axis of the vessel (see the reference). For external walls:

$$\frac{hD_p}{k_g(1-\epsilon_v)(c_s\rho_s/c_g\rho_g)} = f(1 + 7.5e^{-x}) \tag{11-52b}$$

where $x = 0.44L_Hc_s/D_tc_g$ and $f$ is given by Fig. 11-33. An important feature of this equation is inclusion of the ratio of bed depth to vessel diameter $L_H/D_t$.

For **dilute fluidized beds** on the shell side of an unbaffled tubular bundle Genetti and Knudsen [*Inst. Chem. Eng. (London) Symp. Ser.* 3,172 (1968)] obtained the relation:

$$\frac{hD_p}{k} = \frac{5\phi(1-\epsilon_v)}{\left[1 + \frac{580}{N_{Re}}\left(\frac{k_s}{D_p^{1.5}c_s\rho_s g^{0.5}}\right)\left(\frac{\rho_s}{\rho_g}\right)^{1.1}\left(\frac{G_{mf}}{G}\right)^{7/3}\right]^2} \tag{11-53a}$$

where $\phi$ = particle surface area per area of sphere of same diameter. When particle transport occurred through the bundle, the heat-transfer coefficients could be predicted by

$$j_H = 0.14(N_{Re}/\phi)^{-0.68} \tag{11-53b}$$

In Eqs. (11-53*a*) and (11-53*b*), $N_{Re}$ is based on particle diameter and superficial fluid velocity.

Zenz and Othmer (see "Introduction: General References") give an excellent summary of fluidized bed-to-wall heat-transfer investigations.

**Solidification** involves heavy heat loads transferred essentially at a steady temperature difference. It also involves the varying values of liquid- and solid-phase thickness and thermal diffusivity. When these are substantial and/or in the case of a liquid flowing over a changing solid layer interface, Siegel and Savino (ASME Paper 67-WA/Ht-34, November 1967) offer equations and charts for prediction of the layer-growth time. For solidification (or melting) of a slab or a semi-infinite bar, initially at its transition temperature, the position of the interface is given by the one-dimensional Newmann's solution given in Carslaw and Jaeger (*Conduction of Heat in Solids*, Clarendon Press, Oxford, 1959).

Later work by Hashem and Sliepcevich [*Chem. Eng. Prog.*, **63**, *Symp. Ser.* 79, 35, 42 (1967)] offers more accurate second-order finite-difference equations.

The heat-transfer rate is found to be substantially higher under conditions of **agitation.** The heat transfer is usually said to occur by combined conductive and convective modes. A discussion and explanation are given by Holt [*Chem. Eng.*, **69**(1), 110 (1962)]. Prediction of $U_{co}$ by Eq. (11-48) can be accomplished by replacing $\alpha$ by $\alpha_e$, the effective thermal diffusivity of the bed. To date so little work has been performed in evaluating the effect of mixing parameters that few predictions can be made. However, for agitated liquid-phase devices Eq. (18-19) is applicable. Holt (loc. cit.) shows that this equation can be converted for solids heat transfer to yield

$$U_{co} = a'c_sD_t^{-0.3}N^{0.7}(\cos\,\omega)^{0.2} \tag{11-54}$$

where $D_t$ = agitator or vessel diameter; $N$ = turning speed, r/min; $\omega$ = effective angle of repose of the burden; and $a'$ is a proportionality con-

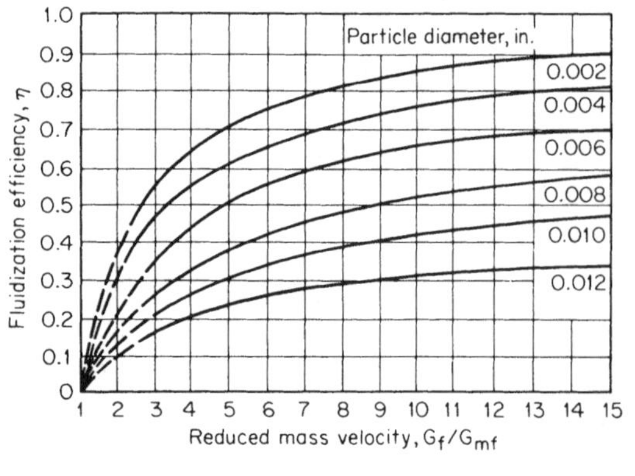


**FIG. 11-31** Fluidization efficiency.

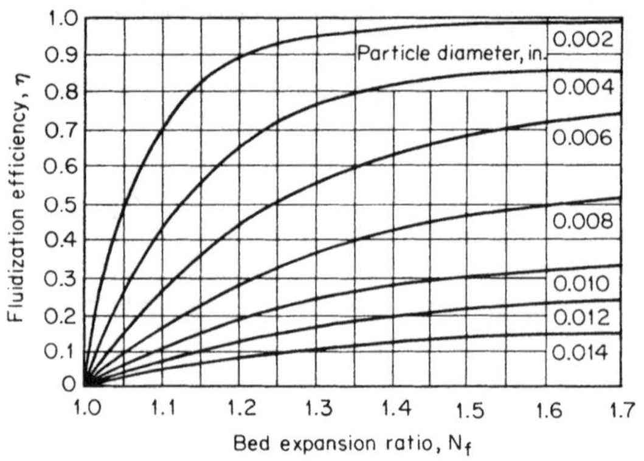


**FIG. 11-32** Bed expansion ratio.

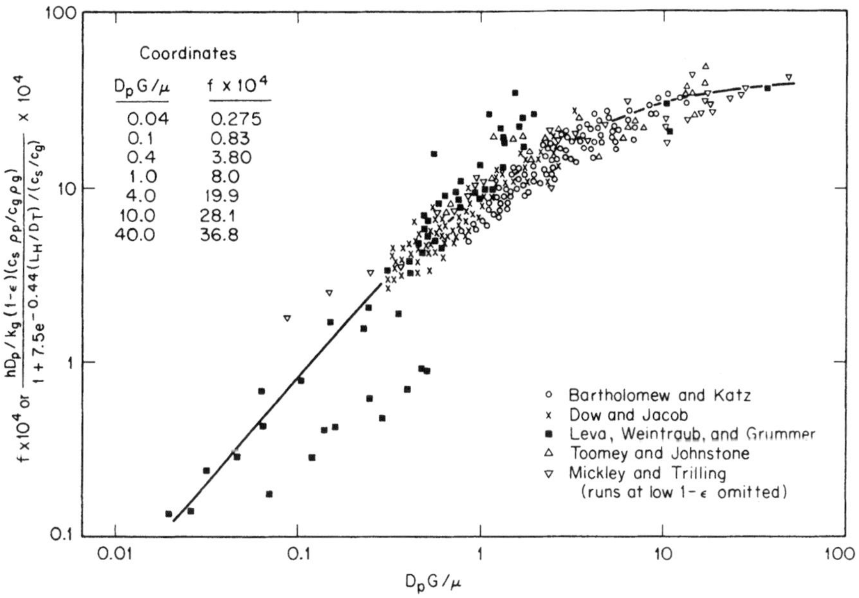


**FIG. 11-33** *f* factor for Eq. (11-52*b*).

stant. This is applicable for such devices as agitated pans, agitated kettles, spiral conveyors, and rotating shells.

The solids passage time through **rotary devices** is given by Saemann [*Chem. Eng. Prog.*, **47,** 508, (1951)]:

$$\theta = 0.318L \sin \omega / S_r N D_t \tag{11-55a}$$

and by Marshall and Friedman [*Chem. Eng. Prog.*, **45,** 482–493, 573–588 (1949)]:

$$\theta = (0.23L/S_r N^{0.9} D_t) \pm (0.6BLG/F_a) \tag{11-55b}$$

where the second term of Eq. (11-55*b*) is positive for counterflow of air, negative for concurrent flow, and zero for indirect rotary shells. From these equations a predictive equation is developed for rotary-shell devices, which is analogous to Eq. (11-54):

$$U_{co} = \frac{b' c_s D_t N^{0.9} Y}{(\Delta t) L \sin \omega} \tag{11-56}$$

where $\theta$ = solids-bed passage time through the shell, min; $S_r$ = shell slope; $L$ = shell length; $Y$ = percent fill; and $b'$ is a proportionality constant.

**Vibratory devices** which constantly agitate the solids bed maintain a relatively constant value for $U_{co}$ such that

$$U_{co} = a' c_s \alpha_e \tag{11-57}$$

with $U_{co}$ having a nominal value of 114 J/(m²·s·K) [20 Btu/(h·ft²·°F)].

**Contactive (Direct) Heat Transfer** Contactive heat-transfer equipment is so constructed that the particulate burden in solid phase is directly exposed to and permeated by the heating or cooling medium (Sec. 20). The carrier may either heat or cool the solids. A large amount of the industrial heat processing of solids is effected by this mechanism. Physically, these can be classified into packed beds and various degrees of agitated beds from dilute to dense fluidized beds.

The temperature difference for heat transfer is the log-mean temperature difference when the particles are large and/or the beds packed, or the difference between the inlet fluid temperature $t_3$ and average exhausting fluid temperature $t_4$, expressed $\Delta_3 t_4$, for small particles. The use of the log mean for packed beds has been confirmed by Thodos and Wilkins (Second American Institute of Chemical Engineers-IIQPR Meeting, Paper 30D, Tampa, May 1968). When fluid and solid flow directions are axially concurrent and particle size is small, as in a vertical-shell fluid bed, the temperature of the exiting solids $t_2$ (which is also that of exiting gas $t_4$) is used as $\Delta_3 t_2$, as shown by Levenspiel, Olson, and Walton [*Ind. Eng. Chem.*, **44,** 1478 (1952)], Marshall [*Chem. Eng. Prog.*, **50,** *Monogr. Ser.* 2, 77 (1954)], Leva (*Fluidization*, McGraw-Hill, New York, 1959), and Holt (Fourth Int. Heat Transfer Conf. Paper 11, American Institute of Chemical Engineers-American Society of Mechanical Engineers, Buffalo, 1960). This temperature difference is also applicable for well-fluidized beds of small particles in cross-flow as in various vibratory carriers.

The **packed-bed-to-fluid heat-transfer coefficient** has been investigated by Baumeister and Bennett [*Am. Inst. Chem. Eng. J.*, **4,** 69 (1958)], who proposed the equation

$$j_H = (h/cG)(c\mu/k)^{2/3} = aN_{Re}^m \tag{11-58}$$

where $N_{Re}$ is based on particle diameter and superficial fluid velocity. Values of $a$ and $m$ are as follows:

| $D_t/D_p$ (dimensionless) | $a$ | $m$ |
|---|---|---|
| 10.7 | 1.58 | −0.40 |
| 16.0 | 0.95 | −0.30 |
| 25.7 | 0.92 | −0.28 |
| >30 | 0.90 | −0.28 |

Glaser and Thodos [*Am. Inst. Chem. Eng. J.*, **4,** 63 (1958)] give a correlation involving individual particle shape and bed porosity. Kunii and Suzuki [*Int. J. Heat Mass Transfer,* **10,** 845 (1967)] discuss heat and mass transfer in packed beds of fine particles.

**Particle-to-fluid** heat-transfer coefficients in gas **fluidized beds** are predicted by the relation (Zenz and Othmer, op. cit.)

$$\frac{hD_p}{k} = 0.017(D_p G_{mf}/\mu)^{1.21} \tag{11-59a}$$

where $G_{mf}$ is the superficial mass velocity at incipient fluidization.

A more general equation is given by Frantz [*Chem. Eng.*, **69**(20), 89 (1962)]:

$$hD_p/k = 0.015(D_p G/\mu)^{1.6}(c\mu/k)^{0.67} \tag{11-59b}$$

where $h$ is based on true gas temperature.

**Bed-to-wall coefficients in dilute-phase transport** generally can be predicted by an equation of the form of Eq. (5-50). For example,

Bonilla et al. (American Institute of Chemical Engineers Heat Transfer Symp., Atlantic City, N.J., December 1951) found for 1- to 2-μm chalk particles in water up to 8 percent by volume that the coefficient on Eq. (5-50) is 0.029 where $k$, $\rho$, and $c$ were arithmetic weighted averages and the viscosity was taken equal to the coefficient of rigidity. Farber and Morley [*Ind. Eng. Chem.*, **49,** 1143 (1957)] found the coefficient on Eq. (5-50) to be 0.025 for the upward flow of air transporting silica-alumina catalyst particles at rates less than 2 kg solids kg air (2 lb solids/lb air). Physical properties used were those of the transporting gas. See Zenz and Othmer (op. cit.) for additional details covering wider porosity ranges.

The thermal performance of **cylindrical rotating shell** units is based upon a volumetric heat-transfer coefficient

$$U_{ct} = \frac{Q}{V_r(\Delta t)} \tag{11-60a}$$

where $V_r$ = volume. This term indirectly includes an area factor so that thermal performance is governed by a cross-sectional area rather than by a heated area. Use of the heated area is possible, however:

$$U_{ct} = \frac{Q}{(\Delta_3 t_2)A} \quad \text{or} \quad \frac{Q}{(\Delta_3 t_4)A} \tag{11-60b}$$

For **heat transfer directly to solids,** predictive equations give directly the volume $V$ or the heat-transfer area $A$, as determined by heat balance and airflow rate. For devices with gas flow normal to a fluidized-solids bed,

$$A = \frac{Q}{\Delta t_p(c\rho_g)(F_g)} \tag{11-61}$$

where $\Delta t_p = \Delta_3 t_4$ as explained above, $c\rho$ = volumetric specific heat, and $F_g$ = gas flow rate. For air, $c\rho$ at normal temperature and pressure is about 1100 J/(m³·K) [0.0167 Btu/(ft³·°F)]; so

$$A = \frac{bQ}{(\Delta_3 t_4)F_g} \tag{11-62}$$

where $b$ = 0.0009 (SI) or 60 (U.S. customary). Another such equation, for stationary vertical-shell and some horizontal rotary-shell and pneumatic-transport devices in which the gas flow is parallel with and directionally concurrent with the fluidized bed, is the same as Eq. (11-62) with $\Delta_3 t_4$ replaced by $\Delta_3 t_2$. If the operation involves drying or chemical reaction, the heat load $Q$ is much greater than for sensible-heat transfer only. Also, the gas flow rate to provide moisture carry-off and stoichiometric requirements must be considered and simultaneously provided. A good treatise on the latter is given by Pinkey and Plint (*Miner. Process.*, June 1968, p. 17).

**Evaporative cooling** is a special patented technique that often can be advantageously employed in cooling solids by contactive heat transfer. The drying operation is terminated *before* the desired final moisture content is reached, and solids temperature is at a moderate value. The cooling operation involves contacting the burden (preferably fluidized) with air at normal temperature and pressure. The air adiabatically absorbs and carries off a large part of the moisture and, in doing so, picks up heat from the warm (or hot) solids particles to supply the latent heat demand of evaporation. For entering solids at temperatures of 180°C (350°F) and less with normal heat-capacity values of 0.85 to 1.0 kJ/(kg·K) [0.2 to 0.25 Btu/(lb·°F)], the effect can be calculated by:

1. Using 285 m³ (1000 ft³) of airflow at normal temperature and pressure at 40 percent relative humidity to carry off 0.45 kg (1 lb) of water [latent heat 2326 kJ/kg (1000 Btu/lb)] and to lower temperature by 22 to 28°C (40 to 50°F).
2. Using the lowered solids temperature as $t_3$ and calculating the remainder of the heat to be removed in the regular manner by Eq. (11-62). The required air quantity for (2) must be equal to or greater than that for (1).

When the solids heat capacity is higher (as is the case for most organic materials), the temperature reduction is inversely proportional to the heat capacity.

A nominal result of this technique is that the required airflow rate and equipment size is about two-thirds of that when evaporative cooling is not used. See Sec. 20 for equipment available.

**Convective Heat Transfer** Equipment using the true convective mechanism when the heated particles are mixed with (and remain with) the cold particles is used so infrequently that performance and sizing equations are not available. Such a device is the pebble heater as described by Norton (*Chem. Metall. Eng.*, July 1946). For operation data, see Sec. 9.

Convective heat transfer is often used as an adjunct to other modes, particularly to the conductive mode. It is often more convenient to consider the agitative effect a performance-improvement influence on the thermal diffusivity factor $\alpha$, modifying it to $\alpha_e$, the effective value.

A *pseudo-convective heat-transfer* operation is one in which the heating gas (generally air) is passed over a bed of solids. Its use is almost exclusively limited to drying operations (see Sec. 12, tray and shelf dryers). The operation, sometimes termed direct, is more akin to the conductive mechanism. For this operation, Tsao and Wheelock [*Chem. Eng.*, **74**(13), 201 (1967)] predict the heat-transfer coefficient when radiative and conductive effects are absent by

$$h = bG^{0.8} \tag{11-63}$$

where $b$ = 14.31 (SI) or 0.0128 (U.S. customary), $h$ = convective heat transfer, and $G$ = gas flow rate.

The **drying rate** is given by

$$K_{cv} = \frac{h(T_d - T_w)}{\lambda} \tag{11-64}$$

where $K_{cv}$ = drying rate, for constant-rate period, kg/(m²·s) [lb/(h·ft²)]; $T_d$ and $T_w$ = respective dry-bulb and wet-bulb temperatures of the air; and $\lambda$ = latent heat of evaporation at temperature $T_w$. Note here that the temperature-difference determination of the operation is a simple linear one and of a steady-state nature. Also note that the operation is a function of the airflow rate. Further, the solids are granular with a fairly uniform size, have reasonable capillary voids, are of a firm texture, and have the particle surface wetted.

The coefficient $h$ is also used to predict (in the constant-rate period) the total overall air-to-solids heat-transfer coefficient $U_{cv}$ by

$$1/U_{cv} = 1/h + x/k \tag{11-65}$$

where $k$ = solids thermal conductivity and $x$ is evaluated from

$$x = \frac{z(X_c - X_o)}{X_c - X_e} \tag{11-65a}$$

where $z$ = bed (or slab) thickness and is the total thickness when drying and/or heat transfer is from one side only but is one-half of the thickness when drying and/or heat transfer is simultaneously from both sides; $X_o$, $X_c$, and $X_e$ are respectively the initial (or feed-stock), critical, and equilibrium (with the drying air) moisture contents of the solids, all in kg $H_2O$/kg dry solids (lb $H_2O$/lb dry solids). This coefficient is used to predict the *instantaneous* drying rate

$$-\frac{W}{A}\frac{dX}{d\theta} = \frac{U_{cv}(T_d - T_w)}{\lambda} \tag{11-66}$$

By rearrangement, this can be made into a design equation as follows:

$$A = -\frac{W\lambda(dX/d\theta)}{U_{cv}(T_d - T_w)} \tag{11-67}$$

where $W$ = weight of dry solids in the equipment, $\lambda$ = latent heat of evaporation, and $\theta$ = drying time. The reader should refer to the full reference article by Tsao and Wheelock (loc. cit.) for other solids conditions qualifying the use of these equations.

**Radiative Heat Transfer** Heat-transfer equipment using the radiative mechanism for divided solids is constructed as a "table" which is stationary, as with trays, or moving, as with a belt, and/or agitated, as with a vibrated pan, to distribute and expose the burden in a plane parallel to (but not in contact with) the plane of the radiant-heat sources. Presence of air is not necessary (see Sec. 12 for vacuum-shelf dryers and Sec. 22 for resublimation). In fact, if air in the intervening space has a high humidity or $CO_2$ content, it acts as an energy absorber, thereby depressing the performance.

For the radiative mechanism, the temperature difference is evaluated as

$$\Delta t = T_e^4 - T_r^4 \tag{11-68}$$

where $T_e$ = absolute temperature of the radiant-heat source, K (°R); and $T_r$ = absolute temperature of the bed of divided solids, K (°R).

Numerical values for $U_{ra}$ for use in the general design equation may be calculated from experimental data by

$$U_{ra} = \frac{Q}{A(T_e^4 - T_r^4)} \quad (11\text{-}69)$$

The literature to date offers practically no such values. However, enough proprietary work has been performed to present a reliable evaluation for the comparison of mechanisms (see "Introduction: Modes of Heat Transfer").

For the radiative mechanism of heat transfer to solids, the rate equation for parallel-surface operations is

$$q_{ra} = b(T_e^4 - T_r^4)i_f \quad (11\text{-}70)$$

where $b = (5.67)(10^{-8})$(SI) or $(0.172)(10^{-8})$(U.S. customary), $q_{ra}$ = radiative heat flux, and $i_f$ = an interchange factor which is evaluated from

$$1/i_f = 1/e_s + 1/e_r - 1 \quad (11\text{-}70a)$$

where $e_s$ = coefficient of emissivity of the source and $e_r$ = "emissivity" (or "absorptivity") of the receiver, which is the divided-solids bed. For the emissivity values, particularly of the heat source $e_s$, an important consideration is the wavelength at which the radiant source emits as well as the flux density of the emission. Data for these values are available from Polentz [*Chem. Eng.*, **65**(7), 137; (8), 151 (1958)] and Adlam (*Radiant Heating,* Industrial Press, New York, p. 40). Both give radiated flux density versus wavelength at varying temperatures. Often, the seemingly cooler but longer wavelength source is the better selection.

Emitting sources are (1) pipes, tubes, and platters carrying steam, 2100 kPa (300 lbf/in²); (2) electrical-conducting glass plates, 150 to 315°C (300 to 600°F) range; (3) light-bulb type (tungsten-filament resistance heater); (4) modules of refractory brick for gas burning at high temperatures and high fluxes; and (5) modules of quartz tubes, also operable at high temperatures and fluxes. For some emissivity values see Table 11-10.

For *predictive work,* where $U_{ra}$ is desired for sizing, this can be obtained by dividing the flux rate $q_{ra}$ by $\Delta t$:

$$U_{ra} = q_{ra}/(T_e^4 - T_r^4) = i_f b \quad (11\text{-}71)$$

where $b = (5.67)(10^{-8})$ (SI) or $(0.172)(10^{-8})$ (U.S. customary). Hence:

$$A = \frac{Q}{U_{ra}(T_e^4 - T_r^4)} \quad (11\text{-}72)$$

where $A$ = bed area of solids in the equipment.

Important considerations in the application of the foregoing equations are:

1. Since the temperature of the emitter is generally known (preselected or readily determined in an actual operation), the absorptivity value $e_r$ is the unknown. This absorptivity is partly a measure of the ability of radiant heat to penetrate the body of a solid particle (or a moisture film) instantly, as compared with diffusional heat transfer by conduction. Such instant penetration greatly reduces processing time and case-hardening effects. Moisture release and other mass transfer, however, still progress by diffusional means.

2. In one of the major applications of radiative devices (drying), the surface-held moisture is a good heat absorber in the 2- to 7-μm wavelength range. Therefore, the absorptivity, color, and nature of the solids are of little importance.

3. For drying, it is important to provide a small amount of venting air to carry away the water vapor. This is needed for two reasons. First, water vapor is a good absorber of 2- to 7-μm energy. Second, water-vapor accumulation depresses further vapor release by the solids. If the air over the solids is kept fairly dry by venting, very little heat is carried off, because dry air does not absorb radiant heat.

4. For some of the devices, when the overall conversion efficiency has been determined, the application is primarily a matter of computing the required heat load. It should be kept in mind, however, that there are two conversion efficiencies that must be differentiated. One measure of efficiency is that with which the source converts input energy to output radiated energy. The other is the overall efficiency that measures the proportion of input energy that is actually absorbed by the solids. This latter is, of course, the one that really matters.

Other applications of radiant-heat processing of solids are the toasting, puffing, and baking of foods and the low-temperature roasting and preheating of plastic powder or pellets. Since the determination of heat loads for these operations is not well established, bench and pilot tests are generally necessary. Such processes require a fast input of heat and higher heat fluxes than can generally be provided by indirect equipment. Because of this, infrared-equipment size and space requirements are often much lower.

Although direct contactive heat transfer can provide high temperatures and heat concentrations and at the same time be small in size, its use may not always be preferable because of undesired side effects such as drying, contamination, case hardening, shrinkage, off color, and dusting.

When radiating and receiving surfaces are not in parallel, as in rotary-kiln devices, and the solids burden bed may be only intermittently exposed and/or agitated, the calculation and procedures become very complex, with photometric methods of optics requiring consideration. The following equation for heat transfer, which allows for convective effects, is commonly used by designers of **high-temperature furnaces:**

$$q_{ra} = Q/A = b\sigma\,[(T_g/100)^4 - (T_s/100)^4] \quad (11\text{-}73)$$

where $b$ = 5.67 (SI) or 0.172 (U.S. customary); $Q$ = total furnace heat transfer; $\sigma$ = an emissivity factor with recommended values of 0.74 for gas, 0.75 for oil, and 0.81 for coal; $A$ = effective area for absorbing heat (here the solids burden exposed area); $T_g$ = exiting-combustion-gas absolute temperature; and $T_s$ = absorbing surface temperature.

In rotary devices, reradiation from the exposed shell surface to the solids bed is a major design consideration. A treatise on furnaces, including radiative heat-transfer effects, is given by Ellwood and Danatos [*Chem. Eng.*, **73**(8), 174 (1966)]. For discussion of radiation heat-transfer computational methods, heat fluxes obtainable, and emissivity values, see Schornshort and Viskanta (ASME Paper 68-H 7-32), Sherman (ASME Paper 56-A-111), and the following subsection.

## SCRAPED-SURFACE EXCHANGERS

Scraped-surface exchangers have a rotating element with spring-loaded scraper blades to scrape the inside surface (Fig. 11-34). Generally a

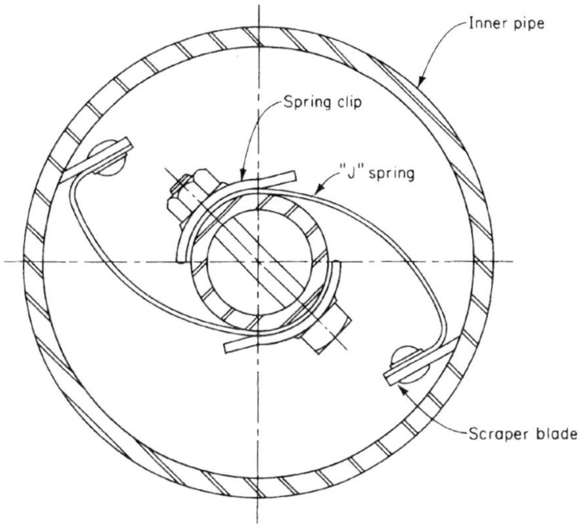


**FIG. 11-34** Scraper blade of scraped-surface exchanger. (*Henry Vogt Machine Co., Inc.*)

**TABLE 11-10 Normal Total Emissivity of Various Surfaces**

A. Metals and Their Oxides

| Surface | t, °F.* | Emissivity* |
|---|---|---|
| Aluminum | | |
| Highly polished plate, 98.3% pure | 440–1070 | 0.039–0.057 |
| Polished plate | 73 | 0.040 |
| Rough plate | 78 | 0.055 |
| Oxidized at 1110°F | 390–1110 | 0.11–0.19 |
| Aluminum-surfaced roofing | 100 | 0.216 |
| Calorized surfaces, heated at 1110°F. | | |
| Copper | 390–1110 | 0.18–0.19 |
| Steel | 390–1110 | 0.52–0.57 |
| Brass | | |
| Highly polished: | | |
| 73.2% Cu, 26.7% Zn | 476–674 | 0.028–0.031 |
| 62.4% Cu, 36.8% Zn, 0.4% Pb, 0.3% Al | 494–710 | 0.033–0.037 |
| 82.9% Cu, 17.0% Zn | 530 | 0.030 |
| Hard rolled, polished: | | |
| But direction of polishing visible | 70 | 0.038 |
| But somewhat attacked | 73 | 0.043 |
| But traces of stearin from polish left on | 75 | 0.053 |
| Polished | 100–600 | 0.096 |
| Rolled plate, natural surface | 72 | 0.06 |
| Rubbed with coarse emery | 72 | 0.20 |
| Dull plate | 120–660 | 0.22 |
| Oxidized by heating at 1110°F | 390–1110 | 0.61–0.59 |
| Chromium; see Nickel Alloys for Ni-Cr steels | 100–1000 | 0.08–0.26 |
| Copper | | |
| Carefully polished electrolytic copper | 176 | 0.018 |
| Commercial, emeried, polished, but pits remaining | 66 | 0.030 |
| Commercial, scraped shiny but not mirror-like | 72 | 0.072 |
| Polished | 242 | 0.023 |
| Plate, heated long time, covered with thick oxide layer | 77 | 0.78 |
| Plate heated at 1110°F | 390–1110 | 0.57 |
| Cuprous oxide | 1470–2010 | 0.66–0.54 |
| Molten copper | 1970–2330 | 0.16–0.13 |
| Gold | | |
| Pure, highly polished | 440–1160 | 0.018–0.035 |
| Iron and steel | | |
| Metallic surfaces (or very thin oxide layer): | | |
| Electrolytic iron, highly polished | 350–440 | 0.052–0.064 |
| Polished iron | 800–1880 | 0.144–0.377 |
| Iron freshly emeried | 68 | 0.242 |
| Cast iron, polished | 392 | 0.21 |
| Wrought iron, highly polished | 100–480 | 0.28 |
| Cast iron, newly turned | 72 | 0.435 |
| Polished steel casting | 1420–1900 | 0.52–0.56 |
| Ground sheet steel | 1720–2010 | 0.55–0.61 |
| Smooth sheet iron | 1650–1900 | 0.55–0.60 |
| Cast iron, turned on lathe | 1620–1810 | 0.60–0.70 |
| Oxidized surfaces: | | |
| Iron plate, pickled, then rusted red | 68 | 0.612 |
| Completely rusted | 67 | 0.685 |
| Rolled sheet steel | 70 | 0.657 |
| Oxidized iron | 212 | 0.736 |
| Cast iron, oxidized at 1100°F | 390–1110 | 0.64–0.78 |
| Steel, oxidized at 1100°F | 390–1110 | 0.79 |
| Smooth oxidized electrolytic iron | 260–980 | 0.78–0.82 |
| Iron oxide | 930–2190 | 0.85–0.89 |
| Rough ingot iron | 1700–2040 | 0.87–0.95 |
| Sheet steel, strong rough oxide layer | 75 | 0.80 |
| Dense shiny oxide layer | 75 | 0.82 |
| Cast plate: | | |
| Smooth | 73 | 0.80 |
| Rough | 73 | 0.82 |
| Cast iron, rough, strongly oxidized | 100–480 | 0.95 |
| Wrought iron, dull oxidized | 70–680 | 0.94 |
| Steel plate, rough | 100–700 | 0.94–0.97 |
| High temperature alloy steels (see Nickel Alloys) | | |
| Molten metal | | |
| Cast iron | 2370–2550 | 0.29 |
| Mild steel | 2910–3270 | 0.28 |
| Lead | | |
| Pure (99.96%), unoxidized | 260–440 | 0.057–0.075 |
| Gray oxidized | 75 | 0.281 |
| Oxidized at 390°F | 390 | 0.63 |
| Mercury | 32–212 | 0.09–0.12 |
| Molybdenum filament | 1340–4700 | 0.096–0.292 |
| Monel metal, oxidized at 1110°F | 390–1110 | 0.41–0.46 |
| Nickel | | |
| Electroplated on polished iron, then polished | 74 | 0.045 |
| Technically pure (98.9% Ni, + Mn), polished | 440–710 | 0.07–0.087 |
| Electropolated on pickled iron, not polished | 68 | 0.11 |
| Wire | 368–1844 | 0.096–0.186 |
| Plate, oxidized by heating at 1110°F | 390–1110 | 0.37–0.48 |
| Nickel oxide | 1200–2290 | 0.59–0.86 |
| Nickel alloys | | |
| Chromnickel | 125–1894 | 0.64–0.76 |
| Nickelin (18–32 Ni; 55–68 Cu; 20 Zn), gray oxidized | 70 | 0.262 |
| KA-2S alloy steel (8% Ni; 18% Cr), light silvery, rough, brown, after heating | 420–914 | 0.44–0.36 |
| After 42 hr. heating at 980°F | 420–980 | 0.62–0.73 |
| NCT-3 alloy (20% Ni; 25% Cr), brown, splotched, oxidized from service | 420–980 | 0.90–0.97 |
| NCT-6 alloy (60% Ni; 12% Cr), smooth, black, firm adhesive oxide coat from service | 520–1045 | 0.89–0.82 |
| Platinum | | |
| Pure, polished plate | 440–1160 | 0.054–0.104 |
| Strip | 1700–2960 | 0.12–0.17 |
| Filament | 80–2240 | 0.036–0.192 |
| Wire | 440–2510 | 0.073–0.182 |
| Silver | | |
| Polished, pure | 440–1160 | 0.0198–0.0324 |
| Polished | 100–700 | 0.0221–0.0312 |
| Steel, see Iron | | |
| Tantalum filament | 2420–5430 | 0.194–0.31 |
| Tin—bright tinned iron sheet | 76 | 0.043 and 0.064 |
| Tungsten | | |
| Filament, aged | 80–6000 | 0.032–0.35 |
| Filament | 6000 | 0.39 |
| Zinc | | |
| Commercial, 99.1% pure, polished | 440–620 | 0.045–0.053 |
| Oxidized by heating at 750°F. | 750 | 0.11 |
| Galvanized sheet iron, fairly bright | 82 | 0.228 |
| Galvanized sheet iron, gray oxidized | 75 | 0.276 |

B. Refractories, Building Materials, Paints, and Miscellaneous

| Surface | t, °F.* | Emissivity* |
|---|---|---|
| Asbestos | | |
| Board | 74 | 0.96 |
| Paper | 100–700 | 0.93–0.945 |
| Brick | | |
| Red, rough, but no gross irregularities | 70 | 0.93 |
| Silica, unglazed, rough | 1832 | 0.80 |
| Silica, glazed, rough | 2012 | 0.85 |
| Grog brick, glazed | 2012 | 0.75 |
| See Refractory Materials below. | | |
| Carbon | | |
| T-carbon (Gebr. Siemens) 0.9% ash (this started with emissivity at 260°F. of 0.72, but on heating changed to values given) | 260–1160 | 0.81–0.79 |
| Carbon filament | 1900–2560 | 0.526 |
| Candle soot | 206–520 | 0.952 |
| Lampblack-waterglass coating | 209–362 | 0.959–0.947 |

**TABLE 11-10 Normal Total Emissivity of Various Surfaces** (*Concluded*)

A. Metals and Their Oxides

| Surface | *t*, °F.* | Emissivity* |
|---|---|---|
| Same | 260–440 | 0.957–0.952 |
| Thin layer on iron plate | 69 | 0.927 |
| Thick coat | 68 | 0.967 |
| Lampblack, 0.003 in. or thicker | 100–700 | 0.945 |
| Enamel, white fused, on iron | 66 | 0.897 |
| Glass, smooth | 72 | 0.937 |
| Gypsum, 0.02 in. thick on smooth or blackened plate | 70 | 0.903 |
| Marble, light gray, polished | 72 | 0.931 |
| Oak, planed | 70 | 0.895 |
| Oil layers on polished nickel (lube oil) | 68 | |
| Polished surface, alone | | 0.045 |
| +0.001-in. oil | | 0.27 |
| +0.002-in. oil | | 0.46 |
| +0.005-in. oil | | 0.72 |
| Infinitely thick oil layer | | 0.82 |
| Oil layers on aluminum foil (linseed oil) | | |
| Al foil | 212 | 0.087† |
| +1 coat oil | 212 | 0.561 |
| +2 coats oil | 212 | 0.574 |
| Paints, lacquers, varnishes | | |
| Snowhite enamel varnish or rough iron plate | 73 | 0.906 |
| Black shiny lacquer, sprayed on iron | 76 | 0.875 |
| Black shiny shellac on tinned iron sheet | 70 | 0.821 |
| Black matte shellac | 170–295 | 0.91 |
| Black lacquer | 100–200 | 0.80–0.95 |
| Flat black lacquer | 100–200 | 0.96–0.98 |
| White lacquer | 100–200 | 0.80–0.95 |

| Surface | *t*, °F.* | Emissivity* |
|---|---|---|
| Oil paints, sixteen different, all colors | 212 | 0.92–0.96 |
| Aluminum paints and lacquers | | |
| 10% Al, 22% lacquer body, on rough or smooth surface | 212 | 0.52 |
| 26% Al, 27% lacquer body, on rough or smooth surface | 212 | 0.3 |
| Other Al paints, varying age and Al content | 212 | 0.27–0.67 |
| Al lacquer, varnish binder, on rough plate | 70 | 0.39 |
| Al paint, after heating to 620°F | 300–600 | 0.35 |
| Paper, thin | | |
| Pasted on tinned iron plate | 66 | 0.924 |
| On rough iron plate | 66 | 0.929 |
| On black lacquered plate | 66 | 0.944 |
| Plaster, rough lime | 50–190 | 0.91 |
| Porcelain, glazed | 72 | 0.924 |
| Quartz, rough, fused | 70 | 0.932 |
| Refractory materials, 40 different | 1110–1830 | |
| poor radiators | | 0.65 – 0.75<br>0.70 |
| good radiators | | 0.80 – 0.85<br>0.85 – 0.90 |
| Roofing paper | 69 | 0.91 |
| Rubber | | |
| Hard, glossy plate | 74 | 0.945 |
| Soft, gray, rough (reclaimed) | 76 | 0.859 |
| Serpentine, polished | 74 | 0.900 |
| Water | 32–212 | 0.95–0.963 |

*When two temperatures and two emissivities are given, they correspond, first to first and second to second, and linear interpolation is permissible. °C = (°F − 32)/1.8.
†Although this value is probably high, it is given for comparison with the data by the same investigator to show the effect of oil layers. See Aluminum, Part A of this table.

double-pipe construction is used; the scraping mechanism is in the inner pipe, where the process fluid flows; and the cooling or heating medium is in the outer pipe. The most common size has 6-in inside and 8-in outside pipes. Also available are 3- by 4-in, 8- by 10-in, and 12- by 14-in sizes (in × 25.4 = mm). These double-pipe units are commonly connected in series and arranged in double stands.

For **chilling** and **crystallizing** with an evaporating refrigerant, a 27-in shell with seven 6-in pipes is available (Henry Vogt Machine Co.). In direct contact with the scraped surface is the process fluid which may deposit crystals upon chilling or be extremely fouling or of very high viscosity. Motors, chain drives, appropriate guards, and so on are required for the rotating element. For chilling service with a refrigerant in the outer shell, an accumulator drum is mounted on top of the unit.

Scraped-surface exchangers are particularly suitable for heat transfer with crystallization, heat transfer with severe fouling of surfaces, heat transfer with solvent extraction, and heat transfer of high-viscosity fluids. They are extensively used in paraffin-wax plants and in petrochemical plants for crystallization.

# TEMA-STYLE SHELL-AND-TUBE HEAT EXCHANGERS

## TYPES AND DEFINITIONS

TEMA-style shell-and-tube-type exchangers constitute the bulk of the unfired heat-transfer equipment in chemical-process plants, although increasing emphasis has been developing in other designs. These exchangers are illustrated in Fig. 11-35, and their features are summarized in Table 11-11.

**TEMA Numbering and Type Designation** Recommended practice for the designation of TEMA-style shell-and-tube heat exchangers by numbers and letters has been established by the Tubular Exchanger Manufacturers Association (TEMA). This information from the sixth edition of the TEMA Standards is reproduced in the following paragraphs.

It is recommended that heat-exchanger size and type be designated by numbers and letters.

1. *Size.* Sizes of shells (and tube bundles) shall be designated by numbers describing shell (and tube-bundle) diameters and tube lengths as follows:

2. *Diameter.* The nominal diameter shall be the inside diameter of the shell in inches, rounded off to the nearest integer. For kettle reboilers the nominal diameter shall be the port diameter followed by the shell diameter, each rounded off to the nearest integer.

3. *Length.* The nominal length shall be the tube length in inches. Tube length for straight tubes shall be taken as the actual overall length. For U tubes the length shall be taken as the straight length from end of tube to bend tangent.

4. *Type.* Type designation shall be by letters describing stationary head, shell (omitted for bundles only), and rear head, in that order, as indicated in Fig. 11-1.

**Typical Examples** (*A*) Split-ring floating-heat exchanger with removable channel and cover, single-pass shell, 591-mm (23¼-in) inside diameter with tubes 4.9 m (16 ft) long. SIZE 23–192 TYPE AES.

(*B*) U-tube exchanger with bonnet-type stationary head, split-flow shell, 483-mm (19-in) inside diameter with tubes 21-m (7-ft) straight length. SIZE 19–84 TYPE GBU.

(*C*) Pull-through floating-heat-kettle-type reboiler having stationary head integral with tube sheet, 584-mm (23-in) port diameter and 940-mm (37-in) inside shell diameter with tubes 4.9-m (16-ft) long. SIZE 23/37–192 TYPE CKT.

(*D*) Fixed-tube sheet exchanger with removable channel and cover, bonnet-type rear head, two-pass shell, 841-mm (33⅛-in) diameter with tubes 2.4 m (8-ft) long. SIZE 33–96 TYPE AFM.

(*E*) Fixed-tube sheet exchanger having stationary and rear heads integral with tube sheets, single-pass shell, 432-mm (17-in) inside diameter with tubes 4.9-m (16-ft) long. SIZE 17–192 TYPE CEN.

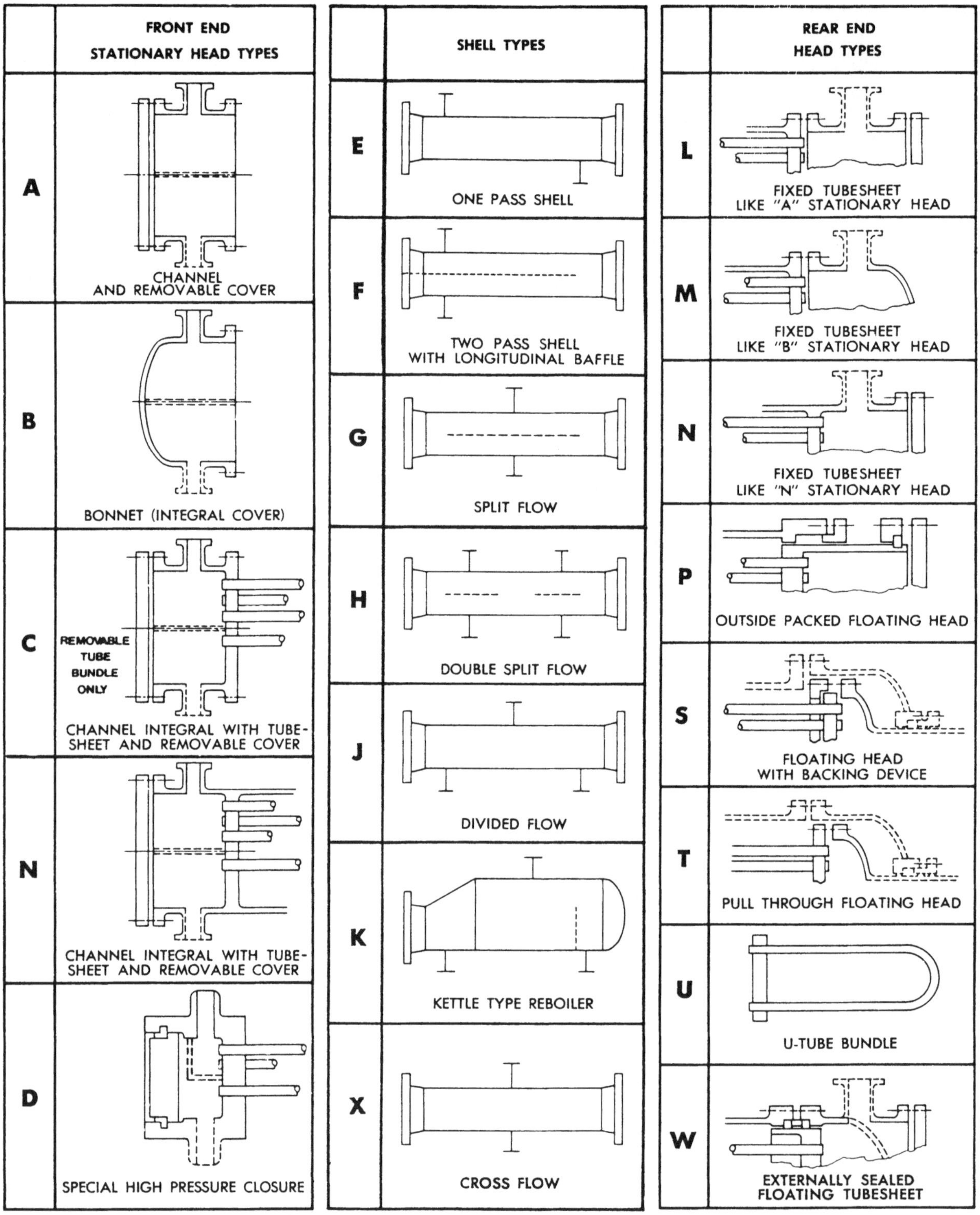


**FIG. 11-35** TEMA-type designations for shell-and-tube heat exchangers. (Standards of Tubular Exchanger Manufacturers Association, 6th ed., 1978.)

**TABLE 11-11 Features of TEMA Shell-and-Tube-Type Exchangers***

| Type of design | Fixed tube sheet | U-tube | Packed lantern-ring floating head | Internal floating head (split backing ring) | Outside-packed floating head | Pull-through floating head |
|---|---|---|---|---|---|---|
| T.E.M.A. rear-head type | L or M or N | U | W | S | P | T |
| Relative cost increases from A (least expensive) through E (most expensive) | B | A | C | E | D | E |
| Provision for differential expansion | Expansion joint in shell | Individual tubes free to expand | Floating head | Floating head | Floating head | Floating head |
| Removable bundle | No | Yes | Yes | Yes | Yes | Yes |
| Replacement bundle possible | No | Yes | Yes | Yes | Yes | Yes |
| Individual tubes replaceable | Yes | Only those in outside row† | Yes | Yes | Yes | Yes |
| Tube cleaning by chemicals inside and outside | Yes | Yes | Yes | Yes | Yes | Yes |
| Interior tube cleaning mechanically | Yes | Special tools required | Yes | Yes | Yes | Yes |
| Exterior tube cleaning mechanically: | | | | | | |
| Triangular pitch | No | No‡ | No‡ | No‡ | No‡ | No‡ |
| Square pitch | No | Yes | Yes | Yes | Yes | Yes |
| Hydraulic-jet cleaning: | | | | | | |
| Tube interior | Yes | Special tools required | Yes | Yes | Yes | Yes |
| Tube exterior | No | Yes | Yes | Yes | Yes | Yes |
| Double tube sheet feasible | Yes | Yes | No | No | Yes | No |
| Number of tube passes | No practical limitations | Any even number possible | Limited to one or two passes | No practical limitations§ | No practical limitations | No practical limitations§ |
| Internal gaskets eliminated | Yes | Yes | Yes | No | Yes | No |

NOTE: Relative costs A and B are not significantly different and interchange for long lengths of tubing.
*Modified from page a-8 of the Patterson-Kelley Co. Manual No. 700A, Heat Exchangers.
†U-tube bundles have been built with tube supports which permit the U-bends to be spread apart and tubes inside of the bundle replaced.
‡Normal triangular pitch does not permit mechanical cleaning. With a wide triangular pitch, which is equal to 2 (tube diameter plus cleaning lane)/$\sqrt{3}$, mechanical cleaning is possible on removable bundles. This wide spacing is infrequently used.
§For odd number of tube side passes, floating head requires packed joint or expansion joint.

**Functional Definitions** Heat-transfer equipment can be designated by type (e.g., fixed tube sheet, outside packed head, etc.) or by function (chiller, condenser, cooler, etc.). Almost any type of unit can be used to perform any or all of the listed functions. Many of these terms have been defined by Donahue [*Pet. Process.,* 103 (March 1956)].

| Equipment | Function |
|---|---|
| Chiller | Cools a fluid to a temperature below that obtainable if water only were used as a coolant. It uses a refrigerant such as ammonia or Freon. |
| Condenser | Condenses a vapor or mixture of vapors, either alone or in the presence of a noncondensable gas. |
| Partial condenser | Condenses vapors at a point high enough to provide a temperature difference sufficient to preheat a cold stream of process fluid. This saves heat and eliminates the need for providing a separate preheater (using flame or steam). |
| Final condenser | Condenses the vapors to a final storage temperature of approximately 37.8°C (100°F). It uses water cooling, which means that the transferred heat is lost to the process. |
| Cooler | Cools liquids or gases by means of water. |
| Exchanger | Performs a double function: (1) heats a cold fluid by (2) using a hot fluid which it cools. None of the transferred heat is lost. |
| Heater | Imparts sensible heat to a liquid or a gas by means of condensing steam or Dowtherm. |
| Reboiler | Connected to the bottom of a fractionating tower, it provides the reboil heat necessary for distillation. The heating medium may be either steam or a hot-process fluid. |
| Thermosiphon reboiler | Natural circulation of the boiling medium is obtained by maintaining sufficient liquid head to provide for circulation. |
| Forced-circulation reboiler | A pump is used to force liquid through the reboiler. |
| Steam generator | Generates steam for use elsewhere in the plant by using the available high-level heat in tar or a heavy oil. |
| Superheater | Heats a vapor above the saturation temperature. |
| Vaporizer | A heater which vaporizes part of the liquid. |
| Waste-heat boiler | Produces steam; similar to steam generator, except that the heating medium is a hot gas or liquid produced in a chemical reaction. |

## GENERAL DESIGN CONSIDERATIONS

**Selection of Flow Path** In selecting the flow path for two fluids through an exchanger, several general approaches are used. The tube-side fluid is more corrosive or dirtier or at a higher pressure. The shell-side fluid is a liquid of high viscosity or a gas.

When alloy construction for one of the two fluids is required, a carbon steel shell combined with alloy tube-side parts is less expensive than alloy in contact with the shell-side fluid combined with carbon steel headers.

Cleaning of the inside of tubes is more readily done than cleaning of exterior surfaces.

For gauge pressures in excess of 2068 kPa (300 lbf/in$^2$) for one of the fluids, the less expensive construction has the high-pressure fluid in the tubes.

For a given pressure drop, higher heat-transfer coefficients are obtained on the shell side than on the tube side.

Heat-exchanger shutdowns are most often caused by fouling, corrosion, and erosion.

**Construction Codes** "Rules for Construction of Pressure Vessels, Division 1," which is part of Section VIII of the ASME Boiler and Pressure Vessel Code (American Society of Mechanical Engineers), serves as a construction code by providing minimum standards. New editions of the code are usually issued every 3 years. Interim revisions are made semiannually in the form of addenda. Compliance with ASME Code requirements is mandatory in much of the United States and Canada. Originally these rules were not prepared for heat exchangers. However, the welded joint between tube sheet and shell of the fixed-tube-sheet heat exchanger is now included. A nonmandatory appendix on tube-to-tube-sheet joints is also included. Additional rules for heat exchangers are being developed.

*Standards of Tubular Exchanger Manufacturers Association,* 6th ed., 1978 (commonly referred to as the TEMA Standards), serve to supplement and define the ASME Code for all shell-and-tube-type heat-exchanger applications (other than double-pipe construction). TEMA Class R design is "for the generally severe requirements of petroleum and related processing applications. Equipment fabricated in accordance with these standards is designed for safety and durability under the rigorous service and maintenance conditions in such applications." TEMA Class C design is "for the generally moderate requirements of commercial and general process applications," while TEMA Class B is "for chemical process service."

The mechanical-design requirements are identical for all three classes of construction. The differences between the TEMA classes are minor and were listed by Rubin [*Hydrocarbon Process.,* **59,** 92 (June 1980)].

Among the topics of the TEMA Standards are nomenclature, fabrication tolerances, inspection, guarantees, tubes, shells, baffles and support plates, floating heads, gaskets, tube sheets, channels, nozzles, end flanges and bolting, material specifications, and fouling resistances.

*Shell and Tube Heat Exchangers for General Refinery Services,* API Standard 660, 4th ed., 1982, is published by the American Petroleum Institute to supplement both the TEMA Standards and the ASME Code. Many companies in the chemical and petroleum processing fields have their own standards to supplement these various requirements. *The Interrelationships between Codes, Standards, and Customer Specifications for Process Heat Transfer Equipment* is a symposium volume which was edited by F. L. Rubin and published by ASME in December 1979. (See discussion of pressure-vessel codes in Sec. 6.)

Design pressures and temperatures for exchangers usually are specified with a margin of safety beyond the conditions expected in service. Design pressure is generally about 172 kPa (25 lbf/in$^2$) greater than the maximum expected during operation or at pump shutoff. Design temperature is commonly 14°C (25°F) greater than the maximum temperature in service.

**Tube Bundle Vibration** Damage from tube vibration has become an increasing problem as plate baffled heat exchangers are designed for higher flow rates and pressure drops. The most effective method of dealing with this problem is the avoidance of cross flow by use of tube support baffles which promote only longitudinal flow. However, even then, strict attention must be given the bundle area under the shell inlet nozzle where flow is introduced through the side of the shell. TEMA has devoted an entire section in its standards to this topic. In general, the mechanisms of tube vibration are as follows:

***Vortex Shedding*** The vortex-shedding frequency of the fluid in cross-flow over the tubes may coincide with a natural frequency of the tubes and excite large resonant vibration amplitudes.

***Fluid-Elastic Coupling*** Fluid flowing over tubes causes them to vibrate with a whirling motion. The mechanism of fluid-elastic coupling occurs when a "critical" velocity is exceeded and the vibration then becomes self-excited and grows in amplitude. This mechanism frequently occurs in process heat exchangers which suffer vibration damage.

***Pressure Fluctuation*** Turbulent pressure fluctuations which develop in the wake of a cylinder or are carried to the cylinder from upstream may provide a potential mechanism for tube vibration. The tubes respond to the portion of the energy spectrum that is close to their natural frequency.

***Acoustic Coupling*** When the shell-side fluid is a low-density gas, acoustic resonance or coupling develops when the standing waves in the shell are in phase with vortex shedding from the tubes. The standing waves are perpendicular to the axis of the tubes and to the direction of cross-flow. Damage to the tubes is rare. However, the noise can be extremely painful.

**Testing** Upon completion of shop fabrication and also during maintenance operations it is desirable hydrostatically to test the shell side of tubular exchangers so that visual examination of tube ends can be made. Leaking tubes can be readily located and serviced. When leaks are determined without access to the tube ends, it is necessary to reroll or reweld all the tube-to-tube-sheet joints with possible damage to the satisfactory joints.

Testing for leaks in heat exchangers was discussed by Rubin [*Chem. Eng.,* **68,** 160–166 (July 24, 1961)].

**Performance** testing of heat exchangers is described in the American Institute of Chemical Engineers' *Standard Testing Procedure for Heat Exchangers,* Sec. 1. "Sensible Heat Transfer in Shell-and-Tube-Type Equipment."

## PRINCIPAL TYPES OF CONSTRUCTION

Figure 11-36 shows details of the construction of the TEMA types of shell-and-tube heat exchangers. These and other types are discussed in the following paragraphs.

**Fixed-Tube-Sheet Heat Exchangers** Fixed-tube-sheet exchangers (Fig. 11-36*b*) are used more often than any other type, and the frequency of use has been increasing in recent years. The tube sheets are welded to the shell. Usually these extend beyond the shell and serve as flanges to which the tube-side headers are bolted. This construction requires that the shell and tube-sheet materials be weldable to each other.

When such welding is not possible, a "blind"-gasket type of construction is utilized. The blind gasket is not accessible for maintenance or replacement once the unit has been constructed. This construction is used for steam surface condensers, which operate under vacuum.

The tube-side header (or channel) may be welded to the tube sheet, as shown in Fig. 11-35 for type C and N heads. This type of construction is less costly than types B and M or A and L and still offers the advantage that tubes may be examined and replaced without disturbing the tube-side piping connections.

There is no limitation on the number of tube-side passes. Shell-side passes can be one or more, although shells with more than two shell-side passes are rarely used.

Tubes can completely fill the heat-exchanger shell. Clearance between the outermost tubes and the shell is only the minimum necessary for fabrication. Between the inside of the shell and the baffles some clearance must be provided so that baffles can slide into the shell. Fabrication tolerances then require some additional clearance between the outside of the baffles and the outermost tubes. The edge distance between the outer tube limit (OTL) and the baffle diameter must be sufficient to prevent vibration of the tubes from breaking through the baffle holes. The outermost tube must be contained within the OTL. Clearances between the inside shell diameter and OTL are 13 mm (½ in) for 635-mm-(25-in-) inside-diameter shells and up, 11 mm (7⁄16 in) for 254- through 610-mm (10- through 24-in) pipe shells, and slightly less for smaller-diameter pipe shells.

Tubes can be replaced. Tube-side headers, channel covers, gaskets, etc., are accessible for maintenance and replacement. Neither the shell-side baffle structure nor the blind gasket is accessible. During tube removal, a tube may break within the shell. When this occurs, it is most difficult to remove or to replace the tube. The usual procedure is to plug the appropriate holes in the tube sheets.

Differential expansion between the shell and the tubes can develop because of differences in length caused by thermal expansion. Various types of expansion joints are used to eliminate excessive stresses caused by expansion. The need for an expansion joint is a function of both the amount of differential expansion and the cycling conditions to be expected during operation. A number of types of expansion joints are available (Fig. 11-37).

*a. Flat plates.* Two concentric flat plates with a bar at the outer edges. The flat plates can flex to make some allowance for differential expansion. This design is generally used for vacuum service and gauge pressures below 103 kPa (15 lbf/in$^2$). All are subject to severe stress during differential expansion.

*b. Flanged-only heads.* The flat plates are flanged (or curved). The diameter of these heads is generally 203 mm (8 in) or more greater than the shell diameter. The welded joint at the shell is subject to the stress referred to before, but the joint connecting the heads is subjected to less stress during expansion because of the curved shape.

*c. Flared shell or pipe segments.* The shell may be flared to connect with a pipe section, or a pipe may be halved and quartered to produce a ring.

*d. Formed heads.* A pair of dished-only or elliptical or flanged and dished heads can be used. These are welded together or connected by a ring. This type of joint is similar to the flanged-only-head type but apparently is subject to less stress.

*e. Flanged and flued heads.* A pair of flanged-only heads is provided with concentric reverse flue holes. These heads are relatively expensive because of the cost of the fluing operation. The curved shape of the heads reduces the amount of stress at the welds to the shell and also connecting the heads.

*f. Toroidal.* The toroidal joint has a mathematically predictable smooth stress pattern of low magnitude, with maximum stresses at sidewalls of the corrugation and minimum stresses at top and bottom.

The foregoing designs were discussed as ring expansion joints by Kopp and Sayre, "Expansion Joints for Heat Exchangers" (ASME Misc. Pap., vol. 6, no. 211). All are statically indeterminate but are subjected to analysis by introducing various simplifying assumptions. Some joints in current industrial use are of lighter wall construction than is indicated by the method of this paper.

*g. Bellows.* Thin-wall bellows joints are produced by various manufacturers. These are designed for differential expansion and are tested for axial and transverse movement as well as for cyclical life. Bellows may be of stainless steel, nickel alloys, or copper. (Aluminum, Monel, phosphor bronze, and titanium bellows have been manufactured.) Welding nipples of the same composition as the heat-exchanger shell are generally furnished. The bellows may be hydraulically formed from a single piece of metal or may consist of welded pieces. External insulation covers of carbon steel are often provided to protect the light-gauge bellows from damage. The cover also prevents insulation from interfering with movement of the bellows (see *h*).

*h. Toroidal bellows.* For high-pressure service the bellows type of joint has been modified so that movement is taken up by thin-wall small-diameter bellows of a toroidal shape. Thickness of parts under high pressure is reduced considerably (see *f*).

Improper handling during manufacture, transit, installation, or maintenance of the heat exchanger equipped with the thin-wall-bellows type or toroidal type of expansion joint can damage the joint. In larger units these light-wall joints are particularly susceptible to damage, and some designers prefer the use of the heavier walls of formed heads.

Chemical-plant exchangers requiring expansion joints most commonly have used the flanged-and-flued-head type. There is a trend toward more common use of the light-wall-bellows type.

**U-Tube Heat Exchanger (Fig. 11-36*d*)** The tube bundle consists of a stationary tube sheet, U tubes (or hairpin tubes), baffles or support plates, and appropriate tie rods and spacers. The tube bundle can be removed from the heat-exchanger shell. A tube-side header (stationary head) and a shell with integral shell cover, which is welded to the shell, are provided. Each tube is free to expand or contract without any limitation being placed upon it by the other tubes.

The U-tube bundle has the advantage of providing minimum clearance between the outer tube limit and the inside of the shell for any of the removable-tube-bundle constructions. Clearances are of the same magnitude as for fixed-tube-sheet heat exchangers.

The number of tube holes in a given shell is less than that for a fixed-tube-sheet exchanger because of limitations on bending tubes of a very short radius.

The U-tube design offers the advantage of reducing the number of joints. In high-pressure construction this feature becomes of considerable importance in reducing both initial and maintenance costs. The use of U-tube construction has increased significantly with the development

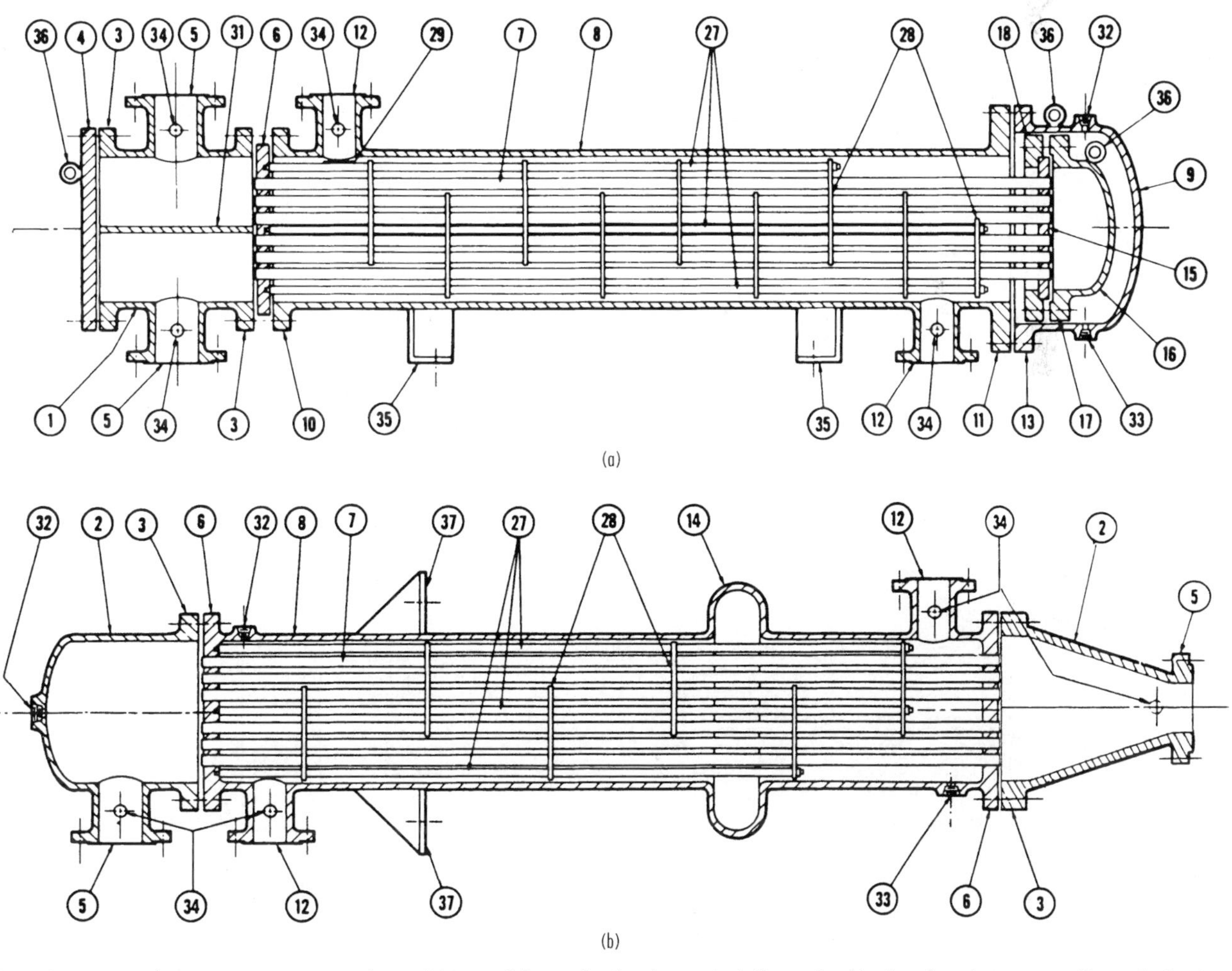


**FIG. 11-36** Heat-exchanger-component nomenclature. (*a*) Internal-floating-head exchanger (with floating-head backing device). Type AES. (*b*) Fixed-tube-sheet exchanger. Type BEM. (Standards of the Tubular Exchanger Manufacturers Association, *6th ed., 1978.*)

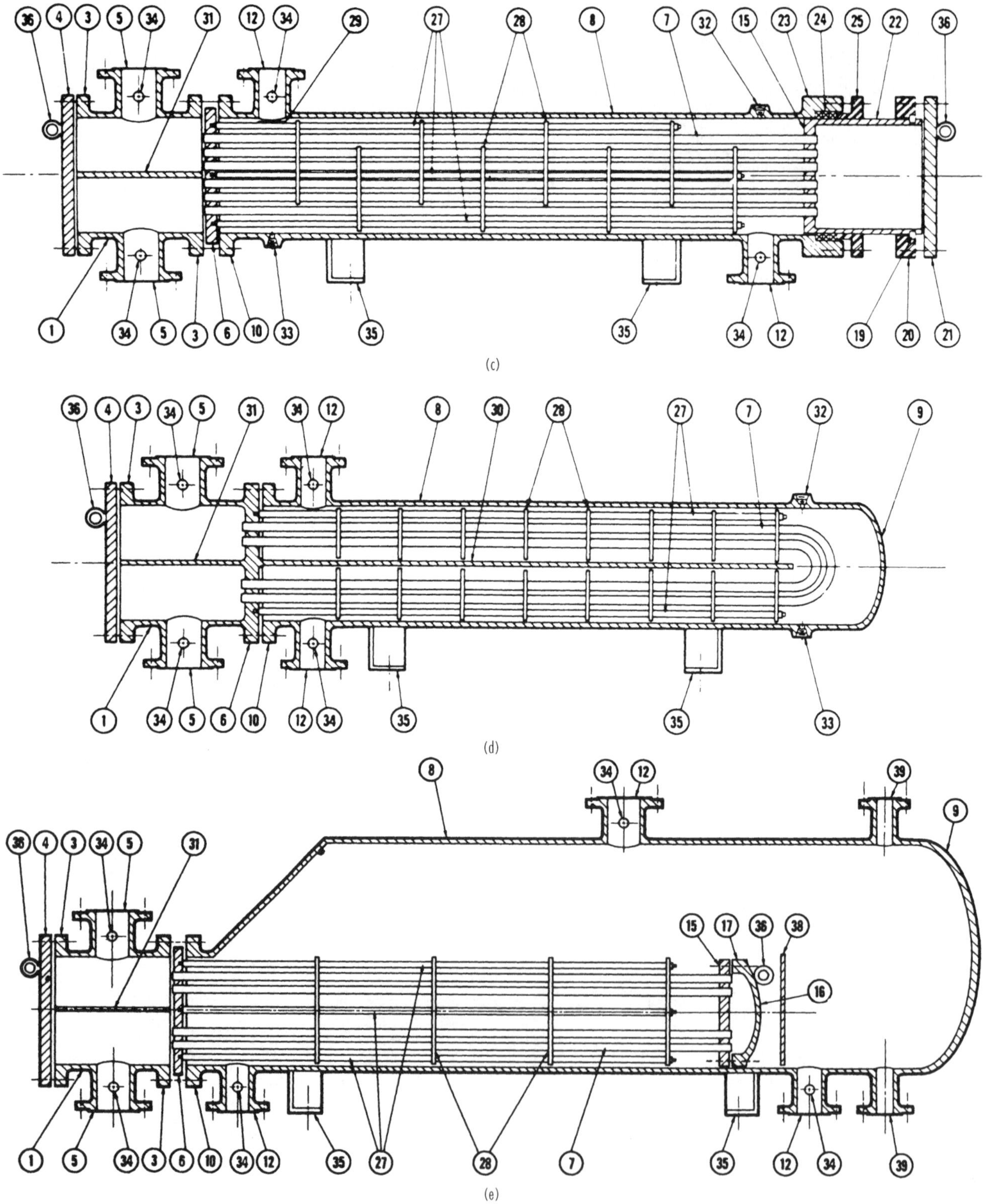


**FIG. 11-36** (*Continued*) Heat-exchanger-component nomenclature. (*c*) Outside-packed floating-head exchanger. Type AEP. (*d*) U-tube heat exchanger. Type CFU. (*e*) Kettle-type floating-head reboiler. Type AKT. (Standards of the Tubular Exchanger Manufacturers Association, *6th ed., 1978.*)

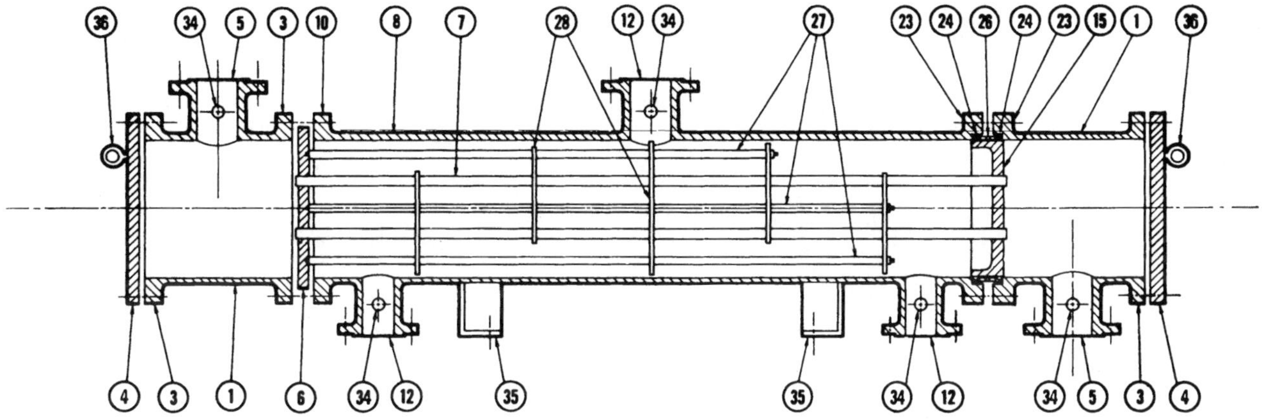


(f)

1. Stationary Head—Channel
2. Stationary Head—Bonnet
3. Stationary Head Flange—Channel or Bonnet
4. Channel Cover
5. Stationary Head Nozzle
6. Stationary Tubesheet
7. Tubes
8. Shell
9. Shell Cover
10. Shell Flange—Stationary Head End
11. Shell Flange—Rear Head End
12. Shell Nozzle
13. Shell Cover Flange
14. Expansion Joint
15. Floating Tubesheet
16. Floating Head Cover
17. Floating Head Flange
18. Floating Head Backing Device
19. Split Shear Ring
20. Slip-on Backing Flange
21. Floating Head Cover—External
22. Floating Tubesheet Skirt
23. Packing Box Flange
24. Packing
25. Packing Gland
26. Lantern Ring
27. Tie Rods and Spacers
28. Transverse Baffles or Support Plates
29. Impingement Plate
30. Longitudinal Baffle
31. Pass Partition
32. Vent Connection
33. Drain Connection
34. Instrument Connection
35. Support Saddle
36. Lifting Lug
37. Support Bracket
38. Weir
39. Liquid Level Connection

**FIG. 11-36** (*Continued*) Heat-exchanger-component nomenclature. (*f*) Exchanger with packed floating tube sheet and lantern ring. Type AJW. (Standards of the Tubular Exchanger Manufacturers Association, *6th ed., 1978.*)

of hydraulic tube cleaners, which can remove fouling residues from both the straight and the U-bend portions of the tubes.

Mechanical cleaning of the inside of the tubes was described by John [*Chem. Eng.*, **66**, 187–192 (Dec. 14, 1959)]. Rods and conventional mechanical tube cleaners cannot pass from one end of the U tube to the other. Power-driven tube cleaners, which can clean both the straight legs of the tubes and the bends, are available.

Hydraulic jetting with water forced through spray nozzles at high pressure for cleaning tube interiors and exteriors of removal bundles is reported by Canaday ("Hydraulic Jetting Tools for Cleaning Heat Exchangers," ASME Pap. 58-A-217, unpublished).

The tank suction heater, as illustrated in Fig. 11-38, contains a U-tube bundle. This design is often used with outdoor storage tanks for heavy fuel oils, tar, molasses, and similar fluids whose viscosity must be lowered to permit easy pumping. Uusally the tube-side heating medium is steam. One end of the heater shell is open, and the liquid being heated passes across the outside of the tubes. Pumping costs can be reduced without heating the entire contents of the tank. Bare tube and integral low-fin tubes are provided with baffles. Longitudinal fin-tube heaters are not baffled. Fins are most often used to minimize the fouling potential in these fluids.

Kettle-type reboilers, evaporators, etc., are often U-tube exchangers with enlarged shell sections for vapor-liquid separation. The U-tube bundle replaces the floating-heat bundle of Fig. 11-36*e*.

The U-tube exchanger with copper tubes, cast-iron header, and other parts of carbon steel is used for water and steam services in office buildings, schools, hospitals, hotels, etc. Nonferrous tube sheets and admiralty or 90-10 copper-nickel tubes are the most frequently used substitute materials. These standard exchangers are available from a number of manufacturers at costs far below those of custom-built process-industry equipment.

**Packed-Lantern-Ring Exchanger (Fig. 11-36*f*)** This construction is the least costly of the straight-tube removable bundle types. The shell- and tube-side fluids are each contained by separate rings of packing separated by a lantern ring and are installed at the floating tube sheet. The lantern ring is provided with weep holes. Any leakage passing the packing goes through the weep holes and then drops to the ground. Leakage at the packing will not result in mixing within the exchanger of the two fluids.

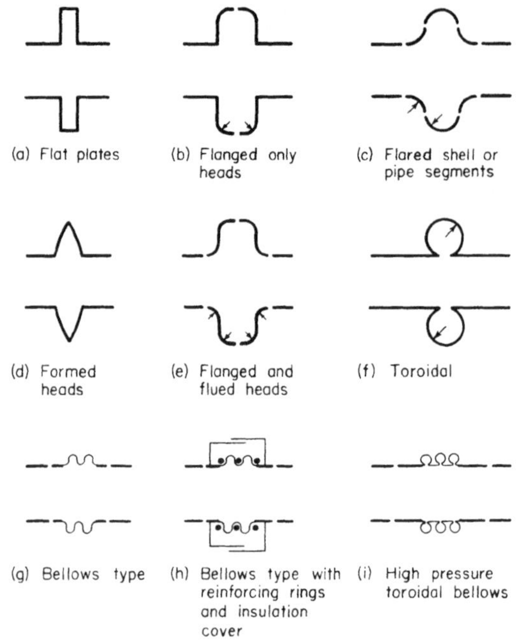


**FIG. 11-37** Expansion joints.

The width of the floating tube sheet must be great enough to allow for the packings, the lantern ring, and differential expansion. Sometimes a small skirt is attached to a thin tube sheet to provide the required bearing surface for packings and lantern ring.

The clearance between the outer tube limit and the inside of the shell is slightly larger than that for fixed-tube-sheet and U-tube exchangers. The use of a floating-tube-sheet skirt increases this clearance. Without the skirt the clearance must make allowance for tube-hole distortion during tube rolling near the outside edge of the tube sheet or for tube-end welding at the floating tube sheet.

The packed-lantern-ring construction is generally limited to design temperatures below 191°C (375°F) and to the mild services of water, steam, air, lubricating oil, etc. Design gauge pressure does not exceed 2068 kPa (300 lbf/in$^2$) for pipe shell exchangers and is limited to 1034 kPa (150 lbf/in$^2$) for 610- to 1067-mm- (24- to 42-in-) diameter shells.

**Outside-Packed Floating-Head Exchanger (Fig. 11-36*c*)** The shell-side fluid is contained by rings of packing, which are compressed within a stuffing box by a packing follower ring. This construction was frequently used in the chemical industry, but in recent years usage has decreased. The removable-bundle construction accommodates differential expansion between shell and tubes and is used for shell-side service up to 4137 kPa gauge pressure (600 lbf/in$^2$) at 316°C (600°F). There are no limitations upon the number of tube-side passes or upon the tube-side design pressure and temperature. The outside-packed floating-head exchanger was the most commonly used type of removable-bundle construction in chemical-plant service.

The floating-tube-sheet skirt, where in contact with the rings of packing, has fine machine finish. A split shear ring is inserted into a groove in the floating-tube-sheet skirt. A slip-on backing flange, which in service is held in place by the shear ring, bolts to the external floating-head cover.

The floating-head cover is usually a circular disk. With an odd number of tube-side passes, an axial nozzle can be installed in such a floating-head cover. If a side nozzle is required, the circular disk is replaced by either a dished head or a channel barrel (similar to Fig. 11-36*f*) bolted between floating-head cover and floating-tube-sheet skirt.

The outer tube limit approaches the inside of the skirt but is farther removed from the inside of the shell than for any of the previously discussed constructions. Clearances between shell diameter and bundle OTL are 22 mm (⅞ in) for small-diameter pipe shells, 44 mm (1¾ in) for large-diameter pipe shells, and 58 mm (2 1⁄16 in) for moderate-diameter plate shells.

**Internal Floating-Head Exchanger (Fig. 11-36*a*)** The internal floating-head design is used extensively in petroleum-refinery service, but in recent years there has been a decline in usage.

The tube bundle is removable, and the floating tube sheet moves (or floats) to accommodate differential expansion between shell and tubes. The outer tube limit approaches the inside diameter of the gasket at the floating tube sheet. Clearances (between shell and OTL) are 29 mm (1⅛ in) for pipe shells and 37 mm (1 7⁄16 in) for moderate-diameter plate shells.

A split backing ring and bolting usually hold the floating-head cover at the floating tube sheet. These are located beyond the end of the shell and within the larger-diameter shell cover. Shell cover, split backing ring, and floating-head cover must be removed before the tube bundle can pass through the exchanger shell.

With an even number of tube-side passes the floating-head cover serves as return cover for the tube-side fluid. With an odd number of passes a nozzle pipe must extend from the floating-head cover through the shell cover. Provision for both differential expansion and tube-bundle removal must be made.

**Pull-Through Floating-Head Exchanger (Fig. 11-36*e*)** Construction is similar to that of the internal-floating-head split-backing-ring exchanger except that the floating-head cover bolts directly to the floating tube sheet. The tube bundle can be withdrawn from the shell without removing either shell cover or floating-head cover. This feature reduces maintenance time during inspection and repair.

The large clearance between the tubes and the shell must provide for both the gasket and the bolting at the floating-head cover. This clearance is about 2 to 2½ times that required by the split-ring design. Sealing strips or dummy tubes are often installed to reduce bypassing of the tube bundle.

**Falling-Film Exchangers** Falling-film shell-and-tube heat exchangers have been developed for a wide variety of services and are described by Sack [*Chem. Eng. Prog.*, **63**, 55 (July 1967)]. The fluid enters at the top of the vertical tubes. Distributors or slotted tubes put the liquid in film flow in the inside surface of the tubes, and the film adheres to the tube surface while falling to the bottom of the tubes. The film can be cooled, heated, evaporated, or frozen by means of the proper heat-transfer medium outside the tubes. Tube distributors have been developed for a wide range of applications. Fixed tube sheets, with or without expansion joints, and outside-packed-head designs are used.

Principal advantages are high rate of heat transfer, no internal pressure drop, short time of contact (very important for heat-sensitive materials), easy accessibility to tubes for cleaning, and, in some cases, prevention of leakage from one side to another.

These falling-film exchangers are used in various services as described in the following paragraphs.

***Liquid Coolers and Condensers*** Dirty water can be used as the cooling medium. The top of the cooler is open to the atmosphere for

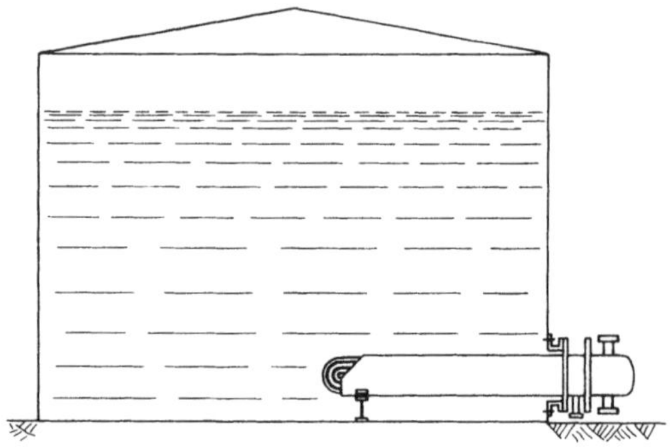

**FIG. 11-38** Tank suction heater.

access to tubes. These can be cleaned without shutting down the cooler by removing the distributors one at a time and scrubbing the tubes.

***Evaporators*** These are used extensively for the concentration of ammonium nitrate, urea, and other chemicals sensitive to heat when minimum contact time is desirable. Air is sometimes introduced in the tubes to lower the partial pressure of liquids whose boiling points are high. These evaporators are built for pressure or vacuum and with top or bottom vapor removal.

***Absorbers*** These have a two-phase flow system. The absorbing medium is put in film flow during its fall downward on the tubes as it is cooled by a cooling medium outside the tubes. The film absorbs the gas which is introduced into the tubes. This operation can be cocurrent or countercurrent.

***Freezers*** By cooling the falling film to its freezing point, these exchangers convert a variety of chemicals to the solid phase. The most common application is the production of sized ice and paradichlorobenzene. Selective freezing is used for isolating isomers. By melting the solid material and refreezing in several stages, a higher degree of purity of product can be obtained.

## TUBE-SIDE CONSTRUCTION

**Tube-Side Header** The tube-side header (or stationary head) contains one or more flow nozzles.

The **bonnet** (Fig. 11-35B) bolts to the shell. It is necessary to remove the bonnet in order to examine the tube ends. The fixed-tube-sheet exchanger of Fig. 11-36*b* has bonnets at both ends of the shell.

The **channel** (Fig. 11-35*A*) has a removable channel cover. The tube ends can be examined by removing this cover without disturbing the piping connections to the channel nozzles. The channel can bolt to the shell as shown in Fig. 11-36*a* and *c*. The Type C and Type N channels of Fig. 11-35 are welded to the tube sheet. This design is comparable in cost with the bonnet but has the advantages of permitting access to the tubes without disturbing the piping connections and of eliminating a gasketed joint.

**Special High-Pressure Closures (Fig. 11-35*D*)** The channel barrel and the tube sheet are generally forged. The removable channel cover is seated in place by hydrostatic pressure, while a shear ring subjected to shearing stress absorbs the end force. For pressures above 6205 kPa (900 lbf/in$^2$) these designs are generally more economical than bolted constructions, which require larger flanges and bolting as pressure increases in order to contain the end force with bolts in tension. Relatively light-gauge internal pass partitions are provided to direct the flow of tube-side fluids but are designed only for the differential pressure across the tube bundle.

**Tube-Side Passes** Most exchangers have an even number of tube-side passes. The fixed-tube-sheet exchanger (which has no shell cover) usually has a return cover without any flow nozzles as shown in Fig. 11-35*M;* Types *L* and *N* are also used. All removable-bundle designs (except for the U tube) have a floating-head cover directing the flow of tube-side fluid at the floating tube sheet.

**Tubes** Standard heat-exchanger tubing is ¼, ⅜, ½, ⅝, ¾, 1, 1¼, and 1½ in in outside diameter (in × 25.4 = mm). Wall thickness is measured in Birmingham wire gauge (BWG) units. A comprehensive list of tubing characteristics and sizes is given in Table 11-12. The most commonly used tubes in chemical plants and petroleum refineries are 19- and 25-mm (¾- and 1-in) outside diameter. Standard tube lengths are 8, 10, 12, 16, and 20 ft, with 20 ft now the most common (ft × 0.3048 = m).

**Manufacturing tolerances** for steel, stainless-steel, and nickel-alloy tubes are such that the tubing is produced to either average or minimum wall thickness. Seamless carbon steel tube of minimum wall thickness may vary from 0 to 20 percent above the nominal wall thickness. Average-wall seamless tubing has an allowable variation of plus or minus 10 percent. Welded carbon steel tube is produced to closer tolerances (0 to plus 18 percent on minimum wall; plus or minus 9 percent on average wall). Tubing of aluminum, copper, and their alloys can be drawn easily and usually is made to minimum wall specifications.

Common practice is to specify **exchanger surface** in terms of total external square feet of tubing. The effective outside heat-transfer surface is based on the length of tubes measured between the inner faces of tube sheets. In most heat exchangers there is little difference between the total and the effective surface. Significant differences are usually found in high-pressure and double-tube-sheet designs.

**Integrally finned tube,** which is available in a variety of alloys and sizes, is being used in shell-and-tube heat exchangers. The fins are radially extruded from thick-walled tube to a height of 1.6 mm (1⁄16 in) spaced at 1.33 mm (19 fins per inch) or to a height of 3.2 mm (⅛ in) spaced at 2.3 mm (11 fins per inch). External surface is approximately 2½ times the outside surface of a bare tube with the same outside diameter. Also available are 0.93-mm- (0.037-in-) high fins spaced 0.91 mm (28 fins per inch) with an external surface about 3.5 times the surface of the bare tube. Bare ends of nominal tube diameter are provided, while the fin height is slightly less than this diameter. The tube can be inserted into a conventional tube bundle and rolled or welded to the tube sheet by the same means, used for bare tubes. An integrally finned tube rolled into a tube sheet with double serrations and flared at the inlet is shown in Fig. 11-39. Internally finned tubes have been manufactured but have limited application.

**Longitudinal fins** are commonly used in double-pipe exchangers upon the outside of the inner tube. U-tube and conventional removable tube bundles are also made from such tubing. The ratio of external to internal surface generally is about 10 or 15:1.

**Transverse fins** upon tubes are used in low-pressure gas services. The primary application is in air-cooled heat exchangers (as discussed under that heading), but shell-and-tube exchangers with these tubes are in service.

**Rolled Tube Joints** Expanded tube-to-tube-sheet joints are standard. Properly rolled joints have uniform tightness to minimize tube fractures, stress corrosion, tube-sheet ligament pushover and enlargement, and dishing of the tube sheet. Tubes are expanded into the tube sheet for a length of two tube diameters, or 50 mm (2 in), or tube-sheet thickness minus 3 mm (⅛ in). Generally tubes are rolled for the last of these alternatives. The expanded portion should never extend beyond the shell-side face of the tube sheet, since removing such a tube is extremely difficult. Methods and tools for tube removal and tube rolling were discussed by John [*Chem. Eng.,* **66,** 77–80 (Dec. 28, 1959)], and rolling techniques by Bach [*Pet. Refiner,* **39,** 8, 104 (1960)].

Tube ends may be projecting, flush, flared, or beaded (listed in order of usage). The flare or bell-mouth tube end is usually restricted to water service in condensers and serves to reduce erosion near the tube inlet.

For moderate general process requirements at gauge pressures less than 2058 kPa (300 lbf/in$^2$) and less than 177°C (350°F), tube-sheet holes without grooves are standard. For all other services with expanded tubes at least two grooves in each tube hole are common. The number of grooves is sometimes changed to one or three in proportion to tube-sheet thickness.

Expanding the tube into the **grooved tube holes** provides a stronger joint but results in greater difficulties during tube removal.

**Welded Tube Joints** When suitable materials of construction are used, the tube ends may be welded to the tube sheets. Welded joints may be seal-welded "for additional tightness beyond that of tube

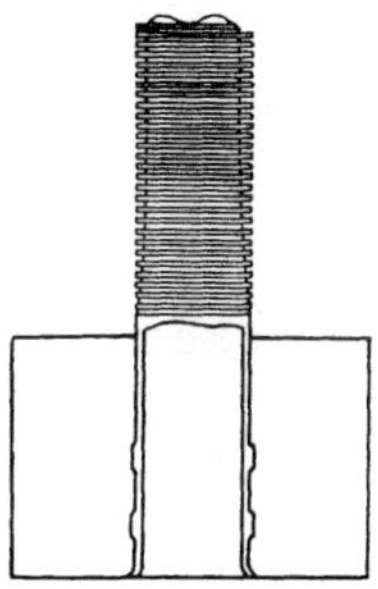

**FIG. 11-39** Integrally finned tube rolled into tube sheet with double serrations and flared inlet. (*Woverine Division, UOP, Inc.*)

**TABLE 11-12 Characterstics of Tubing (From *Standards of the Tubular Exchanger Manufacturers Association*, 8th Ed., 1999; 25 North Broadway, Tarrytown, N.Y.)**

| Tube O.D., in. | B.W.G. gage | Thickness, in. | Internal area, in.$^2$ | Sq. ft. external surface per foot length | Sq. ft. internal surface per foot length | Weight per ft. length steel, lb* | Tube I.D., in. | Moment of inertia, in.$^4$ | Section modulus, in.$^3$ | Radius of gyration, in. | Constant $C$† | O.D. / I.D. | Transverse metal area, in.$^3$ |
|---|---|---|---|---|---|---|---|---|---|---|---|---|---|
| 1/4 | 22 | 0.028 | 0.0296 | 0.0654 | 0.0508 | 0.066 | 0.194 | 0.00012 | 0.00098 | 0.0791 | 46 | 1.289 | 0.0195 |
| | 24 | 0.022 | 0.0333 | 0.0654 | 0.0539 | 0.054 | 0.206 | 0.00010 | 0.00083 | 0.0810 | 52 | 1.214 | 0.0158 |
| | 26 | 0.018 | 0.0360 | 0.0654 | 0.0560 | 0.045 | 0.214 | 0.00009 | 0.00071 | 0.0823 | 56 | 1.168 | 0.0131 |
| | 27 | 0.016 | 0.0373 | 0.0654 | 0.0571 | 0.040 | 0.218 | 0.00008 | 0.00065 | 0.0829 | 58 | 1.147 | 0.0118 |
| 3/8 | 18 | 0.049 | 0.0603 | 0.0982 | 0.0725 | 0.171 | 0.277 | 0.00068 | 0.0036 | 0.1166 | 94 | 1.354 | 0.0502 |
| | 20 | 0.035 | 0.0731 | 0.0982 | 0.0798 | 0.127 | 0.305 | 0.00055 | 0.0029 | 0.1208 | 114 | 1.230 | 0.0374 |
| | 22 | 0.028 | 0.0799 | 0.0982 | 0.0835 | 0.104 | 0.319 | 0.00046 | 0.0025 | 0.1231 | 125 | 1.176 | 0.0305 |
| | 24 | 0.022 | 0.0860 | 0.0982 | 0.0867 | 0.083 | 0.331 | 0.00038 | 0.0020 | 0.1250 | 134 | 1.133 | 0.0244 |
| 1/2 | 16 | 0.065 | 0.1075 | 0.1309 | 0.0969 | 0.302 | 0.370 | 0.0021 | 0.0086 | 0.1555 | 168 | 1.351 | 0.0888 |
| | 18 | 0.049. | 0.1269 | 0.1309 | 0.1052 | 0.236 | 0.402 | 0.0018 | 0.0071 | 0.1604 | 198 | 1.244 | 0.0694 |
| | 20 | 0.035 | 0.1452 | 0.1309 | 0.1126 | 0.174 | 0.430 | 0.0014 | 0.0056 | 0.1649 | 227 | 1.163 | 0.0511 |
| | 22 | 0.028 | 0.1548 | 0.1309 | 0.1162 | 0.141 | 0.444 | 0.0012 | 0.0046 | 0.1672 | 241 | 1.126 | 0.0415 |
| 5/8 | 12 | 0.109 | 0.1301 | 0.1636 | 0.1066 | 0.601 | 0.407 | 0.0061 | 0.0197 | 0.1865 | 203 | 1.536 | 0.177 |
| | 13 | 0.095 | 0.1486 | 0.1636 | 0.1139 | 0.538 | 0.435 | 0.0057 | 0.0183 | 0.1904 | 232 | 1.437 | 0.158 |
| | 14 | 0.083 | 0.1655 | 0.1636 | 0.1202 | 0.481 | 0.459 | 0.0053 | 0.0170 | 0.1939 | 258 | 1.362 | 0.141 |
| | 15 | 0.072 | 0.1817 | 0.1636 | 0.1259 | 0.426 | 0.481 | 0.0049 | 0.0156 | 0.1972 | 283 | 1.299 | 0.125 |
| | 16 | 0.065 | 0.1924 | 0.1636 | 0.1296 | 0.389 | 0.495 | 0.0045 | 0.0145 | 0.1993 | 300 | 1.263 | 0.114 |
| | 17 | 0.058 | 0.2035 | 0.1636 | 0.1333 | 0.352 | 0.509 | 0.0042 | 0.0134 | 0.2015 | 317 | 1.228 | 0.103 |
| | 18 | 0.049 | 0.2181 | 0.1636 | 0.1380 | 0.302 | 0.527 | 0.0037 | 0.0119 | 0.2044 | 340 | 1.186 | 0.089 |
| | 19 | 0.042 | 0.2299 | 0.1636 | 0.1416 | 0.262 | 0.541 | 0.0033 | 0.0105 | 0.2067 | 359 | 1.155 | 0.077 |
| | 20 | 0.035 | 0.2419 | 0.1636 | 0.1453 | 0.221 | 0.555 | 0.0028 | 0.0091 | 0.2090 | 377 | 1.126 | 0.065 |
| 3/4 | 10 | 0.134 | 0.1825 | 0.1963 | 0.1262 | 0.833 | 0.482 | 0.0129 | 0.0344 | 0.2229 | 285 | 1.556 | 0.259 |
| | 11 | 0.120 | 0.2043 | 0.1963 | 0.1335 | 0.808 | 0.510 | 0.0122 | 0.0326 | 0.2267 | 319 | 1.471 | 0.238 |
| | 12 | 0.109 | 0.2223 | 0.1963 | 0.1393 | 0.747 | 0.532 | 0.0116 | 0.0309 | 0.2299 | 347 | 1.410 | 0.219 |
| | 13 | 0.095 | 0.2463 | 0.1963 | 0.1466 | 0.665 | 0.560 | 0.0107 | 0.0285 | 0.2340 | 384 | 1.339 | 0.195 |
| | 14 | 0.083 | 0.2679 | 0.1963 | 0.1529 | 0.592 | 0.584 | 0.0098 | 0.0262 | 0.2376 | 418 | 1.284 | 0.174 |
| | 15 | 0.072 | 0.2884 | 0.1963 | 0.1587 | 0.522 | 0.606 | 0.0089 | 0.0238 | 0.2411 | 450 | 1.238 | 0.153 |
| | 16 | 0.065 | 0.3019 | 0.1963 | 0.1623 | 0.476 | 0.620 | 0.0083 | 0.0221 | 0.2433 | 471 | 1.210 | 0.140 |
| | 17 | 0.058 | 0.3157 | 0.1963 | 0.1660 | 0.429 | 0.634 | 0.0076 | 0.0203 | 0.2455 | 492 | 1.183 | 0.126 |
| | 18 | 0.049 | 0.3339 | 0.1963 | 0.1707 | 0.367 | 0.652 | 0.0067 | 0.0178 | 0.2484 | 521 | 1.150 | 0.108 |
| | 20 | 0.035 | 0.3632 | 0.1963 | 0.1780 | 0.268 | 0.680 | 0.0050 | 0.0134 | 0.2531 | 567 | 1.103 | 0.079 |
| 7/8 | 10 | 0.134 | 0.2894 | 0.2291 | 0.1589 | 1.062 | 0.607 | 0.0221 | 0.0505 | 0.2662 | 451 | 1.442 | 0.312 |
| | 11 | 0.120 | 0.3167 | 0.2291 | 0.1662 | 0.969 | 0.635 | 0.0208 | 0.0475 | 0.2703 | 494 | 1.378 | 0.285 |
| | 12 | 0.109 | 0.3390 | 0.2291 | 0.1720 | 0.893 | 0.657 | 0.0196 | 0.0449 | 0.2736 | 529 | 1.332 | 0.262 |
| | 13 | 0.095 | 0.3685 | 0.2291 | 0.1793 | 0.792 | 0.685 | 0.0180 | 0.0411 | 0.2778 | 575 | 1.277 | 0.233 |
| | 14 | 0.083 | 0.3948 | 0.2291 | 0.1856 | 0.703 | 0.709 | 0.0164 | 0.0374 | 0.2815 | 616 | 1.234 | 0.207 |
| | 15 | 0.072 | 0.4197 | 0.2291 | 0.1914 | 0.618 | 0.731 | 0.0148 | 0.0337 | 0.2850 | 655 | 1.197 | 0.182 |
| | 16 | 0.065 | 0.4359 | 0.2291 | 0.1950 | 0.563 | 0.745 | 0.0137 | 0.0312 | 0.2873 | 680 | 1.174 | 0.165 |
| | 17 | 0.058 | 0.4525 | 0.2291 | 0.1987 | 0.507 | 0.759 | 0.0125 | 0.0285 | 0.2896 | 706 | 1.153 | 0.149 |
| | 18 | 0.049 | 0.4742 | 0.2291 | 0.2034 | 0.433 | 0.777 | 0.0109 | 0.0249 | 0.2925 | 740 | 1.126 | 0.127 |
| | 20 | 0.035 | 0.5090 | 0.2291 | 0.2107 | 0.314 | 0.805 | 0.0082 | 0.0187 | 0.2972 | 794 | 1.087 | 0.092 |
| 1 | 8 | 0.165 | 0.3526 | 0.2618 | 0.1754 | 1.473 | 0.670 | 0.0392 | 0.0784 | 0.3009 | 550 | 1.493 | 0.433 |
| | 10 | 0.134 | 0.4208 | 0.2618 | 0.1916 | 1.241 | 0.732 | 0.0350 | 0.0700 | 0.3098 | 656 | 1.366 | 0.365 |
| | 11 | 0.120 | 0.4536 | 0.2618 | 0.1990 | 1.129 | 0.760 | 0.0327 | 0.0654 | 0.3140 | 708 | 1.316 | 0.332 |
| | 12 | 0.109 | 0.4803 | 0.2618 | 0.2047 | 1.038 | 0.782 | 0.0307 | 0.0615 | 0.3174 | 749 | 1.279 | 0.305 |
| | 13 | 0.095 | 0.5153 | 0.2618 | 0.2121 | 0.919 | 0.810 | 0.0280 | 0.0559 | 0.3217 | 804 | 1.235 | 0.270 |
| | 14 | 0.083 | 0.5463 | 0.2618 | 0.2183 | 0.814 | 0.834 | 0.0253 | 0.0507 | 0.3255 | 852 | 1.199 | 0.239 |
| | 15 | 0.072 | 0.5755 | 0.2618 | 0.2241 | 0.714 | 0.856 | 0.0227 | 0.0455 | 0.3291 | 898 | 1.168 | 0.210 |
| | 16 | 0.065 | 0.5945 | 0.2618 | 0.2278 | 0.650 | 0.870 | 0.0210 | 0.0419 | 0.3314 | 927 | 1.149 | 0.191 |
| | 18 | 0.049 | 0.6390 | 0.2618 | 0.2361 | 0.498 | 0.902 | 0.0166 | 0.0332 | 0.3367 | 997 | 1.109 | 0.146 |
| | 20 | 0.035 | 0.6793 | 0.2618 | 0.2435 | 0.361 | 0.930 | 0.0124 | 0.0247 | 0.3414 | 1060 | 1.075 | 0.106 |
| 1-1/4 | 7 | 0.180 | 0.6221 | 0.3272 | 0.2330 | 2.059 | 0.890 | 0.0890 | 0.1425 | 0.3836 | 970 | 1.404 | 0.605 |
| | 8 | 0.165 | 0.6648 | 0.3272 | 0.2409 | 1.914 | 0.920 | 0.0847 | 0.1355 | 0.3880 | 1037 | 1.359 | 0.562 |
| | 10 | 0.134 | 0.7574 | 0.3272 | 0.2571 | 1.599 | 0.982 | 0.0742 | 0.1187 | 0.3974 | 1182 | 1.273 | 0.470 |
| | 11 | 0.120 | 0.8012 | 0.3272 | 0.2644 | 1.450 | 1.010 | 0.0688 | 0.1100 | 0.4018 | 1250 | 1.238 | 0.426 |
| | 12 | 0.109 | 0.8365 | 0.3272 | 0.2702 | 1.330 | 1.032 | 0.0642 | 0.1027 | 0.4052 | 1305 | 1.211 | 0.391 |
| | 13 | 0.095 | 0.8825 | 0.3272 | 0.2775 | 1.173 | 1.060 | 0.0579 | 0.0926 | 0.4097 | 1377 | 1.179 | 0.345 |
| | 14 | 0.083 | 0.9229 | 0.3272 | 0.2838 | 1.036 | 1.084 | 0.0521 | 0.0833 | 0.4136 | 1440 | 1.153 | 0.304 |
| | 16 | 0.065 | 0.9852 | 0.3272 | 0.2932 | 0.824 | 1.120 | 0.0426 | 0.0682 | 0.4196 | 1537 | 1.116 | 0.242 |
| | 18 | 0.049 | 1.0423 | 0.3272 | 0.3016 | 0.629 | 1.152 | 0.0334 | 0.0534 | 0.4250 | 1626 | 1.085 | 0.185 |
| | 20 | 0.035 | 1.0936 | 0.3272 | 0.3089 | 0.455 | 1.180 | 0.0247 | 0.0395 | 0.4297 | 1706 | 1.059 | 0.134 |
| 1-1/2 | 10 | 0.134 | 1.1921 | 0.3927 | 0.3225 | 1.957 | 1.232 | 0.1354 | 0.1806 | 0.4853 | 1860 | 1.218 | 0.575 |
| | 12 | 0.109 | 1.2908 | 0.3927 | 0.3356 | 1.621 | 1.282 | 0.1159 | 0.1545 | 0.4933 | 2014 | 1.170 | 0.476 |
| | 14 | 0.083 | 1.3977 | 0.3927 | 0.3492 | 1.257 | 1.334 | 0.0931 | 0.1241 | 0.5018 | 2180 | 1.124 | 0.369 |
| | 16 | 0.065 | 1.4741 | 0.3927 | 0.3587 | 0.997 | 1.370 | 0.0756 | 0.1008 | 0.5079 | 2300 | 1.095 | 0.293 |
| 2 | 11 | 0.120 | 2.4328 | 0.5236 | 0.4608 | 2.412 | 1.760 | 0.3144 | 0.3144 | 0.6660 | 3795 | 1.136 | 0.709 |
| | 12 | 0.109 | 2.4941 | 0.5236 | 0.4665 | 2.204 | 1.782 | 0.2904 | 0.2904 | 0.6697 | 3891 | 1.122 | 0.648 |
| | 13 | 0.095 | 2.5730 | 0.5236 | 0.4739 | 1.935 | 1.810 | 0.2586 | 0.2588 | 0.6744 | 4014 | 1.105 | 0.569 |
| | 14 | 0.083 | 2.6417 | 0.5236 | 0.4801 | 1.701 | 1.834 | 0.2300 | 0.2300 | 0.6784 | 4121 | 1.091 | 0.500 |

*Weights are based on low-carbon steel with a density of 0.2836 lb/cu. in. For other metals multiply by the following factors: aluminum, 0.35; titanium, 0.58; A.I.S.I. 400 Series S/steels, 0.99; A.I.S.I. 300 Series S/steels, 1.02; aluminum bronze, 1.04; aluminum brass, 1.06; nickel-chrome-iron, 1.07; Admiralty, 1.09; nickel, 1.13; nickel-copper, 1.12; copper and cupro-nickels, 1.14.

† $\text{Liquid velocity} = \dfrac{\text{lb per tube hour}}{C \times \text{sp gr of liquid}}$ ft/s (sp gr of water at 60°F = 1.0)

rolling" or may be strength-welded. Strength-welded joints have been found satisfactory in very severe services. Welded joints may or may not be rolled before or after welding.

The variables in tube-end welding were discussed in two unpublished papers (Emhardt, "Heat Exchanger Tube-to-Tubesheet Joints," ASME Pap. 69-WA/HT-47; and Reynolds, "Tube Welding for Conventional and Nuclear Power Plant Heat Exchangers," ASME Pap. 69-WA/HT-24), which were presented at the November 1969 meeting of the American Society of Mechanical Engineers.

Tube-end rolling before welding may leave lubricant from the tube expander in the tube hole. Fouling during normal operation followed by maintenance operations will leave various impurities in and near the tube ends. Satisfactory welds are rarely possible under such conditions, since tube-end welding requires extreme cleanliness in the area to be welded.

Tube **expansion after welding** has been found useful for low and moderate pressures. In high-pressure service tube rolling has not been able to prevent leakage after weld failure.

**Double-Tube-Sheet Joints** This design prevents the passage of either fluid into the other because of leakage at the tube-to-tube-sheet joints, which are generally the weakest points in heat exchangers. Any leakage at these joints admits the fluid to the gap between the tube sheets. Mechanical design, fabrication, and maintenance of double-tube-sheet designs require special consideration.

## SHELL-SIDE CONSTRUCTION

**Shell Sizes** Heat-exchanger shells are generally made from standard-wall steel pipe in sizes up to 305-mm (12-in) diameter; from 9.5-mm (⅜-in) wall pipe in sizes from 356 to 610 mm (14 to 24 in); and from steel plate rolled at discrete intervals in larger sizes. Clearances between the outer tube limit and the shell are discussed elsewhere in connection with the different types of construction.

The following formulae may be used to estimate tube counts for various bundle sizes and tube passes. The estimated values include the removal of tubes to provide an entrance area for shell nozzle sizes of one-fifth the shell diameter. Due to the large effect from other parameters such as design pressure/corrosion allowance, baffle cuts, seal strips, and so on, these are to be used as estimates only. Exact tube counts are part of the design package of most reputable exchanger design software and are normally used for the final design.

**Triangular** tube layouts with pitch equal to 1.25 times the tube outside diameter:

$C = 0.75\,(D/d) - 36$; where $D$ = Bundle O.D. $d$ = Tube O.D.

Range of accuracy: $-24 \le C \le 24$.

$$\text{1 Tube Pass: } N_t = 1298. + 74.86C + 1.283C^2 - .0078C^3 - .0006C^4 \qquad (11\text{-}74a)$$

$$\text{2 Tube Pass: } N_t = 1266. + 73.58C + 1.234C^2 - .0071C^3 - .0005C^4 \qquad (11\text{-}74b)$$

$$\text{4 Tube Pass: } N_t = 1196. + 70.79C + 1.180C^2 - .0059C^3 - .0004C^4 \qquad (11\text{-}74c)$$

$$\text{6 Tube Pass: } N_t = 1166. + 70.72C + 1.269C^2 - .0074C^3 - .0006C^4 \qquad (11\text{-}74d)$$

**Square** tube layouts with pitch equal to 1.25 times the tube outside diameter:

$C = (D/d) - 36.$; where $D$ = Bundle O.D. $d$ = Tube O.D.

Range of accuracy: $-24 \le C \le 24$.

$$\text{1 Tube Pass: } N_t = 593.6 + 33.52C + .3782C^2 - .0012C^3 + .0001C^4 \qquad (11\text{-}75a)$$

$$\text{2 Tube Pass: } N_t = 578.8 + 33.36C + .3847C^2 - .0013C^3 + .0001C^4 \qquad (11\text{-}75b)$$

$$\text{4 Tube Pass: } N_t = 562.0 + 33.04C + .3661C^2 - .0016C^3 + .0002C^4 \qquad (11\text{-}75c)$$

$$\text{6 Tube Pass: } N_t = 550.4 + 32.49C + .3873C^2 - .0013C^3 + .0001C^4 \qquad (11\text{-}75d)$$

**Shell-Side Arrangements** The **one-pass shell** (Fig. 11-35*E*) is the most commonly used arrangement. Condensers from single component vapors often have the nozzles moved to the center of the shell for vacuum and steam services.

A solid longitudinal baffle is provided to form a two-pass shell (Fig. 11-35*F*). It may be insulated to improve thermal efficiency. (See further discussion on baffles). A two-pass shell can improve thermal effectiveness at a cost lower than for two shells in series.

For **split flow** (Fig. 11-35*G*), the longitudinal baffle may be solid or perforated. The latter feature is used with condensing vapors.

A **double-split-flow** design is shown in Fig. 11-35*H*. The longitudinal baffles may be solid or perforated.

The **divided flow** design (Fig. 11-35*J*), mechanically is like the one-pass shell except for the addition of a nozzle. Divided flow is used to meet low-pressure-drop requirements.

The **kettle reboiler** is shown in Fig. 11-35*K*. When nucleate boiling is to be done on the shell-side, this common design provides adequate dome space for separation of vapor and liquid above the tube bundle and surge capacity beyond the weir near the shell cover.

## BAFFLES AND TUBE BUNDLES

The **tube bundle** is the most important part of a tubular heat exchanger. The tubes generally constitute the most expensive component of the exchanger and are the one most likely to corrode. Tube sheets, baffles, or support plates, tie rods, and usually spacers complete the bundle.

Minimum **baffle spacing** is generally one-fifth of the shell diameter and not less than 50.8 mm (2 in). Maximum baffle spacing is limited by the requirement to provide adequate support for the tubes. The maximum unsupported tube span in inches equals $74\,d^{0.75}$ (where $d$ is the outside tube diameter in inches). The unsupported tube span is reduced by about 12 percent for aluminum, copper, and their alloys.

Baffles are provided for heat-transfer purposes. When shell-side baffles are not required for heat-transfer purposes, as may be the case in condensers or reboilers, tube supports are installed.

**Segmental Baffles** Segmental or cross-flow baffles are standard. Single, double, and triple segmental baffles are used. Baffle cuts are illustrated in Fig. 11-40. The double segmental baffle reduces cross-flow velocity for a given baffle spacing. The triple segmental baffle reduces both cross-flow and long-flow velocities and has been identified as the "window-cut" baffle.

Baffle cuts are expressed as the ratio of segment opening height to shell inside diameter. Cross-flow baffles with horizontal cut are shown in Fig. 11-36*a, c,* and *f.* This arrangement is not satisfactory for horizontal condensers, since the condensate can be trapped between baffles, or for dirty fluids in which the dirt might settle out. Vertical-cut baffles are used for side-to-side flow in horizontal exchangers with condensing fluids or dirty fluids. Baffles are notched to assure complete drainage when the units are taken out of service. (These notches permit some bypassing of the tube bundle during normal operation.)

Tubes are most commonly arranged on an equilateral triangular pitch. Tubes are arranged on a square pitch primarily for mechanical cleaning purposes in removable-bundle exchangers.

**Maximum baffle cut** is limited to about 45 percent for single segmental baffles so that every pair of baffles will support each tube. Tube bundles are generally provided with baffles cut so that at least one row of tubes passes through all the baffles or support plates. These tubes hold the entire bundle together. In pipe-shell exchangers with a horizontal baffle cut and a horizontal pass rib for directing tube-side flow in the channel, the maximum baffle cut, which permits a minimum of one row of tubes to pass through all baffles, is approximately 33 percent in small shells and 40 percent in larger pipe shells.

Maximum shell-side heat-transfer rates in forced convection are apparently obtained by cross-flow of the fluid at right angles to the tubes. In order to maximize this type of flow some heat exchangers are built with segmental-cut baffles and with "no tubes in the window" (or the baffle cutout). Maximum baffle spacing may thus equal maximum unsupported-tube span, while conventional baffle spacing is limited to one-half of this span.

The maximum baffle spacing for no tubes in the window of single segmental baffles is unlimited when intermediate supports are provided. These are cut on both sides of the baffle and therefore do not affect the flow of the shell-side fluid. Each support engages all the tubes; the supports are spaced to provide adequate support for the tubes.

**Rod Baffles** Rod or bar baffles have either rods or bars extending through the lanes between rows of tubes. A baffle set can consist of a baffle with rods in all the vertical lanes and another baffle with rods in all the horizontal lanes between the tubes. The shell-side flow is uniform and parallel to the tubes. Stagnant areas do not exist.

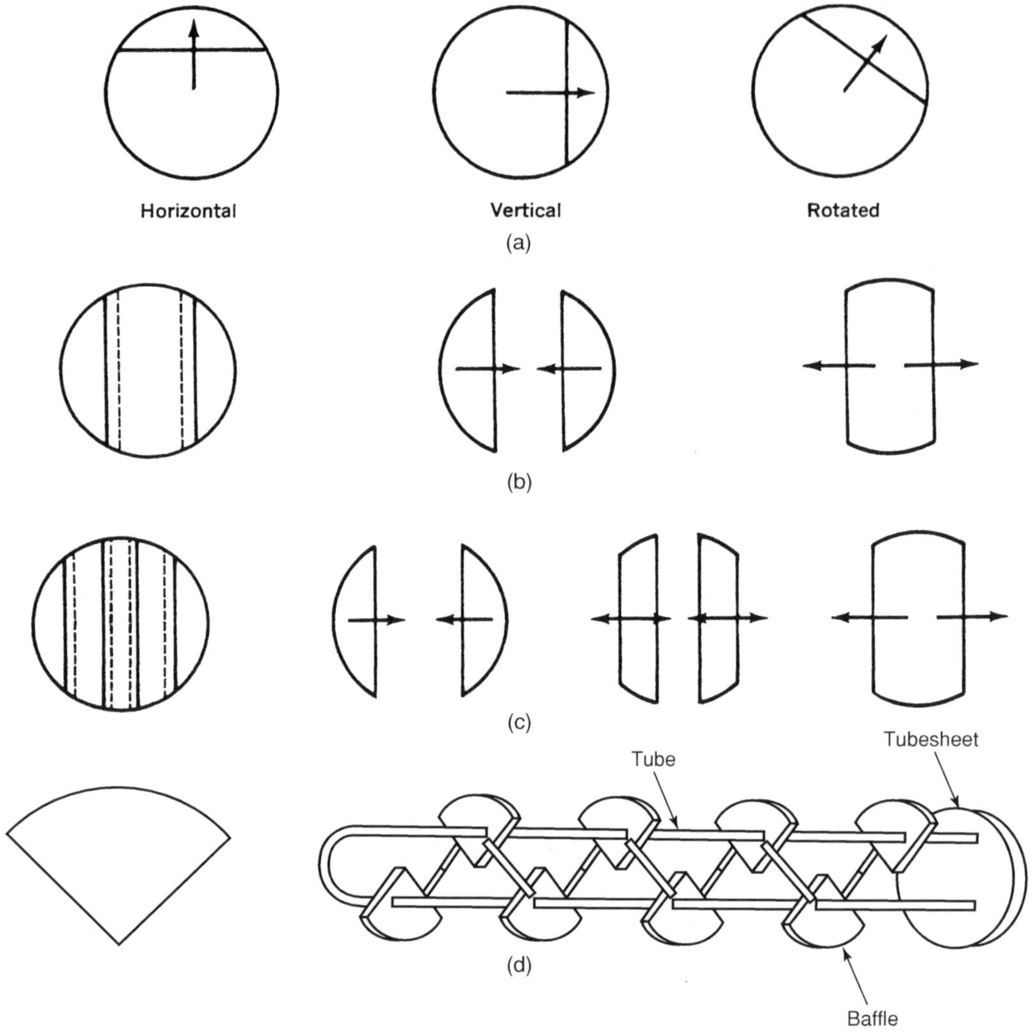


**FIG. 11-40** Plate baffles. (*a*) Baffle cuts for single segmental baffles. (*b*) Baffle cuts for double segmental baffles. (*c*) Baffle cuts for triple segmental baffles. (*d*) Helical baffle construction.

One device uses four baffles in a baffle set. Only half of either the vertical or the horizontal tube lanes in a baffle have rods. The new design apparently provides a maximum shell-side heat-transfer coefficient for a given pressure drop.

**Tie Rods and Spacers** Tie rods are used to hold the baffles in place with spacers, which are pieces of tubing or pipe placed on the rods to locate the baffles. Occasionally baffles are welded to the tie rods, and spacers are eliminated. Properly located tie rods and spacers serve both to hold the bundle together and to reduce bypassing of the tubes.

In very large fixed-tube-sheet units, in which concentricity of shells decreases, baffles are occasionally welded to the shell to eliminate bypassing between the baffle and the shell.

Metal baffles are standard. Occasionally plastic baffles are used either to reduce corrosion or in vibratory service, in which metal baffles may cut the tubes.

**Impingement Baffle** The tube bundle is customarily protected against impingement by the incoming fluid at the shell inlet nozzle when the shell-side fluid is at a high velocity, is condensing, or is a two-phase fluid. Minimum entrance area about the nozzle is generally equal to the inlet nozzle area. Exit nozzles also require adequate area between the tubes and the nozzles. A full bundle without any provision for shell inlet nozzle area can increase the velocity of the inlet fluid by as much as 300 percent with a consequent loss in pressure.

Impingement baffles are generally made of rectangular plate, although circular plates are more desirable. Rods and other devices are sometimes used to protect the tubes from impingement. In order to maintain a maximum tube count the impingement plate is often placed in a conical nozzle opening or in a dome cap above the shell.

Impingement baffles or flow-distribution devices are recommended for axial tube-side nozzles when entrance velocity is high.

**Vapor Distribution** Relatively large shell inlet nozzles, which may be used in condensers under low pressure or vacuum, require provision for uniform vapor distribution.

**Tube-Bundle Bypassing** Shell-side heat-transfer rates are maximized when bypassing of the tube bundle is at a minimum. The most significant bypass stream is generally between the outer tube limit and the inside of the shell. The clearance between tubes and shell is at a minimum for fixed-tube-sheet construction and is greatest for straight-tube removable bundles.

Arrangements to reduce tube-bundle bypassing include:

*Dummy tubes.* These tubes do not pass through the tube sheets and can be located close to the inside of the shell.

*Tie rods with spacers.* These hold the baffles in place but can be located to prevent bypassing.

*Sealing strips.* These longitudinal strips either extend from baffle to baffle or may be inserted in slots cut into the baffles.

Dummy tubes or tie rods with spacers may be located within the pass partition lanes (and between the baffle cuts) in order to ensure maximum bundle penetration by the shell-side fluid.

When tubes are omitted from the tube layout to provide entrance area about an impingement plate, the need for sealing strips or other devices to cause proper bundle penetration by the shell-side fluid is increased.

**Helical Baffles** An increasingly popular variant to the segmental baffle is the helical baffle. These are quadrant-shaped plate baffles installed at an angle to the axial bundle centerline to produce a pseudo-spiraling flow down the length of the tube bundle (Fig. 11-40*d*). This baffle has the advantage of producing shell-side heat-transfer coefficients similar to those of the segmental baffle with much less shell-side pressure loss for the same size of shell. In the case of equal pressure drops, the helical baffle exchanger will be smaller than the segmental baffle exchanger; or, for identical shell sizes, the helical baffle exchanger; will permit a much higher throughput of flow for the same process inlet/outlet temperatures.

A great amount of proprietary research has been conducted by a few companies into the workings of helical baffled heat exchangers. The only known open literature method for estimating helical baffle performance has been "Comparison of Correction Factors for Shell-and-Tube Heat Exchangers with Segmental or Helical Baffles" by Stehlik, Nemcansky, Kral, and Swanson [*Heat Transfer Engineering*, **15**(1), 55–65].

Unique design variables for helical baffles include the baffle angle, adjacent baffle contact diameter (which sets the baffle spacing and is usually about half of the shell I.D.), and the number of baffle starts (i.e., number of intermediate baffle starts). Of course, consideration is also given to the tube layout, tube pitch, use of seal strips, and all the other configuration characteristics common to any plate baffled bundle.

A helical baffle bundle built in this way produces two distinct flow regions. The area outside of the adjacent baffle contact diameter tends to produce a stable helical cross flow. However, inside the diameter where adjacent baffles touch is a second region where vortical flow is induced but in which the intensity of the rotational component tends to decrease as one approaches the center of the bundle. For a fixed flow rate and helix angle, this tendency may be minimized by the proper selection of the baffle contact diameter. With the correct selection, stream temperatures may be made to be close to uniform across the bundle cross section through the shell. However, below a critical velocity (for the baffle configuration and fluid state), the tendency for nonuniformity of temperatures increases as velocity decreases until ever-increasing portions of the central core surface area pinch out with respect to temperature and become ineffective for further heat transfer.

The design approach involves varying the baffle spacing for the primary purpose of balancing the flows in the two regions and maximizing the effectiveness of the total surface area. In many cases, a shallower helix angle is chosen in conjunction with the baffle spacing in order to minimize the central core component while still achieving a reduced overall bundle pressure drop.

**Longitudinal Flow Baffles** In fixed-tube-sheet construction with multipass shells, the baffle is usually welded to the shell and positive assurance against bypassing results. Removable tube bundles have a sealing device between the shell and the longitudinal baffle. Flexible light-gauge sealing strips and various packing devices have been used. Removable U-tube bundles with four tube-side passes and two shell-side passes can be installed in shells with the longitudinal baffle welded in place.

In split-flow shells the longitudinal baffle may be installed without a positive seal at the edges if design conditions are not seriously affected by a limited amount of bypassing.

Fouling in petroleum-refinery service has necessitated rough treatment of tube bundles during cleaning operations. Many refineries avoid the use of longitudinal baffles, since the sealing devices are subject to damage during cleaning and maintenance operations.

## CORROSION IN HEAT EXCHANGERS

Some of the special considerations in regard to heat-exchanger corrosion are discussed in this subsection. A more extended presentation in Sec. 23 covers corrosion and its various forms as well as materials of construction.

**Materials of Construction** The most common material of construction for heat exchangers is carbon steel. Stainless-steel construction throughout is sometimes used in chemical-plant service and on rare occasions in petroleum refining. Many exchangers are constructed from dissimilar metals. Such combinations are functioning satisfactorily in certain services. Extreme care in their selection is required since electrolytic attack can develop.

Carbon steel and alloy combinations appear in Table 11-13 "Alloys" in chemical- and petrochemical-plant service in approximate order of use are stainless-steel series 300, nickel, Monel, copper alloy, aluminum, Inconel, stainless steel series 400, and other alloys. In petroleum-refinery service the frequency order shifts, with copper alloy (for water-cooled units) in first place and low-alloy steel in second place. In some segments of the petroleum industry copper alloy, stainless series 400, low-alloy steel, and aluminum are becoming the most commonly used alloys.

Copper-alloy tubing, particularly inhibited admiralty, is generally used with cooling water. Copper-alloy tube sheets and baffles are generally of naval brass.

Aluminum alloy (and in particular alclad aluminum) tubing is sometimes used in water service. The alclad alloy has a sacrificial aluminum-alloy layer metallurgically bonded to a core alloy.

Tube-side headers for water service are made in a wide variety of materials: carbon steel, copper alloy, cast iron, and lead-lined or plastic-lined or specially painted carbon steel.

**Bimetallic Tubes** When corrosive requirements or temperature conditions do not permit the use of a single alloy for the tubes, bimetallic (or duplex) tubes may be used. These can be made from almost any possible combination of metals. Tube sizes and gauges can be varied. For thin gauges the wall thickness is generally divided equally between the two components. In heavier gauges the more expensive component may comprise from a fifth to a third of the total thickness.

The component materials comply with applicable ASTM specifications, but after manufacture the outer component may increase in hardness beyond specification limits, and special care is required during the tube-rolling operation. When the harder material is on the outside, precautions must be exercised to expand the tube properly. When the inner material is considerably softer, rolling may not be practical unless ferrules of the soft material are used.

In order to eliminate galvanic action the outer tube material may be stripped from the tube ends and replaced with ferrules of the inner tube material. When the end of a tube with a ferrule is expanded or welded to a tube sheet, the tube-side fluid can contact only the inner tube material, while the outer material is exposed to the shell-side fluid.

Bimetallic tubes are available from a small number of tube mills and are manufactured only on special order and in large quantities.

**TABLE 11-13 Dissimilar Materials in Heat-Exchanger Construction**

| | Relative use | 1 | 2 | 3 | 4 | 5 | 6 |
|---|---|---|---|---|---|---|---|
| Part | Relative cost | A | B | C | D | C | E |
| Tubes | | ● | ● | ● | ● | ● | ● |
| Tube sheets | | | ● | ● | ● | ● | ● |
| Tube-side headers | | | | ● | ● | | |
| Baffles | | | | | ● | ● | ● |
| Shell | | | | | | | ● |

Carbon steel replaced by an alloy when ● appears.
Relative use: from 1 (most popular) through 6 (least popular) combinations.
Relative cost: from A (least expensive) to E (most expensive).

**Clad Tube Sheets** Usually tube sheets and other exchanger parts are of a solid metal. Clad or bimetallic tube sheets are used to reduce costs or because no single metal is satisfactory for the corrosive conditions. The alloy material (e.g., stainless steel, Monel) is generally bonded or clad to a carbon steel backing material. In fixed-tube-sheet construction a copper-alloy-clad tube sheet can be welded to a steel shell, while most copper-alloy tube sheets cannot be welded to steel in a manner acceptable to ASME Code authorities.

Clad tube sheets in service with carbon steel backer material include stainless-steel types 304, 304L, 316, 316L, and 317, Monel, Inconel, nickel, naval rolled brass, copper, admiralty, silicon bronze, and titanium. Naval rolled brass and Monel clad on stainless steel are also in service.

Ferrous-alloy-clad tube sheets are generally prepared by a weld overlay process in which the alloy material is deposited by welding upon the face of the tube sheet. Precautions are required to produce a weld deposit free of defects, since these may permit the process fluid to attack the base metal below the alloy. Copper-alloy-clad tube sheets are prepared by brazing the alloy to the carbon steel backing material.

Clad materials can be prepared by bonding techniques, which involve rolling, heat treatment, explosive bonding, etc. When properly manufactured, the two metals do not separate because of thermal-expansion differences encountered in service. Applied tube-sheet facings prepared by tack welding at the outer edges of alloy and base metal or by bolting together the two metals are in limited use.

**Nonmetallic Construction** Shell-and-tube exchangers with glass tubes 14 mm (0.551 in) in diameter and 1 mm (0.039 in) thick with tube lengths from 2.015 m (79.3 in) to 4.015 m (158 in) are available. Steel shell exchangers have a maximum design pressure of 517 kPa (75 lbf/in$^2$). Glass shell exchangers have a maximum design gauge pressure of 103 kPa (15 lbf/in$^2$). Shell diameters are 229 mm (9 in), 305 mm (12 in), and 457 mm (18 in). Heat-transfer surface ranges from 3.16 to 51 m$^2$ (34 to 550 ft$^2$). Each tube is free to expand, since a Teflon sealer sheet is used at the tube-to-tube-sheet joint.

**Impervious graphite** heat-exchanger equipment is made in a variety of forms, including outside-packed-head shell-and-tube exchangers. They are fabricated with impervious graphite tubes and tube-side headers and metallic shells. Single units containing up to 1300 m$^2$ (14,000 ft$^2$) of heat-transfer surface are available.

Teflon heat exchangers of special construction are described later in this section.

**Fabrication** Expanding the tube into the tube sheet reduces the tube wall thickness and work-hardens the metal. The induced stresses can lead to **stress corrosion.** Differential expansion between tubes and shell in fixed-tube-sheet exchangers can develop stresses, which lead to stress corrosion.

When austenitic stainless-steel tubes are used for corrosion resistance, a close fit between the tube and the tube hole is recommended in order to minimize work hardening and the resulting loss of corrosion resistance.

In order to facilitate removal and replacement of tubes it is customary to roller-expand the tubes to within 3 mm (⅛ in) of the shell-side face of the tube sheet. A 3-mm- (⅛-in-) long gap is thus created between the tube and the tube hole at this tube-sheet face. In some services this gap has been found to be a focal point for corrosion.

It is standard practice to provide a chamfer at the inside edges of tube holes in tube sheets to prevent cutting of the tubes and to remove burrs produced by drilling or reaming the tube sheet. In the lower tube sheet of vertical units this chamfer serves as a pocket to collect material, dirt, etc., and to serve as a corrosion center.

Adequate venting of exchangers is required both for proper operation and to reduce corrosion. Improper venting of the water side of exchangers can cause alternate wetting and drying and accompanying chloride concentration, which is particularly destructive to the series 300 stainless steels.

Certain corrosive conditions require that special consideration be given to complete drainage when the unit is taken out of service. Particular consideration is required for the upper surfaces of tube sheets in vertical heat exchangers, for sagging tubes, and for shell-side baffles in horizontal units.

## SHELL-AND-TUBE EXCHANGER COSTS

Basic costs of shell-and-tube heat exchangers made in the United States of carbon steel construction in 1958 are shown in Fig. 11-41.

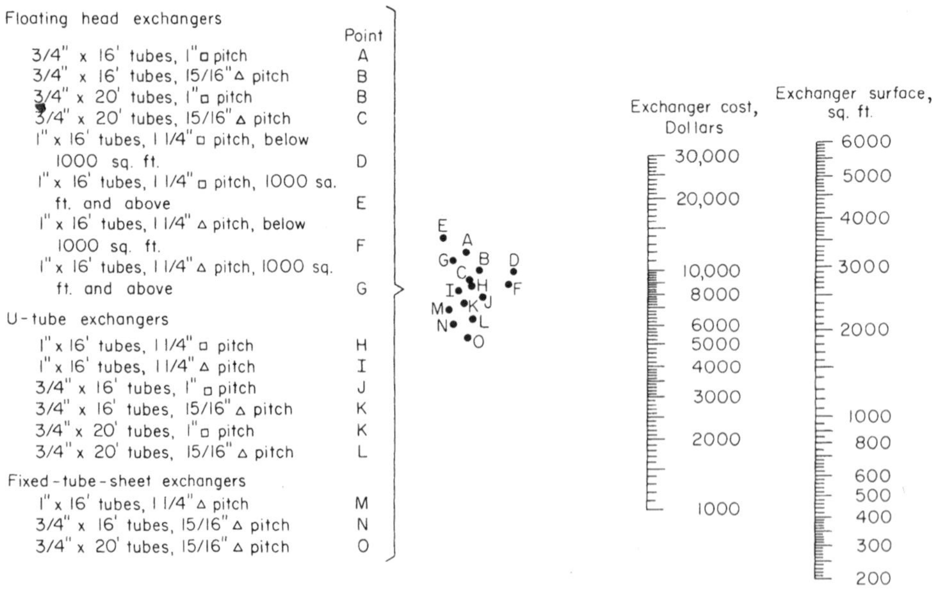


**FIG. 11-41** Costs of basic exchangers—all steel, TEMA Class R, 150 lbf/in$^2$, 1958. To convert pounds-force per square inch to kilopascals, multiply by 6.895; to convert square feet to square meters, multiply by 0.0929; to convert inches to millimeters, multiply by 25.4; and to convert feet to meters, multiply by 0.3048.

**TABLE 11-14 Extras for Pressure and Alloy Construction and Surface and Weights***

Percent of steel base price, 1500-lbf/in$^2$ working pressure

| | Shell diameters, in | | | | | | | | | | | | |
|---|---|---|---|---|---|---|---|---|---|---|---|---|---|
| **Pressure**† | 12 | 14 | 16 | 18 | 20 | 22 | 24 | 27 | 30 | 33 | 36 | 39 | 42 |
| 300 lbf/in$^2$ | 7 | 7 | 8 | 8 | 9 | 9 | 10 | 11 | 11 | 12 | 13 | 14 | 15 |
| 450 lbf/in$^2$ | 18 | 19 | 20 | 21 | 22 | 23 | 24 | 27 | 29 | 31 | 32 | 33 | 35 |
| 600 lbf/in$^2$ | 28 | 29 | 31 | 33 | 35 | 37 | 39 | 40 | 41 | 32 | 44 | 45 | 50 |
| **Alloy** | | | | | | | | | | | | | |
| All-steel heat exchanger | 100 | 100 | 100 | 100 | 100 | 100 | 100 | 100 | 100 | 100 | 100 | 100 | 100 |
| Tube sheets and baffles | | | | | | | | | | | | | |
| Naval rolled brass | 14 | 17 | 19 | 21 | 22 | 22 | 22 | 22 | 22 | 23 | 24 | 24 | 25 |
| Monel | 24 | 31 | 35 | 37 | 39 | 39 | 40 | 40 | 41 | 41 | 41 | 41 | 42 |
| 1¼ Cr, ½ Mo | 6 | 7 | 7 | 7 | 8 | 8 | 8 | 8 | 9 | 10 | 10 | 10 | 11 |
| 4–6 Cr, ½ Mo | 19 | 22 | 24 | 25 | 26 | 26 | 26 | 25 | 25 | 25 | 26 | 26 | 26 |
| 11–13 Cr (stainless 410) | 21 | 24 | 26 | 27 | 27 | 27 | 27 | 27 | 27 | 27 | 27 | 27 | 28 |
| Stainless 304 | 22 | 27 | 29 | 30 | 31 | 31 | 31 | 31 | 30 | 30 | 30 | 31 | 31 |
| Shell and shell cover | | | | | | | | | | | | | |
| Monel | 45 | 48 | 51 | 52 | 53 | 52 | 52 | 51 | 49 | 47 | 45 | 44 | 44 |
| 1¼ Cr, ½ Mo | 20 | 22 | 24 | 25 | 25 | 25 | 24 | 22 | 20 | 19 | 18 | 17 | 17 |
| 4–6 Cr, ½ Mo | 28 | 31 | 33 | 35 | 35 | 35 | 34 | 32 | 30 | 28 | 27 | 26 | 26 |
| 11–13 Cr (stainless 410) | 29 | 33 | 35 | 36 | 36 | 36 | 35 | 34 | 32 | 30 | 29 | 27 | 27 |
| Stainless 304 | 32 | 34 | 36 | 37 | 38 | 37 | 37 | 35 | 33 | 31 | 30 | 29 | 28 |
| Channel and floating-head cover | | | | | | | | | | | | | |
| Monel | 40 | 42 | 42 | 43 | 42 | 41 | 40 | 37 | 34 | 32 | 31 | 40 | 30 |
| 1¼ Cr, ½ Mo | 23 | 24 | 24 | 25 | 24 | 24 | 23 | 22 | 21 | 21 | 21 | 20 | 20 |
| 4–6 Cr, ½ Mo | 36 | 37 | 38 | 38 | 37 | 36 | 34 | 31 | 29 | 27 | 26 | 25 | 24 |
| 11–13 Cr (stainless 410) | 37 | 38 | 39 | 39 | 38 | 37 | 35 | 32 | 30 | 28 | 27 | 26 | 25 |
| Stainless 304 | 37 | 39 | 39 | 39 | 38 | 37 | 36 | 33 | 31 | 29 | 28 | 26 | 26 |
| **Surface** | | | | | | | | | | | | | |
| Surface, ft$^2$, internal floating head, ¾-in OD by 1-in square pitch, 16 ft 0 in, tube‡ | 251 | 302 | 438 | 565 | 726 | 890 | 1040 | 1470 | 1820 | 2270 | 2740 | 3220 | 3700 |
| 1-in OD by 1¼-in square pitch, 16-ft 0-in tube§ | 218 | 252 | 352 | 470 | 620 | 755 | 876 | 1260 | 1560 | 1860 | 2360 | 2770 | 3200 |
| Weight, lb, internal floating head, 1-in OD, 14 BWG tube | 2750 | 3150 | 4200 | 5300 | 6600 | 7800 | 9400 | 11,500 | 14,300 | 17,600 | 20,500 | 24,000 | 29,000 |

*Modified from E. N. Sieder and G. H. Elliot, *Pet. Refiner,* **39**(5), 223 (1960).

†Total extra is 0.7 × pressure extra on shell side plus 0.3 × pressure extra on tube side.

‡Fixed-tube-sheet construction with ¾-in OD tube on 15⁄16-in triangular pitch provides 36 percent more surface.

§Fixed-tube-sheet construction with 1-in OD tube on 1¼-in triangular pitch provides 18 percent more surface.

For an all-steel heat exchanger with mixed design pressures the total extra for pressure is 0.7 × pressure extra on shell side plus 0.3 × pressure extra tube side.

For an exchanger with alloy parts and a design pressure of 150 lbf/in$^2$, the alloy extras are added. For shell and shell cover the combined alloy-pressure extra is the alloy extra times the shell-side pressure extra/100. For channel and floating-head cover the combined alloy-pressure extra is the alloy extra times the tube-side pressure extra/100. For tube sheets and baffles the combined alloy-pressure extra is the alloy extra times the higher-pressure extra times 0.9/100. (The 0.9 factor is included since baffle thickness does not increase because of pressure.)

NOTE: To convert pounds-force per square inch to kilopascals, multiply by 6.895; to convert square feet to square meters, multiply by 0.0929; and to convert inches to millimeters, multiply by 25.4.

Cost data for shell-and-tube exchangers from 15 sources were correlated and found to be consistent when scaled by the Marshall and Swift index [Woods et al., *Can. J. Chem. Eng.,* **54,** 469–489 (December 1976)].

Costs of shell-and-tube heat exchangers can be estimated from Fig. 11-41 and Tables 11-14 and 11-15. These 1960 costs should be updated by use of the Marshall and Swift Index, which appears in each issue of *Chemical Engineering.* Note that during periods of high and low demand for heat exchangers the prices in the marketplace may vary significantly from those determined by this method.

Small heat exchangers and exchangers bought in small quantities are likely to be more costly than indicated.

Standard heat exchangers (which are in some instances off-the-shelf items) are available in sizes ranging from 1.9 to 37 m$^2$ (20 to 400 ft$^2$) at costs lower than for custom-built units. Steel costs are approximately one-half, admiralty tube-side costs are two-thirds, and stainless costs are three-fourths of those for equivalent custom-built exchangers.

**Kettle-type-reboiler** costs are 15 to 25 percent greater than for equivalent internal-floating-head or U-tube exchangers. The higher extra is applicable with relatively large kettle-to-port-diameter ratios and with increased internals (e.g., vapor-liquid separators, foam breakers, sight glasses).

To estimate exchanger costs for varying construction details and alloys, first determine the base cost of a similar heat exchanger of basic construction (carbon steel, Class R, 150 lbf/in$^2$) from Fig. 11-41. From Table 11-14, select appropriate extras for higher pressure rating and for alloy construction of tube sheets and baffles, shell and shell cover, and channel and floating-head cover. Compute these extras in accordance with the notes below the table. For tubes other than welded carbon steel, compute the extra by multiplying the exchanger surface by the appropriate cost per square foot from Table 11-15.

When points for 20-ft-long tubes do not appear in Fig. 11-41, use 0.95 times the cost of the equivalent 16-ft-long exchanger. Length variation of steel heat exchangers affects costs by approximately $1 per square foot. Shell diameters for a given surface are approximately equal for U-tube and floating-head construction.

Low-fin tubes (1⁄16-in-high fins) provide 2.5 times the surface per lineal foot. Surface required should be divided by 2.5; then use Fig. 11-41 to determine basic cost of the heat exchanger. Actual surface times extra costs (from Table 11-15) should then be added to determine cost of fin-tube exchanger.

**TABLE 11-15 Base Quantity Extra Cost for Tube Gauge and Alloy**
Dollars per square foot

| | ¾-in OD tubes | | | 1-in OD tubes | | |
|---|---|---|---|---|---|---|
| | 16 BWG | 14 BWG | 12 BWG | 16 BWG | 14 BWG | 12 BWG |
| Carbon steel | 0 | 0.02 | 0.06 | 0 | 0.01 | 0.07 |
| Admiralty | 0.78 | 1.20 | 1.81 | 0.94 | 1.39 | 2.03 |
| (T-11) 1¼ Cr, ½ Mo | 1.01 | 1.04 | 1.11 | 0.79 | 0.82 | 0.95 |
| (T-5) 4–6 Cr | 1.61 | 1.65 | 1.74 | 1.28 | 1.32 | 1.48 |
| Stainless 410 welded | 2.62 | 3.16 | 4.12 | 2.40 | 2.89 | 3.96 |
| Stainless 410 seamless | 3.10 | 3.58 | 4.63 | 2.84 | 3.31 | 4.47 |
| Stainless 304 welded | 2.50 | 3.05 | 3.99 | 2.32 | 2.83 | 3.88 |
| Stainless 304 seamless | 3.86 | 4.43 | 5.69 | 3.53 | 4.08 | 5.46 |
| Stainless 316 welded | 3.40 | 4.17 | 5.41 | 3.25 | 3.99 | 5.36 |
| Stainless 316 seamless | 7.02 | 7.95 | 10.01 | 6.37 | 7.27 | 9.53 |
| 90-10 cupronickel | 1.33 | 1.89 | 2.67 | 1.50 | 2.09 | 2.90 |
| Monel | 4.25 | 5.22 | 6.68 | 4.01 | 4.97 | 6.47 |
| Low fin | | | | | | |
| Carbon steel | 0.22 | 0.23 | | 0.18 | 0.19 | |
| Admiralty | 0.58 | 0.75 | | 0.70 | 0.87 | |
| 90-10 cupronickel | 0.72 | 0.96 | | 0.86 | 1.06 | |

NOTE: To convert inches to millimeters, multiply by 25.4.

# HAIRPIN/DOUBLE-PIPE HEAT EXCHANGERS

## PRINCIPLES OF CONSTRUCTION

Hairpin heat exchangers (often also referred to as "double pipes") are characterized by a construction form which imparts a U-shaped appearance to the heat exchanger. In its classical sense, the term *double pipe* refers to a heat exchanger consisting of a pipe within a pipe, usually of a straight-leg construction with no bends. However, due to the need for removable bundle construction and the ability to handle differential thermal expansion while avoiding the use of expansion joints (often the weak point of the exchanger), the current U-shaped configuration has become the standard in the industry (Fig. 11-42). A further departure from the classical definition comes when more than one pipe or tube is used to make a tube bundle, complete with tubesheets and tube supports similar to the TEMA style exchanger.

Hairpin heat exchangers consist of two shell assemblies housing a common set of tubes and interconnected by a return-bend cover referred to as the *bonnet*. The shell is supported by means of bracket assemblies designed to cradle both shells simultaneously. These brackets are configured to permit the modular assembly of many hairpin sections into an exchanger bank for inexpensive future-expansion capability and for providing the very long thermal lengths demanded by special process applications.

The bracket construction permits support of the exchanger without fixing the supports to the shell. This provides for thermal movement of the shells within the brackets and prevents the transfer of thermal stresses into the process piping. In special cases the brackets may be welded to the shell. However, this is usually avoided due to the resulting loss of flexibility in field installation and equipment reuse at other sites and an increase in piping stresses.

The hairpin heat exchanger, unlike the removable bundle TEMA styles, is designed for bundle insertion and removal from the return end rather than the tubesheet end. This is accomplished by means of removable split rings which slide into grooves machined around the outside of each tubesheet and lock the tubesheets to the external closure flanges. This provides a distinct advantage in maintenance since bundle removal takes place at the exchanger end furthest from the plant process piping without disturbing any gasketed joints of this piping.

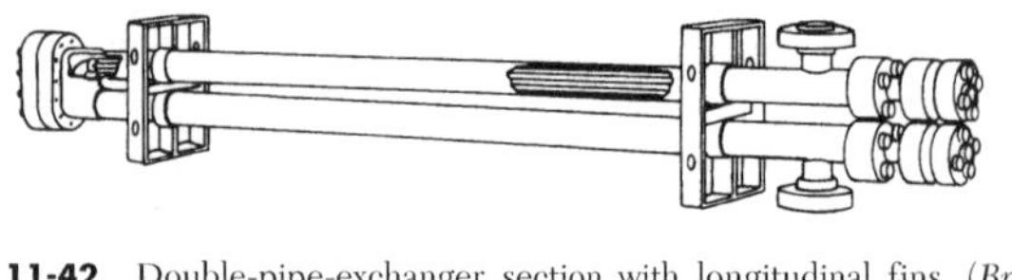

**FIG. 11-42** Double-pipe-exchanger section with longitudinal fins. (*Brown Fin-tube Co.*)

## FINNED DOUBLE PIPES

The design of the classical single-tube double-pipe heat exchanger is an exercise in pure longitudinal flow with the shellside and tubeside coefficients differing primarily due to variations in flow areas. Adding longitudinal fins gives the more common double-pipe configuration (Table 11-16). Increasing the number of tubes yields the *multitube* hairpin.

## MULTITUBE HAIRPINS

For years, the slightly higher mechanical-design complexity of the hairpin heat exchanger relegated it to only the smallest process requirements with shell sizes not exceeding 100 mm. In the early 1970s the maximum available sizes were increased to between 300 and 400 mm depending upon the manufacturer. At the present time, due to recent advances in design technology, hairpin exchangers are routinely produced in shell sizes between 50 (2 in) and 800 mm (30 in) for a wide range of pressures and temperatures and have been made in larger sizes as well. Table 11-17 gives common hairpin tube counts and areas for 19 mm (¾ in) O.D. tubes arranged on a 24 mm (¹⁵⁄₁₆ in) triangular tube layout.

The hairpin width and the centerline distance of the two legs (shells) of the hairpin heat exchanger are limited by the outside diameter of the closure flanges at the tubesheets. This diameter, in turn, is a function of the design pressures. As a general rule, for low-to-moderate design pressures (less than 15 bar), the center-to-center distance is approximately

**TABLE 11-16 Double-Pipe Hairpin Section Data**

| Shell pipe O.D. | | Inner pipe O.D. | | Fin height | | Fin count | Surface-area-per-unit length | |
|---|---|---|---|---|---|---|---|---|
| mm | in | mm | in | mm | in | (max) | sq m/m | sq ft/ft |
| 60.33 | 2.375 | 25.4 | 1.000 | 12.7 | 0.50 | 24 | 0.692 | 2.27 |
| 88.9 | 3.500 | 48.26 | 1.900 | 12.7 | 0.50 | 36 | 1.07 | 3.51 |
| 114.3 | 4.500 | 48.26 | 1.900 | 25.4 | 1.00 | 36 | 1.98 | 6.51 |
| 114.3 | 4.500 | 60.33 | 2.375 | 19.05 | 0.75 | 40 | 1.72 | 5.63 |
| 114.3 | 4.500 | 73.03 | 2.875 | 12.70 | 0.50 | 48 | 1.45 | 4.76 |
| 141.3 | 5.563 | 88.9 | 3.500 | 17.46 | 0.6875 | 56 | 2.24 | 7.34 |
| 168.3 | 6.625 | 114.3 | 4.500 | 17.46 | 0.6875 | 72 | 2.88 | 9.44 |

**TABLE 11-17 Multitube Hairpin Section Data**

| | Shell O.D. | | Shell thickness | | Tube count | Surface area for 6.1 m (20 ft.) nominal length | |
|---|---|---|---|---|---|---|---|
| Size | mm | in | mm | in | 19 mm | sq m | sq ft |
| 03-MT | 88.9 | 3.500 | 5.49 | 0.216 | 5 | 3.75 | 40.4 |
| 04-MT | 114.3 | 4.500 | 6.02 | 0.237 | 9 | 6.73 | 72.4 |
| 05-MT | 141.3 | 5.563 | 6.55 | 0.258 | 14 | 10.5 | 113.2 |
| 06-MT | 168.3 | 6.625 | 7.11 | 0.280 | 22 | 16.7 | 179.6 |
| 08-MT | 219.1 | 8.625 | 8.18 | 0.322 | 42 | 32.0 | 344.3 |
| 10-MT | 273.1 | 10.75 | 9.27 | 0.365 | 68 | 52.5 | 564.7 |
| 12-MT | 323.9 | 12.75 | 9.53 | 0.375 | 109 | 84.7 | 912.1 |
| 14-MT | 355.6 | 14.00 | 9.53 | 0.375 | 136 | 107. | 1159. |
| 16-MT | 406.4 | 16.00 | 9.53 | 0.375 | 187 | 148. | 1594. |
| 18-MT | 457.2 | 18.00 | 9.53 | 0.375 | 241 | 191. | 2054. |
| 20-MT | 508.0 | 20.00 | 9.53 | 0.375 | 304 | 244. | 2622. |
| 22-MT | 558.8 | 22.00 | 9.53 | 0.375 | 380 | 307. | 3307. |
| 24-MT | 609.6 | 24.00 | 9.53 | 0.375 | 463 | 378 | 4065. |
| 26-MT | 660.4 | 26.00 | 9.53 | 0.375 | 559 | 453. | 4879. |
| 28-MT | 711.2 | 28.00 | 9.53 | 0.375 | 649 | 529. | 5698. |
| 30-MT | 762.0 | 30.00 | 11.11 | 0.4375 | 752 | 630. | 6776. |

1.5 to 1.8 times the shell outside diameter, with this ratio decreasing slightly for the larger sizes.

One interesting consequence of this fact is the inability to construct a hairpin tube bundle having the smallest radius bends common to a conventional U-tube, TEMA shell, and tube bundle. In fact, in the larger hairpin sizes the tubes might be better described as curved rather than bent. The smallest U-bend diameters are greater than the outside diameter of shells less than 300 mm in size. The U-bend diameters are greater than 300 mm in larger shells. As a general rule, mechanical tube cleaning around the radius of a U-bend may be accomplished with a flexible shaft-cleaning tool for bend diameters greater than ten times the tube's inside diameter. This permits the tool to pass around the curve of the tube bend without binding.

In all of these configurations, maintaining longitudinal flow on both the shellside and tubeside allows the decision for placement of a fluid stream on either one side or the other to be based upon design efficiency (mass flow rates, fluid properties, pressure drops, and velocities) and not because there is any greater tendency to foul on one side than the other. Experience has shown that, in cases where fouling is influenced by flow velocity, overall fouling in tube bundles is less in properly designed longitudinal flow bundles where areas of low velocity can be avoided without flow-induced tube vibration.

This same freedom of stream choice is not as readily applied when a segmental baffle is used. In those designs, the baffle's creation of low velocities and stagnant flow areas on the outside of the bundle can result in increased shellside fouling at various locations of the bundle. The basis for choosing the stream side in those cases will be similar to the common shell and tube heat exchanger. At times a specific selection of stream side must be made regardless of tube-support mechanism in expectation of an unresolvable fouling problem. However, this is often the exception rather than the rule.

## DESIGN APPLICATIONS

One benefit of the hairpin exchanger is its ability to handle high tubeside pressures at a lower cost than other removable-bundle exchangers. This is due in part to the lack of pass partitions at the tubesheets which complicate the gasketing design process. Present mechanical design technology has allowed the building of dependable, removable-bundle, hairpin multitubes at tubeside pressures of 825 bar (12,000 psi).

The best known use of the hairpin is its operation in true countercurrent flow which yields the most efficient design for processes that have a close temperature approach or temperature cross. However, maintaining countercurrent flow in a tubular heat exchanger usually implies one tube pass for each shell pass. As recently as 30 years ago, the lack of inexpensive, multiple-tube pass capability often diluted the advantages gained from countercurrent flow.

The early attempts to solve this problem led to investigations into the area of heat transfer augmentation. This familiarity with augmentation techniques inevitably led to improvements in the efficiency and capacity of the small heat exchangers. The result has been the application of the hairpin heat exchanger to the solution of unique process problems, such as dependable, once-through, convective boilers offering high-exit qualities, especially in cases of process-temperature crosses.

# AIR-COOLED HEAT EXCHANGERS

## AIR-COOLED HEAT EXCHANGERS

Atmospheric air has been used for many years to cool and condense fluids in areas of water scarcity. During the 1960s the use of air-cooled heat exchangers grew rapidly in the United States and elsewhere. In Europe, where seasonal variations in ambient temperatures are relatively small, air-cooled exchangers are used for the greater part of process cooling. In some new plants all cooling is done with air. Increased use of air-cooled heat exchangers has resulted from lack of available water, significant increases in water costs, and concern for water pollution.

Air-cooled heat exchangers include a tube bundle, which generally has spiral-wound fins upon the tubes, and a fan, which moves air across the tubes and is provided with a driver. Electric motors are the most commonly used drivers; typical drive arrangements require a V belt or a direct right-angle gear. A plenum and structural supports are basic components. Louvers are often used:

A bay generally has two tube bundles installed in parallel. These may be in the same or different services. Each bay is usually served by two (or more) fans and is furnished with a structure, a plenum, and other attendant equipment.

The location of air-cooled heat exchangers must consider the large space requirements and the possible recirculation of heated air because of the effect of prevailing winds upon buildings, fired heaters, towers, various items of equipment, and other air-cooled exchangers. Inlet air temperature at the exchanger can be significantly higher than the ambient air temperature at a nearby weather station. See *Air-Cooled Heat Exchangers for General Refinery Services,* API Standard 661, 2d ed., January 1978, for information on refinery-process air-cooled heat exchangers.

**Forced and Induced Draft** The forced-draft unit, which is illustrated in Fig. 11-43 pushes air across the finned tube surface. The fans are located below the tube bundles. The induced-draft design has the fan above the bundle, and the air is pulled across the finned tube surface. In theory, a primary advantage of the forced-draft unit is that less power is required. This is true when the air-temperature rise exceeds 30°C (54°F).

Air-cooled heat exchangers are generally arranged in banks with several exchangers installed side by side. The height of the bundle aboveground must be one-half of the tube length to produce an inlet velocity equal to the face velocity. This requirement applies both to ground-mounted exchangers and to those pipe-rack-installed exchangers which have a fire deck above the pipe rack.

The forced-draft design offers better accessibility to the fan for on-stream maintenance and fan-blade adjustment. The design also provides a fan and V-belt assembly, which are not exposed to the hot-air stream that exits from the unit. Structural costs are less, and mechanical life is longer.

Induced-draft design provides more even distribution of air across the bundle, since air velocity approaching the bundle is relatively low.

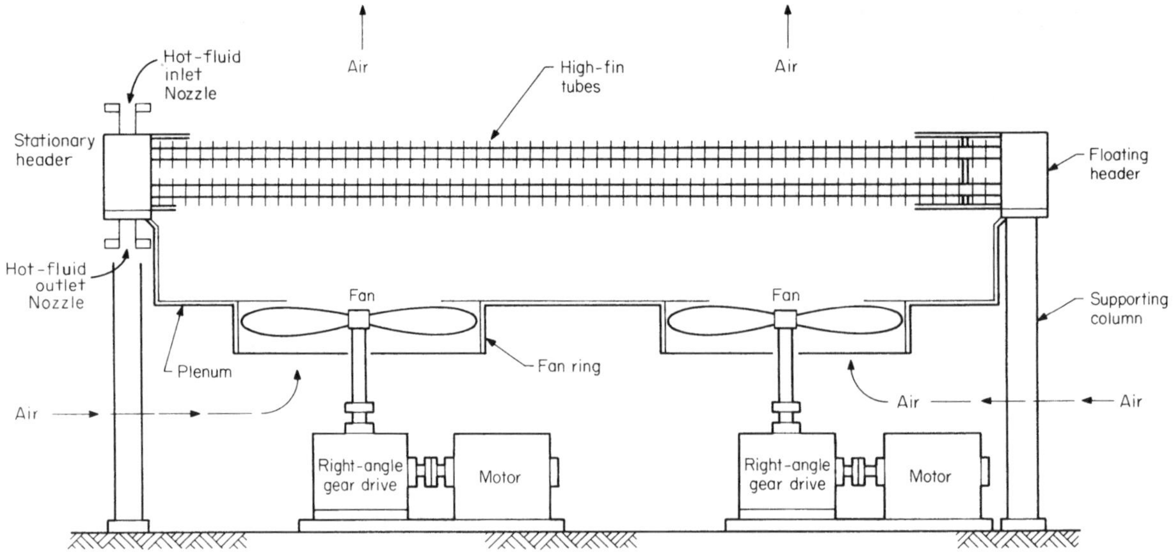


**FIG. 11-43** Forced-draft air-cooled heat exchanger. [*Chem. Eng.*, **114** (Mar. 27, 1978).]

This design is better suited for exchangers designed for a close approach of product outlet temperature to ambient-air temperature.

Induced-draft units are less likely to recirculate the hot exhaust air, since the exit air velocity is several times that of the forced-draft unit. Induced-draft design more readily permits the installation of the air-cooled equipment above other mechanical equipment such as pipe racks or shell-and-tube exchangers.

In a service in which sudden temperature change would cause upset and loss of product, the induced-draft unit gives more protection in that only a fraction of the surface (as compared with the forced-draft unit) is exposed to rainfall, sleet, or snow.

**Tube Bundle** The principal parts of the tube bundle are the finned tubes and the header. Most commonly used is the plug header, which is a welded box that is illustrated in Fig. 11-44. The finned tubes are described in a subsequent paragraph. The components of a tube bundle are identified in the figure.

The second most commonly used header is a cover-plate header. The cover plate is bolted to the top, bottom, and end plates of the header. Removing the cover plate provides direct access to the tubes without the necessity of removing individual threaded plugs.

Other types of headers include the bonnet-type header, which is constructed similarly to the bonnet construction of shell-and-tube heat exchangers; manifold-type headers, which are made from pipe and have tubes welded into the manifold; and billet-type headers, made from a solid piece of material with machined channels for distributing the fluid. Serpentine-type tube bundles are sometimes used for very viscous fluids. A single continuous flow path through pipe is provided.

Tube bundles are designed to be rigid and self-contained and are mounted so that they expand independently of the supporting structure.

The face area of the tube bundle is its length times width. The net free area for air flow through the bundle is about 50 percent of the face area of the bundle.

The standard air face velocity (FV) is the velocity of standard air passing through the tube bundle and generally ranges from 1.5 to 3.6 m/s (300 to 700 ft/min).

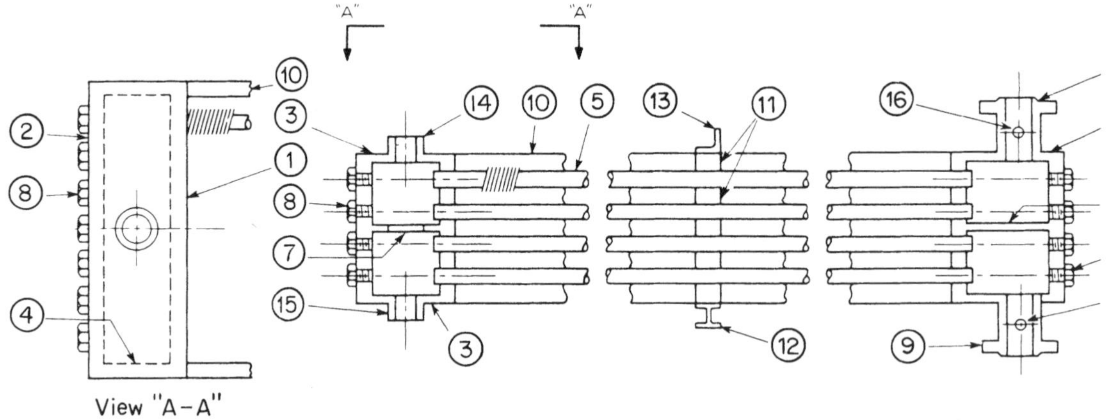


**FIG. 11-44** Typical construction of a tube bundle with plug headers: (1) tube sheet; (2) plug sheet; (3) top and bottom plates; (4) end plate; (5) tube; (6) pass partition; (7) stiffener; (8) plug; (9) nozzle; (10) side frame; (11) tube spacer; (12) tube-support cross member; (13) tube keeper; (14) vent; (15) drain; (16) instrument connection. (API Standard 661.)

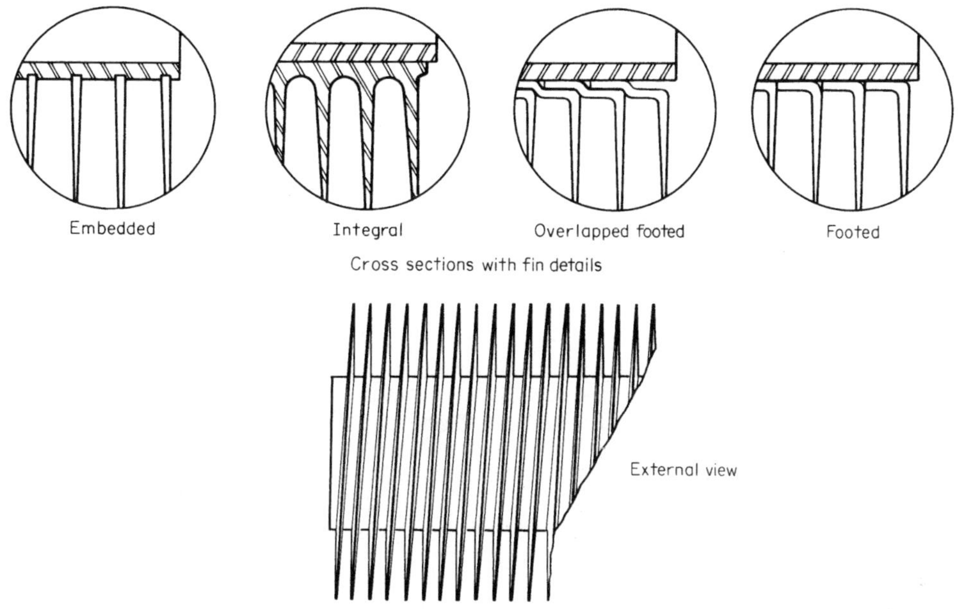


**FIG. 11-45** Finned-tube construction.

**Tubing** The 25.4-mm (1-in) outside-diameter tube is most commonly used. Fin heights vary from 12.7 to 15.9 mm (0.5 to 0.625 in), fin spacing from 3.6 to 2.3 mm (7 to 11 per linear inch), and tube triangular pitch from 50.8 to 63.5 mm (2.0 to 2.5 in). Ratio of extended surface to bare-tube outside surface varies from about 7 to 20. The 38-mm (1½-in) tube has been used for flue-gas and viscous-oil service. Tube size, fin heights, and fin spacing can be further varied.

Tube lengths vary and may be as great as 18.3 m (60 ft). When tube length exceeds 12.2 m (40 ft), three fans are generally installed in each bay. Frequently used tube lengths vary from 6.1 to 12.2 m (20 to 40 ft).

**Finned-Tube Construction** The following are descriptions of commonly used finned-tube constructions (Fig. 11-45).

1. *Embedded.* Rectangular-cross-section aluminum fin which is wrapped under tension and mechanically embedded in a groove 0.25 ± 0.05 mm (0.010 ± 0.002 in) deep, spirally cut into the outside surface of a tube.
2. *Integral (or extruded).* An aluminum outer tube from which fins have been formed by extrusion, mechanically bonded to an inner tube or liner.
3. *Overlapped footed.* L-shaped aluminum fin wrapped under tension over the outside surface of a tube, with the tube fully covered by the overlapped feet under and between the fins.
4. *Footed.* L-shaped aluminum fin wrapped under tension over the outside surface of a tube with the tube fully covered by the feet between the fins.
5. *Bonded.* Tubes on which fins are bonded to the outside surface by hot-dip galvanizing, brazing, or welding.

Typical metal design temperatures for these finned-tube constructions are 399°C (750°F) embedded, 288°C (550°F) integral, 232°C (450°F) overlapped footed, and 177°C (350°F) footed.

Tube ends are left bare to permit insertion of the tubes into appropriate holes in the headers or tube sheets. Tube ends are usually roller-expanded into these tube holes.

**Fans** Axial-flow fans are large-volume, low-pressure devices. Fan diameters are selected to give velocity pressures of approximately 2.5 mm (0.1 in) of water. Total fan efficiency (fan, driver, and transmission device) is about 75 percent, and fan drives usually have a minimum of 95 percent mechanical efficiency.

Usually fans are provided with four or six blades. Larger fans may have more blades. Fan diameter is generally slightly less than the width of the bay.

At the fan-tip speeds required for economical performance, a large amount of noise is produced. The predominant source of noise is vortex shedding at the trailing edge of the fan blade. Noise control of air-cooled exchangers is required by the Occupational Safety and Health Act (OSHA). API Standard 661 (*Air-Cooled Heat Exchangers for General Refinery Services,* 2d ed., January 1978) has the purchaser specifying sound-pressure-level (SPL) values per fan at a location designated by the purchaser and also specifying sound-power-level (PWL) values per fan. These are designated at the following octave-band-center frequencies: 63, 125, 250, 1000, 2000, 4000, 8000, and also the dBa value (the dBa is a weighted single-value sound-pressure level).

Reducing the fan-tip speed results in a straight-line reduction in air flow while the noise level decreases. The API Standard limits fan-tip speed to 61 m/s (12,000 ft/min) for typical constructions. Fan-design changes which reduce noise include increasing the number of fan blades, increasing the width of the fan blades, and reducing the clearance between fan tip and fan ring.

Both the quantity of air and the developed static pressure of fans in air-cooled heat exchangers are lower than indicated by fan manufacturers' test data, which are applicable to testing-facility tolerances and not to heat-exchanger constructions.

The axial-flow fan is inherently a device for moving a consistent volume of air when blade setting and speed of rotation are constant. Variation in the amount of air flow can be obtained by adjusting the blade angle of the fan and the speed of rotation. The blade angle can be either (1) permanently fixed, (2) hand-adjustable, or (3) automatically adjusted. Air delivery and power are a direct function of blade pitch angle.

Fan mounting should provide a minimum of one-half to three-fourths diameter between fan and ground on a forced-draft heat exchanger and one-half diameter between tubes and fan on an induced-draft cooler.

Fan blades can be made of aluminum, molded plastic, laminated plastic, carbon steel, stainless steel, and Monel.

**Fan Drivers** Electric motors or steam turbines are most commonly used. These connect with gears or V belts. (Gas engines connected through gears and hydraulic motors either direct-connected or connected through gears are in use. Fans may be driven by a prime mover such as a compressor with a V-belt takeoff from the flywheel to a jack shaft and then through a gear or V belt to the fan. Direct motor drive is generally limited to small-diameter fans.

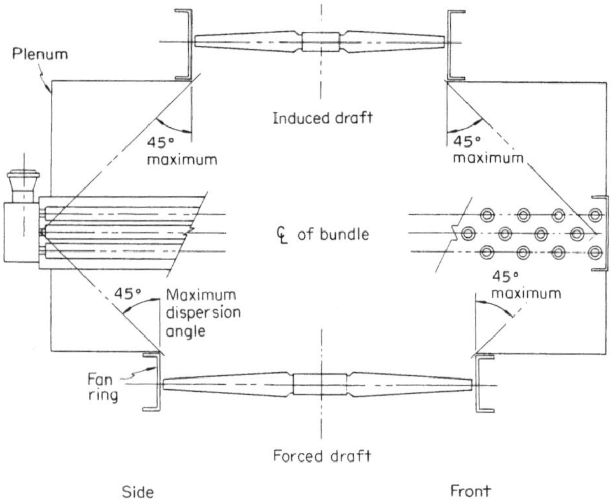


**FIG. 11-46** Fan dispersion angle. (API Standard 661.)

V-belt drive assemblies are generally used with fans 3 m (10 ft) and less in diameter and motors of 22.4 kW (30 hp) and less.

Right-angle gear drive is preferred for fans over 3 m (10 ft) in diameter, for electric motors over 22.4 kW (30 hp), and with steam-turbine drives.

**Fan Ring and Plenum Chambers** The air must be distributed from the circular fan to the rectangular face of the tube bundle. The air velocity at the fan is between 3.8 and 10.2 m/s (750 and 2000 ft/in). The plenum-chamber depth (from fan to tube bundle) is dependent upon the fan dispersion angle (Fig. 11-46), which should have a maximum value of 45°.

The fan ring is made to commercial tolerances for the relatively large diameter fan. These tolerances are greater than those upon closely machined fan rings used for small-diameter laboratory-performance testing. Fan performance is directly affected by this increased clearance between the blade tip and the ring, and adequate provision in design must be made for the reduction in air flow. API Standard 661 requires that fan-tip clearance be a maximum of 0.5 percent of the fan diameter for diameters between 1.9 and 3.8 m (6.25 and 12.5 ft). Maximum clearance is 9.5 mm (⅜ in) for smaller fans and 19 mm (¾ in) for larger fans.

The depth of the fan ring is critical. Worsham (ASME Pap. 59-PET-27, Petroleum Mechanical Engineering Conference, Houston, 1959) reports an increase in flow varying from 5 to 15 percent with the same power consumption when the depth of a fan ring was doubled. The percentage increase was proportional to the volume of air and static pressure against which the fan was operating.

When making a selection, the stall-out condition, which develops when the fan cannot produce any more air regardless of power input, should be considered.

**Air-Flow Control** Process operating requirements and weather conditions are considered in determining the method of controlling air flow. The most common methods include simple on-off control, on-off step control (in the case of multiple-driver units), two-speed-motor control, variable-speed drivers, controllable fan pitch, manually or automatically adjustable louvers, and air recirculation.

Winterization is the provision of design features, procedures, or systems for air-cooled heat exchangers to avoid process-fluid operating problems resulting from low-temperature inlet air. These include fluid freezing, pour point, wax formation, hydrate formation, laminar flow, and condensation at the dew point (which may initiate corrosion). Freezing points for some commonly encountered fluids in refinery service include: benzene, 5.6°C (42°F); *p*-xylene 15.5°C (55.9°F); cyclohexane, 6.6°C (43.8°F); phenol, 40.9°C (105.6°F); monoethanolamine, 10.3°C (50.5°F); and diethanolamine, 25.1°C (77.2°F). Water solutions

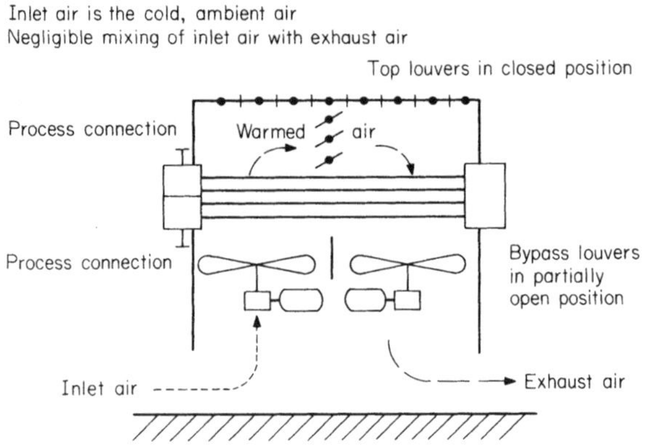


**FIG. 11-47** Contained internal recirculation (with internal louvers). [*Hydrocarbon Process,* **59,** 148–149 (October 1980).]

of these organic compounds are likely to freeze in air-cooled exchangers during winter service. Paraffinic and olefinic gases ($C_1$ through $C_4$) saturated with water vapor form hydrates when cooled. These hydrates are solid crystals which can collect and plug exchanger tubes.

Air-flow control in some services can prevent these problems. Cocurrent flow of air and process fluid during winter may be adequate to prevent problems. (Normal design has countercurrent flow of air and process fluid.) In some services when the hottest process fluid is in the bottom tubes, which are exposed to the lowest-temperature air, winterization problems may be eliminated.

Following are references which deal with problems in low-temperature environments: Brown and Benkley, "Heat Exchangers in Cold Service—A Contractor's View," *Chem. Eng. Prog.,* **70,** 59–62 (July 1974); Franklin and Munn, "Problems with Heat Exchangers in Low Temperature Environments," *Chem. Eng. Prog.,* **70,** 63–67 (July 1974); Newell, "Air-Cooled Heat Exchangers in Low Temperature Environments: A Critique," *Chem. Eng. Prog.,* **70,** 86–91 (October 1974); Rubin, "Winterizing Air Cooled Heat Exchangers," *Hydrocarbon Process.,* **59,** 147–149 (October 1980); Shipes, "Air-Cooled Heat Exchangers in Cold Climates," *Chem. Eng. Prog.,* **70,** 53–58 (July 1974).

**Air Recirculation** Recirculation of air which has been heated as it crosses the tube bundle provides the best means of preventing operating problems due to low-temperature inlet air. Internal recirculation is the movement of air within a bay so that the heated air which has crossed the bundle is directed by a fan with reverse flow across another part of the bundle. Wind skirts and louvers are generally provided to minimize the entry of low-temperature air from the surroundings. Contained internal recirculation uses louvers within the bay to control the flow of warm air in the bay as illustrated in Fig. 11-47. Note that low-temperature inlet air has access to the tube bundle.

External recirculation is the movement of the heated air within the bay to an external duct, where this air mixes with inlet air, and the mixture serves as the cooling fluid within the bay. Inlet air does not have direct access to the tube bundle; an adequate mixing chamber is essential. Recirculation over the end of the exchanger is illustrated in Fig. 11-48. Over-the-side recirculation also is used. External recirculation systems maintain the desired low temperature of the air crossing the tube bundle.

**Trim Coolers** Conventional air-cooled heat exchangers can cool the process fluid to within 8.3°C (15°F) of the design dry-bulb temperature. When a lower process outlet temperature is required, a trim cooler is installed in series with the air-cooled heat exchanger. The water-cooled trim cooler can be designed for a 5.6 to 11.1°C (10 to 20°F) approach to the wet-bulb temperature (which in the United States is about 8.3°C (15°F) less than the dry-bulb temperature). In arid areas the difference between dry- and wet-bulb temperatures is much greater.

**Humidification Chambers** The air-cooled heat exchanger is provided with humidification chambers in which the air is cooled to a

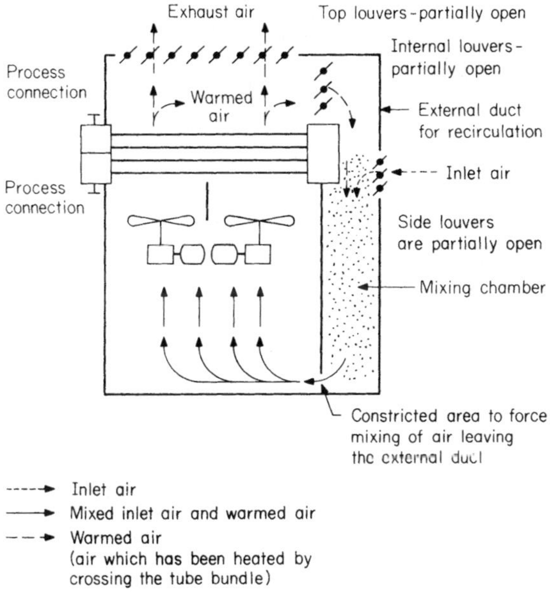


**FIG. 11-48** External recirculation with adequate mixing chamber. [*Hydrocarbon Process,* **59,** 148–149 (October 1980).]

close approach to the wet-bulb temperature before entering the finned-tube bundle of the heat exchanger.

**Evaporative Cooling** The process fluid can be cooled by using evaporative cooling with the sink temperature approaching the wet-bulb temperature.

**Steam Condensers** Air-cooled steam condensers have been fabricated with a single tube-side pass and several rows of tubes. The bottom row has a higher temperature difference than the top row, since the air has been heated as it crosses the rows of tubes. The bottom row condenses all the entering steam before the steam has traversed the length of the tube. The top row, with a lower temperature driving force, does not condense all the entering steam. At the exit header, uncondensed steam flows from the top row into the bottom row. Since noncondensable gases are always present in steam, these accumulate within the bottom row because steam is entering from both ends of the tube. Performance suffers.

Various solutions have been used. These include orifices to regulate the flow into each tube, a "blow-through steam" technique with a vent condenser, complete separation of each row of tubes, and inclined tubes.

**Air-Cooled Overhead Condensers** Air-cooled overhead condensers (AOC) have been designed and installed above distillation columns as integral parts of distillation systems. The condensers generally have inclined tubes, with air flow over the finned surfaces induced by a fan. Prevailing wind affects both structural design and performance.

AOC provide the additional advantages of reducing ground-space requirements and piping and pumping requirements and of providing smoother column operation.

The downflow condenser is used mainly for nonisothermal condensation. Vapors enter through a header at the top and flow downward. The reflux condenser is used for isothermal and small-temperature-change conditions. Vapors enter at the bottom of the tubes.

AOC usage first developed in Europe but became more prevalent in the United States during the 1960s. A state-of-the-art article was published by Dehne [*Chem. Eng. Prog.,* **64,** 51 (July 1969)].

**Air-Cooled Heat-Exchanger Costs** The cost data that appear in Table 11-18 are unchanged from those published in the 1963 edition of this *Handbook.* In 1969 Guthrie [*Chem. Eng.,* **75,** 114 (Mar. 24, 1969)] presented cost data for field-erected air-cooled exchangers. These costs are only 25 percent greater than those of Table 11-18 and include the costs of steel stairways, indirect subcontractor charges, and field-erection charges. Since minimal field costs would be this high (i.e., 25 percent of purchase price), the basic costs appear to be unchanged. (Guthrie indicated a cost band of plus or minus 25 percent.) Preliminary design and cost estimating of air-cooled heat exchangers have been discussed by J. E. Lerner ["Simplified Air Cooler Estimating," *Hydrocarbon Process.,* **52,** 93–100 (February 1972)].

**TABLE 11-18 Air-Cooled Heat-Exchanger Costs (1970)**

| Surface (bare tube), sq. ft. | 500 | 1000 | 2000 | 3000 | 5000 |
|---|---|---|---|---|---|
| Cost for 12-row-deep bundle, dollars/square foot | 9.0 | 7.6 | 6.8 | 5.7 | 5.3 |
| Factor for bundle depth: | | | | | |
| 6 rows | 1.07 | 1.07 | 1.07 | 1.12 | 1.12 |
| 4 rows | 1.2 | 1.2 | 1.2 | 1.3 | 1.3 |
| 3 rows | 1.25 | 1.25 | 1.25 | 1.5 | 1.5 |

Base: Bare-tube external surface 1 in. o.d. by 12 B.W.G. by 24 ft. 0 in. steel tube with 8 aluminum fins per inch ⅝-in. high. Steel headers. 150 lb./sq. in. design pressure. V-belt drive and explosion-proof motor. Bare-tube surface 0.262 sq. ft./ft. Fin-tube surface/bare-tube surface ratio is 16.9.

| Factors: | | |
|---|---|---|
| | 20 ft. tube length | 1.05 |
| | 30 ft. tube length | 0.95 |
| | 18 B.W.G. admiralty tube | 1.04 |
| | 16 B.W.G. admiralty tube | 1.12 |

NOTE: To convert feet to meters, multiply by 0.3048; to convert square feet to square meters, multiply by 0.0929; and to convert inches to millimeters, multiply by 25.4.

## Design Considerations

1. *Design dry-bulb temperature.* The typically selected value is the temperature which is equaled or exceeded 2½ percent of the time during the warmest consecutive 4 months. Since air temperatures at industrial sites are frequently higher than those used for these weather-data reports, it is good practice to add 1 to 3°C (2 to 6°F) to the tabulated value.

2. *Air recirculation.* Prevailing winds and the locations and elevations of buildings, equipment, fired heaters, etc., require consideration. All air-cooled heat exchangers in a bank are of one type, i.e., all forced-draft or all induced-draft. Banks of air-cooled exchangers must be placed far enough apart to minimize air recirculation.

3. *Wintertime operations.* In addition to the previously discussed problems of winterization, provision must be made for heavy rain, strong winds, freezing of moisture upon the fins, etc.

4. *Noise.* Two identical fans have a noise level 3 dBa higher than one fan, while eight identical fans have a noise level 9 dBa higher than a single fan. Noise level at the plant site is affected by the exchanger position, the reflective surfaces near the fan, the hardness of these surfaces, and noise from adjacent equipment. The extensive use of air-cooled heat exchangers contributes significantly to plant noise level.

5. *Ground area and space requirements.* Comparisons of the overall space requirements for plants using air cooling versus water cooling are not consistent. Some air-cooled units are installed above other equipment—pipe racks, shell-and-tube exchangers, etc. Some plants avoid such installations because of safety considerations, as discussed later.

6. *Safety.* Leaks in air-cooled units are directly to the atmosphere and can cause fire hazards or toxic-fume hazards. However, the large air flow through an air-cooled exchanger greatly reduces any concentration of toxic fluids. Segal [*Pet. Refiner,* **38,** 106 (April 1959)] reports that air-fin coolers "are not located over pumps, compressors, electrical switchgear, control houses and, in general, the amount of equipment such as drums and shell-and-tube exchangers located beneath them are minimized."

Pipe-rack-mounted air-cooled heat exchangers with flammable fluids generally have concrete fire decks which isolate the exchangers from the piping.

7. *Atmospheric corrosion.* Air-cooled heat exchangers should not be located where corrosive vapors and fumes from vent stacks will pass through them.

8. *Air-side fouling.* Air-side fouling is generally negligible.
9. *Process-side cleaning.* Either chemical or mechanical cleaning on the inside of the tubes can readily be accomplished.
10. *Process-side design pressure.* The high-pressure process fluid is always in the tubes. Tube-side headers are relatively small as compared with water-cooled units when the high pressure is generally on the shell side. High-pressure design of rectangular headers is complicated. The plug-type header is normally used for design gauge pressures to 13,790 kPa (2000 lbf/in$^2$) and has been used to 62,000 kPa (9000 lbf/in$^2$). The use of threaded plugs at these pressures creates problems. Removable cover plate headers are generally limited to gauge pressures of 2068 kPa (300 lbf/in$^2$). The expensive billet-type header is used for high-pressure service.
11. *Bond resistance.* Vibration and thermal cycling affect the bond resistance of the various types of tubes in different manners and thus affect the amount of heat transfer through the fin tube.
12. *Approach temperature.* The approach temperature, which is the difference between the process-fluid outlet temperature and the design dry-bulb air temperature, has a practical minimum of 8 to 14°C (15 to 25°F). When a lower process-fluid outlet temperature is required, an air-humidification chamber can be provided to reduce the inlet air temperature toward the wet-bulb temperature. A 5.6°C (10°F) approach is feasible. Since typical summer wet-bulb design temperatures in the United States are 8.3°C (15°F) lower than dry-bulb temperatures, the outlet process-fluid temperature can be 3°C (5°F) below the dry-bulb temperature.
13. *Mean-temperature-difference (MTD) correction factor.* When the outlet temperatures of both fluids are identical, the MTD correction factor for a 1:2 shell-and-tube exchanger (one pass shell side, two or more passes tube side) is approximately 0.8. For a single-pass air-cooled heat exchanger the factor is 0.91. A two-pass exchanger has a factor of 0.96, while a three-pass exchanger has a factor of 0.99 when passes are arranged for counterflow.
14. *Maintenance cost.* Maintenance for air-cooled equipment as compared with shell-and-tube coolers (complete with cooling-tower costs) indicates that air-cooling maintenance costs are approximately 0.3 to 0.5 those for water-cooled equipment.
15. *Operating costs.* Power requirements for air-cooled heat exchangers can be lower than at the summer design condition provided that an adequate means of air-flow control is used. The annual power requirement for an exchanger is a function of the means of air-flow control, the exchanger service, the air-temperature rise, and the approach temperature.

When the mean annual temperature is 16.7°C (30°F) lower than the design dry-bulb temperature and when both fans in a bay have automatically controllable pitch of fan blades, annual power required has been found to be 22, 36, and 54 percent respectively of that needed at the design condition for three process services [Frank L. Rubin, "Power Requirements Are Lower for Air-Cooled Heat Exchangers with AV Fans," *Oil Gas J.*, 165–167 (Oct. 11, 1982)]. Alternatively, when fans have two-speed motors, these deliver one-half of the design flow of air at half speed and use only one-eighth of the power of the full-speed condition.

# COMPACT AND NONTUBULAR HEAT EXCHANGERS

## COMPACT HEAT EXCHANGERS

With equipment costs rising and limited available plot space, compact heat exchangers are gaining a larger portion of the heat exchange market. Numerous types use special enhancement techniques to achieve the required heat transfer in smaller plot areas and, in many cases, less initial investment. As with all items that afford a benefit there is a series of restrictions that limit the effectiveness or application of these special heat exchanger products. In most products discussed some of these considerations are presented, but a thorough review with reputable suppliers of these products is the only positive way to select a compact heat exchanger. The following guidelines will assist in prequalifying one of these.

## PLATE-AND-FRAME EXCHANGERS

There are two major types gasketed and welded-plate heat exchangers. Each shall be discussed individually.

## GASKETED-PLATE EXCHANGERS (G. PHE)

**Description** This type is the fastest growing of the compact exchangers and the most recognized (see Fig. 11-49). A series of corrugated alloy material channel plates, bounded by elastomeric gaskets are hung off and guided by longitudinal carrying bars, then compressed by large-diameter tightening bolts between two pressure retaining frame plates (cover plates). The frame and channel plates have portholes which allow the process fluids to enter alternating flow passages (the space between two adjacent-channel plates). Gaskets around the periphery of the channel plate prevent leakage to the atmosphere and also prevent process fluids from coming in contact with the frame plates. No interfluid leakage is possible in the port area due to a dual-gasket seal.

The frame plates are typically epoxy-painted carbon-steel material and can be designed per most pressure vessel codes. Design limitations are in the Table 11-19. The channel plates are always an alloy material with 304SS as a minimum (see Table 11-19 for other materials).

Channel plates are typically 0.4 to 0.8 mm thick and have corrugation depths of 2 to 10 mm. Special Wide Gap (WG PHE) plates are available, in limited sizes, for slurry applications with depths of approximately 16 mm. The channel plates are compressed to achieve metal-to-metal contact for pressure-retaining integrity. These narrow gaps and high number of contact points which change fluid flow direction, combine to create a very high turbulence between the plates. This means high individual-heat-transfer coefficients (up to 14,200 W/m$^2$ °C), but also very high pressure drops per length as well. To compensate, the channel plate lengths are usually short, most under 2 and few over 3 meters in length. In general, the same pressure drops as conventional exchangers are used without loss of the enhanced heat transfer.

Expansion of the initial unit is easily performed in the field without special considerations. The original frame length typically has an additional capacity of 15–20 percent more channel plates (i.e., surface area). In fact, if a known future capacity is available during fabrication stages, a longer carrying bar could be installed, and later, increasing the surface area would be easily handled. When the expansion is needed, simply untighten the carrying bolts, pull back the frame plate, add the additional channel plates, and tighten the frame plate.

**Applications** Most PHE applications are liquid-liquid services but there are numerous steam heater and evaporator uses from their heritage in the food industry. Industrial users typically have chevron style channel plates while some food applications are washboard style.

Fine particulate slurries in concentrations up to 70 percent by weight are possible with standard channel spacings. Wide-gap units are used with larger particle sizes. Typical particle size should not exceed 75 percent of the *single* plate (not total channel) gap.

Close temperature approaches and tight temperature control possible with PHE's and the ability to sanitize the entire heat transfer surface easily were a major benefit in the food industry.

Multiple services in a single frame are possible.

Gasket selection is one of the most critical and limiting factors in PHE usage. Table 11-20 gives some guidelines for fluid compatibility. Even trace fluid components need to be considered. The higher the operating temperature and pressure, the shorter the anticipated gasket life. *Always* consult the supplier on gasket selection and obtain an estimated or guaranteed lifetime.

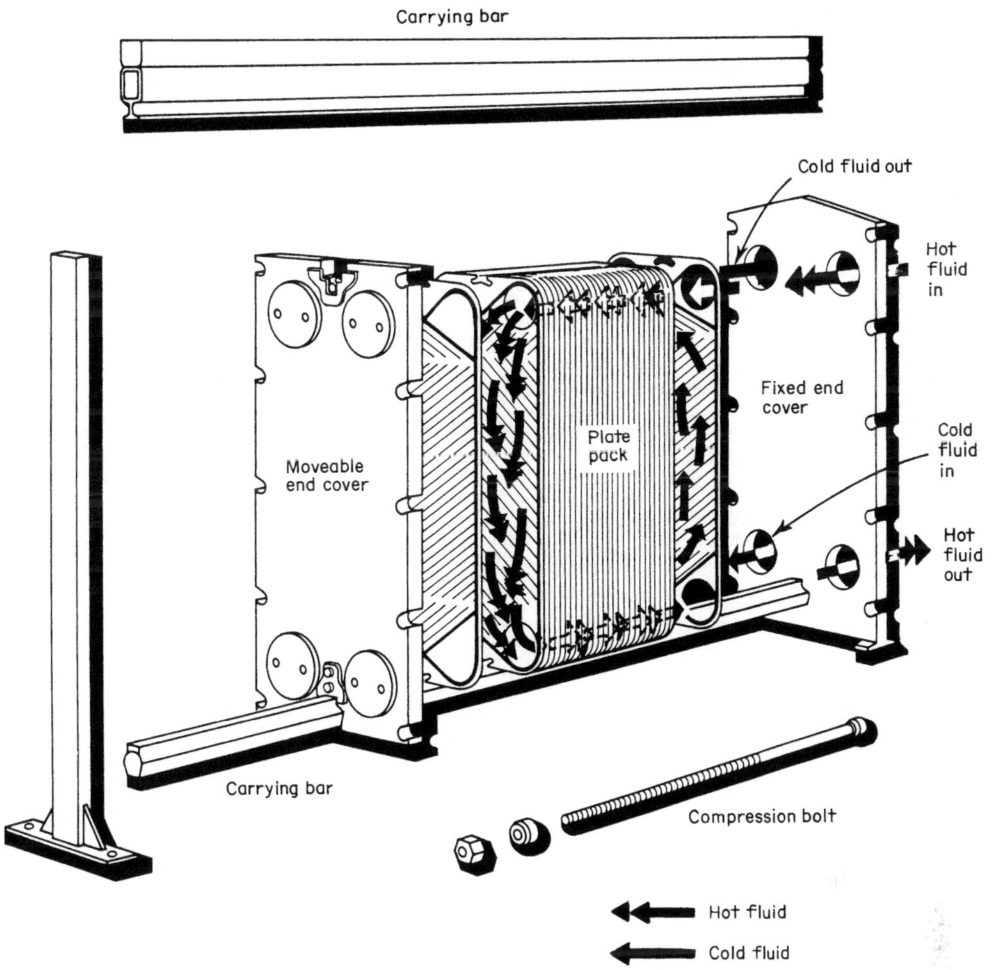


**FIG. 11-49** Plate-and-frame heat exchanger. Hot fluid flows down between alternate plates, and cold fluid flows up between alternate plates. (*Thermal Division, Alfa-Laval, Inc.*)

The major applications are, but not limited to, as follows:

| | |
|---|---|
| Temperature cross applications | (lean/rich solvent) |
| Close approaches | (fresh water/seawater) |
| Viscous fluids | (emulsions) |
| Sterilized surface required | (food, pharmaceutical) |
| Polished surface required | (latex, pharmaceutical) |
| Future expansion required | |
| Space restrictions | |
| Barrier coolant services | (closed-loop coolers) |
| Slurry applications | ($TiO_2$, Kaolin, precipitated calcium carbonate, and beet sugar raw juice) |

**Design** Plate exchangers are becoming so commonplace that there is now an API 662 document available for the specification of these products. In addition, commercial computer programs are available from HTRI among others. Standard channel-plate designs, unique to each manufacturer, are developed with limited modifications of each plates' corrugation depths and included angles. Manufacturers combine their different style plates to custom-fit each service. Due to the possible combinations, it is impossible to present a way to exactly size PHEs. However, it is possible to estimate areas for new units and to predict performance of existing units with different conditions (chevron-type channel plates are presented).

The fixed length and limited corrugation included angles on channel plates makes the NTU method of sizing practical. (Waterlike fluids are assumed for the following examples).

$$\text{NTU} = \frac{\Delta t \text{ of either side}}{\text{LMTD}} \tag{11-76}$$

Most plates have NTU values of 0.5 to 4.0, with 2.0 to 3.0 as the most common, (multipass shell and tube exchangers are typically less than 0.75). The more closely the fluid profile matches that of the channel plate, the smaller the required surface area. Attempting to increase the service NTU beyond the plate's NTU capability causes oversurfacing (inefficiency).

True sizing from scratch is impractical since a pressure balance on a channel-to-channel basis, from channel closest to inlet to furthest, must be achieved and when mixed plate angles are used; this is quite a challenge. Computer sizing is not just a benefit, it is a necessity for supplier's selection. Averaging methods are recommended to perform any sizing calculations.

From the APV heat-transfer handbook—*Design & Application of Paraflow-Plate Heat Exchangers* and J. Marriott's article, "Where and How To Use Plate Heat Exchangers," Chemical Engineering, April 5,

**TABLE 11-19 Compact Exchanger Applications Guide**

| Design conditions | G. PHE | W. PHE | WG. PHE | BHE | DBL | MLT | STE | CP | SHE | THE |
|---|---|---|---|---|---|---|---|---|---|---|
| Design temperature °C | 165 | 150 | 150 | 185 | +500 | +500 | +500 | 450 | +400 | +500 |
| Minimum metal temp °C | −30 | −30 | −30 | −160 | −160 | −160 | −160 | −160 | −160 | −160 |
| Design pressure MPa | 2.5 | 2.5 | 0.7 | 3.1 | +20 | +20 | +20 | 3.1 | 2.0 | +20 |
| Inspect for leakage | Yes | Partial | Yes | No | Yes | Yes | Yes | Partial | Yes | Yes |
| Mechanical cleaning | Yes | Yes/no | Yes | No | Yes | Yes | Yes/no | Yes | Yes | Yes |
| Chemical cleaning | Yes | Yes | Yes | Yes | Yes | Yes | Yes | Yes | Yes | Yes |
| Expansion capability | Yes | Yes | Yes | No | No | No | No | No | No | No |
| Repair | Yes | Yes/no | Yes | No | Yes | Yes | Partial | Partial | Partial | Yes |
| Temperature cross | Yes | Yes | Yes | Yes | Yes | Yes | Yes | Yes | Yes | No* |
| Surface area/unit $m^2$ | 1850 | 900 | 250 | 50 | 10 | 150 | 60 | 275 | 450 | High |
| Holdup volume | Low | Low | Low | Low | Med | Med | Low | Low | Med | High |
| **Materials** | **G. PHE** | **W. PHE** | **WG. PHE** | **BHE** | **DBL** | **MLT** | **STE** | **CP** | **SHE** | **THE** |
| Mild steel | No | No | No | No | Yes | Yes | Yes | Yes | Yes | Yes |
| Stainless | Yes | Yes | Yes | Yes | Yes | Yes | Yes | Yes | Yes | Yes |
| Titanium | Yes | Yes | Yes | No | Yes | Yes | Yes | Yes | Yes | Yes |
| Hastalloy | Yes | Yes | No | No | Yes | Yes | Yes | Yes | Yes | Yes |
| Nickel | Yes | Yes | No | No | Yes | Yes | Yes | Yes | Yes | Yes |
| Alloy 20 | Yes | Yes | No | No | Yes | Yes | Yes | Yes | Yes | Yes |
| Incoloy 825 | Yes | Yes | No | No | Yes | Yes | Yes | Yes | Yes | Yes |
| Monel | Yes | Yes | No | No | Yes | Yes | Yes | Yes | Yes | Yes |
| Impervious graphite | Yes | No | No | No | No | No | No | No | No | Yes |
| **Service** | **G. PHE** | **W. PHE** | **WG. PHE** | **BHE** | **DBL** | **MLT** | **STE** | **CP** | **SHE** | **THE** |
| Clean fluids | A | A | A | A | A | A | A | A | A | A |
| Gasket incompatibility | D | A/D | D | A | A | A | A | A | A | A |
| Medium viscosity | A/B | A/B | A/B | B | A | A | A/B | A/B | A | A |
| High viscosity | A/B | A/B | A/B | D | A | A | A/B | A/B | A | A |
| Slurries & pulp (fine) | B/D | D | A/B | C | A | A/B | C | B | A | A/D |
| Slurries & pulp (coarse) | D | D | B | D | A | B/C | D | B | A | A/D |
| Refrigerants | D | A | D | A | A | A | B/C | A | A | A |
| Thermal fluids | D | A/B | D | A/B | A | A | C | A | A | A |
| Vent condensers | D | D | D | D | A/D | A | A | B/C | A | A |
| Process condenser | D | C | D | D | A/D | A | A | B/C | B | A |
| Vacuum reboil/cond | D | D | B | D | A/D | B | A | B/C | B | A/C |
| Evaporator | D | C | C | A | B | B | A | B/C | C | A |
| Tight temp control | A | A | A | A | A | A | A | B | A | C |
| High scaling | B | B | A | D | A | A/B | B/C | B | B | A/D |

*Multipass
Adapted from Alfa-Laval and Vicarb literature
A—Very good C—Fair
B—Good D—Poor

**TABLE 11-20 Elastomer Selection Guide**

| | Uses | Avoid |
|---|---|---|
| Nitrile (NBR) | Oil resistant<br>Fat resistant<br>Food stuffs<br>Mineral oil<br>Water | Oxidants<br>Acids<br>Aromatics<br>Alkalies<br>Alcohols |
| Resin cured butyl (IIR) | Acids<br>Lyes<br>Strong alkalies<br>Strong phosphoric acid<br>Dilute mineral acids<br>Ketones<br>Amines<br>Water | Fats and fatty acids<br>Petroleum oils<br>Chlorinated hydrocarbons<br>Liquids with dissolved chlorine<br>Mineral oil<br>Oxygen rich demin. water<br>Strong oxidants |
| Ethylene-propylene (EPDM) | Oxidizing agents<br>Dilute acids<br>Amines<br>Water<br>(Mostly any IIR fluid) | Oils<br>Hot & conc. acids<br>Very strong oxidants<br>Fats & fatty acids<br>Chlorinated hydrocarbons |
| Viton (FKM, FPM) | Water<br>Petroleum oils<br>Many inorganic acids<br>(Most all NBR fluids) | Amines<br>Ketones<br>Esters<br>Organic acids<br>Liquid ammonia |

1971, there are the following equations for plate heat transfer.

$$\mathrm{Nu} = \frac{h\mathrm{De}}{k} = 0.28 * (\mathrm{Re})^{0.65} * (\mathrm{Pr})^{0.4} \qquad (11\text{-}77)$$

where De = 2 × depth of single-plate corrugation

$$G = \frac{W}{\mathrm{Np} * w * \mathrm{De}} \qquad (11\text{-}78)$$

Width of the plate ($w$) is measured from inside to inside of the channel gasket. If not available, use the tear-sheet drawing width and subtract two times the bolt diameter and subtract another 50 mm. For depth of corrugation ask supplier, or take the compressed plate pack dimension, divide by the number of plates and subtract the plate thickness from the result. The number of passages (Np) is the number of plates minus 1 then divided by 2.

Typical overall coefficients to start a rough sizing are as below. Use these in conjunction with the NTU calculated for the process. The closer the NTU matches the plate (say between 2.0 and 3.0), the higher the range of listed coefficients can be used. The narrower (smaller) the depth of corrugation, the higher the coefficient (and pressure drop), but also the lower the ability to carry through any particulate.

| | |
|---|---|
| Water-water | 5700–7400 W/($m^2$ °C) |
| Steam-water | 5700–7400 W/($m^2$ °C) |
| Glycol/Glycol | 2300–4000 W/($m^2$ °C) |
| Amine/Amine | 3400–5000 W/($m^2$ °C) |
| Crude/Emulsion | 400–1700 W/($m^2$ °C) |

Pressure drops typically can match conventional tubular exchangers. Again from the APV handbook an average correlation is as follows:

$$\Delta P = \frac{2fG^2L}{g\rho De} \tag{11-79}$$

where $f = 2.5\ (GDe/\mu)^{-0.3}$
$g$ = gravitational constant

Fouling factors are typically 1/10 of TEMA values or a percent oversurfacing of 10–20 percent is used (J. Kerner, "Sizing Plate Exchangers," *Chemical Engineering,* November 1993).

LMTD is calculated like a 1 pass-1 pass shell and tube with no $F$ correction factor required in most cases.

Overall coefficients are determined like shell and tube exchangers; that is, sum all the resistances, then invert. The resistances include the hot-side coefficient, the cold-side coefficient, the fouling factor (usually only a total value not individual values per fluid side) and the wall resistance.

## WELDED- AND BRAZED-PLATE EXCHANGERS (W. PHE & BHE)

The title of this group of plate exchangers has been used for a great variety of designs for various applications from normal gasketed-plate exchanger services to air-preheater services on fired heaters or boilers. The intent here is to discuss more traditional heat-exchanger designs, not the heat-recovery designs on fired equipment flue-gas streams. Many similarities exist between these products but the manufacturing techniques are quite different due to the normal operating conditions these units experience.

To overcome the gasket limitations, PHE manufacturers have developed welded-plate exchangers. There are numerous approaches to this solution: weld plate pairs together with the other fluid-side conventionally gasketed, weld up both sides but use a horizonal stacking of plates method of assembly, entirely braze the plates together with copper or nickel brazing, diffusion bond then pressure form plates and bond etched, passage plates.

Most methods of welded-plate manufacturing do not allow for inspection of the heat-transfer surface, mechanical cleaning of that surface, and have limited ability to repair or plug off damage channels. Consider these limitations when the fluid is heavily fouling, has solids, or in general the repair or plugging ability for severe services.

One of the previous types has an additional consideration of the brazing material to consider for fluid compatibility. The brazing compound entirely coats both fluid's heat-transfer surfaces.

The second type, a Compabloc (CP) from Alpha-Laval Thermal, has the advantage of removable cover plates, similar to air-cooled exchanger headers, to observe both fluids surface area. The fluids flow at 90° angles to each other on a horizonal plane. LMTD correction factors approach 1.0 for Compabloc just like the other welded and gasketed PHEs. Hydroblasting of Compabloc surfaces is also possible. The Compabloc has higher operating conditions than PHE's or W-PHE.

The performances and estimating methods of welded PHEs match those of gasketed PHEs in most cases, but normally the Compabloc, with larger depth of corrugations, can be lower in overall coefficient. Some extensions of the design operating conditions are possible with welded PHEs, most notably is that cryogenic applications are possible. Pressure vessel code acceptance is available on most units.

## COMBINATION WELDED-PLATE EXCHANGERS

Plate exchangers are well known for their high efficiency but suffer from limitations on operating pressure. Several companies have rectified this limitation by placing the welded plate exchanger inside a pressure vessel to withstand the pressure. One popular application is the feed effluent exchange in a catalytic reforming plant for oil refineries. Large volumes of gases with some liquids require cross-exchange to feed a reactor system. Close temperature approaches and lower pressure drops are required. These combined units provide an economic alternative to shell-and-tube exchangers.

## SPIRAL-PLATE EXCHANGERS (SHE)

**Description** The *spiral-plate heat exchanger* (SHE) may be one exchanger selected primarily on its virtues and not on its initial cost. SHEs offer high reliability and on-line performance in many severely fouling services such as slurries.

The SHE is formed by rolling two strips of plate, with welded-on spacer studs, upon each other into clock-spring shape. This forms two passages. Passages are sealed off on one end of the SHE by welding a bar to the plates; hot and cold fluid passages are sealed off on opposite ends of the SHE. A single rectangular flow passage is now formed for each fluid, producing very high shear rates compared to tubular designs. Removable covers are provided on each end to access and clean the entire heat transfer surface. Pure countercurrent flow is achieved and LMTD correction factor is essentially = 1.0.

Since there are no dead spaces in a SHE, the helical flow pattern combines to entrain any solids and create high turbulence creating a self-cleaning flow passage.

There are no thermal-expansion problems in spirals. Since the center of the unit is not fixed, it can torque to relieve stress.

The SHE can be expensive when only one fluid requires a high-alloy material. Since the heat-transfer plate contacts both fluids, it is required to be fabricated out of the higher alloy. SHEs can be fabricated out of any material that can be cold-worked and welded.

The channel spacings can be different on each side to match the flow rates and pressure drops of the process design. The spacer studs are also adjusted in their pitch to match the fluid characteristics.

As the coiled plate spirals outward, the plate thickness increases from a minimum of 2 mm to a maximum (as required by pressure) up to 10 mm. This means relatively thick material separates the two fluids compared to tubing of conventional exchangers. Pressure vessel code conformance is a common request.

**Applications** The most common applications that fit SHE are slurries. The rectangular channel provides high shear and turbulence to sweep the surface clear of blockage and causes no distribution problems associated with other exchanger types. A localized restriction causes an increase in local velocity which aids in keeping the unit free flowing. Only fibers that are long and stringy cause SHE to have a blockage it cannot clear itself.

As an additional antifoulant measure, SHEs have been coated with a phenolic lining. This provides some degree of corrosion protection as well, but this is not guaranteed due to pinholes in the lining process.

There are three types of SHE to fit different applications:

*Type I* is the spiral-spiral flow pattern. It is used for all heating and cooling services and can accommodate temperature crosses such as lean/rich services in one unit. The removable covers on each end allow access to one side at a time to perform maintenance on that fluid side. Never remove a cover with one side under pressure as the unit will telescope out like a collapsible cup.

*Type II* units are the condenser and reboiler designs. One side is spiral flow and the other side is in cross flow. These SHEs provide very stable designs for vacuum condensing and reboiling services. A SHE can be fitted with special mounting connections for reflux-type vent-condenser applications. The vertically mounted SHE directly attaches on the column or tank.

*Type III* units are a combination of the Type I and Type II where part is in spiral flow and part is in cross flow. This SHE can condense and subcool in a single unit.

The unique channel arrangement has been used to provide on-line cleaning, by switching fluid sides to clean the fouling (caused by the fluid that previously flowed there) off the surface. Phosphoric acid coolers use pond water for cooling and both sides foul; water, as you expect, and phosphoric acid deposit crystals. By reversing the flow sides, the water dissolves the acid crystals and the acid clears up the organic fouling. SHEs are also used as oleum coolers, sludge coolers/heaters, slop oil heaters, and in other services where multiple-flow-passage designs have not performed well.

**Design** A thorough article by P. E. Minton of Union Carbide called "Designing Spiral-Plate Heat Exchangers," appeared in *Chemical Engineering,* May 4, 1970. It covers the design in detail. Also an article in *Chemical Engineering Progress* titled "Applications of Spiral Plate Heat Exchangers" by A. Hargis, A. Beckman, and J. Loicano appeared in July 1967, provides formulae for heat-transfer and pressure-drop calculations.

Spacings are from 6.35 to 31.75 mm (in 6.35 mm increments) with 9.5 mm the most common. Stud densities are 60 × 60 to 110 × 110 mm, the former the most common. The width (measured to the spiral flow passage), is from 150 to 2500 mm (in 150 mm increments). By varying the spacing and the width, separately for each fluid, velocities can be maintained at optimum rates to reduce fouling tendencies or utilize the allowable pressure drop most effectively. Diameters can reach 1500 mm. The total surface areas exceed 465 sqm. Materials that work harder are not suitable for spirals since hot-forming is not possible and heat treatment after forming is impractical.

$$Nu = \frac{HDe}{k} = 0.0315\,(Re)^{0.8}\,(Pr)^{0.25}\,(\mu/\mu_w)^{0.17} \qquad (11\text{-}80)$$

where De = 2 × spacing
Flow area = width × spacing

$$\Delta P = \frac{LV^2\rho}{1.705E03} * 1.45 \quad (1.45 \text{ for } 60 \times 60 \text{ mm studs}) \qquad (11\text{-}81)$$

LMTD and overall coefficient are calculated like in PHE section above.

## BRAZED-PLATE-FIN HEAT EXCHANGER

Brazed-aluminum-plate-fin heat exchangers (or core exchangers or cold boxes) as they are sometimes called, were first manufactured for the aircraft industry during World War II. In 1950, the first tonnage air-separation plant with these compact, lightweight, reversing heat exchangers began producing oxygen for a steel mill. Aluminum-plate-fin exchangers are used in the process and gas-separation industries, particularly for services below −45°C.

Core exchangers are made up of a stack of rectangular sheets of aluminum separated by a wavy, usually perforated, aluminum fin. Two ends are sealed off to form a passage (see Fig. 11-50). The layers have the wavy fins and sealed ends alternating at 90° to each. Aluminum half-pipe-type headers are attached to the open ends to route the fluids into the alternating passages. Fluids usually flow at this same 90° angle to each other. Variations in the fin height, number of passages, and the length and width of the prime sheet allow for the core exchanger to match the needs of the intended service.

Design conditions range in pressures from full vacuum to 96.5 bar g and in temperatures from −269°C to 200°C. This is accomplished meeting the quality standards of most pressure vessel codes.

**Design and Application** Brazed plate heat exchangers have two design standards that are available. One is ALPEMA, the Brazed Aluminum Plate-Fin Heat Exchanger Manufacturers' Association, and the other is the API 662 document for plate heat exchangers.

Applications are varied for this highly efficient, compact exchanger. Mainly it is seen in the cryogenic fluid services of air-separation plants, refrigeration trains like in ethylene plants, and in natural-gas processing plants. Fluids can be all vapor, liquid, condensing, or vaporizing. Multifluid exchangers and multiservice cores, that is one exchanger with up to 10 different fluids, are common for this type of product.

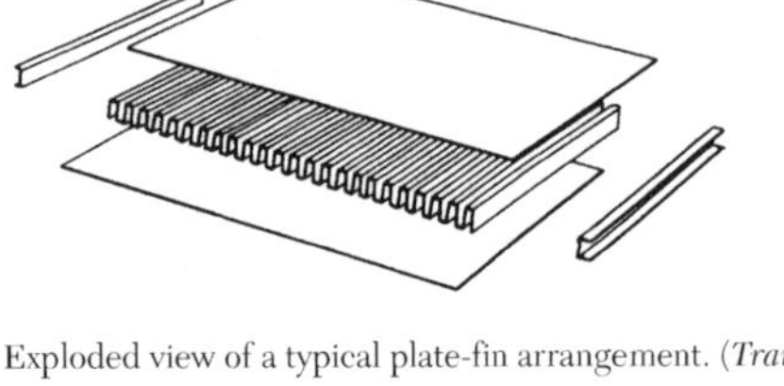

**FIG. 11-50** Exploded view of a typical plate-fin arrangement. (*Trane Co.*)

Cold boxes are a group of cores assembled into a single structure or module, prepiped for minimum field connections. (Data obtained from ALTEC INTERNATIONAL, now Chart Industries. For detailed information refer to *GPSA Engineering Handbook* Section 9.)

## PLATE-FIN TUBULAR EXCHANGERS (PFE)

**Description** These shell and tube exchangers are designed to use a group of tightly spaced plate fins to increase the shellside heat transfer performance as fins do on double-pipe exchangers. In this design, a series of very thin plates (fins), usually of copper or aluminum material, are punched to the same pattern as the tube layout, spaced very close together, and mechanically bonded to the tube. Fin spacing is 315–785 FPM (Fins Per Meter) with 550 FPM most common. The fin thicknesses are 0.24 mm for aluminum and 0.19 mm for copper. Surface-area ratios over bare prime-tube units can be 20:1 to 30:1. The cost of the additional plate-fin material, without a reduction in shell diameter in many cases, and increased fabrication has to be offset by the total reduction of plot space and prime tube-surface area. The more costly the prime tube or plot space cost, the better the payout for this design. A rectangular tube layout is normally used, *no tubes in the window* (NTIW). The window area (where no tubes are) of the plate-fins are cut out. This causes a larger shell diameter for a given tube count compared to conventional tubular units. A dome area on top and bottom of the inside of the shell has been created for the fluid to flow along the tube length. In order to exit the unit the fluid must flow across the plate-finned tube bundle with extremely low pressure loss. The units from the outside and from the tubeside appear like any conventional shell and tube exchanger.

**Applications** Two principal applications are rotating equipment oil coolers and compressor inter- and after-coolers. Although seemingly different applications, both rely on the shellside finning to enhance the heat transfer of low heat-transfer characteristic fluids, viscous oils, and gases. By nature of the fluids and their applications, both are clean servicing. The tightly spaced fins would be a maintenance problem otherwise.

**Design** The economics usually work out in the favor of gas coolers when the centrifugal machine's flow rate reaches about 5000 scfm. The pressure loss can be kept to 7.0 kPa in most cases. When the ratio of $A_t h_t$ to $A_s h_s$ is 20:1, is another point to consider these plate-fin designs. Vibration is practically impossible with this design, and uses in reciprocating compressors are possible due to this.

Marine and hydraulic-oil coolers use these characteristics to enhance the coefficient of otherwise poorly performing fluids. The higher metallurgies in marine applications like 90/10 Cu-Ni afford the higher cost of plate-fin design to be offset by the less amount of alloy material being used. On small hydraulic coolers, these fins usually allow one to two size smaller coolers for the package and save skid space and initial cost.

Always check on metallurgy compatibility and cleanliness of the shellside fluid! (Data provided by Bos-Hatten and ITT-Standard.)

## PRINTED-CIRCUIT HEAT EXCHANGERS

These are a variation of the welded or brazed plate heat exchangers but using a chemical etching process to form the flow channels and diffusion bonding technique to secure the plates together. These units have the high heat-transfer characteristics and extended operating conditions that welded or brazed units have, but the diffusion process makes the bond the same strength as that of the prime plate material. The chemical etching, similar to that used in printed circuitry, allows greater flexibility in flow channel patterns than any other heat exchanger. This type of heat exchanger is perhaps the most compact design of all due to the infinite variations in passage size, layout, and direction.

Headers are welded on the core block to direct the fluids into the appropriate passages. The all-metal design allows very high operating conditions for both temperature and pressure. The diffusion bonding provides a near-homogeneous material for fluids that are corrosive or require high purity.

These exchangers can handle gases, liquids, and two-phase applications. It has the greatest potential in cryogenic, refrigeration, gas processing, and corrosive chemical applications. Other applications are possible with the exception of fluids containing solids: the narrow passages, like most plate exchangers, are conducive to plugging.

## SPIRAL-TUBE EXCHANGERS (STE)

**Description** These exchangers are typically a series of stacked helical-coiled tubes connected to manifolds, then inserted into a casing or shell. They have many advantages like spiral-plate designs, such as avoiding differential expansion problems, acceleration effects of the helical flow increasing the heat transfer coefficient, and compactness of plot area. They are typically selected because of their economical design.

The most common form has both sides in helical flow patterns, pure countercurrent flow is followed and the LMTD correction factor approaches 1.0. Temperature crosses are possible in single units. Like the spiral-plate unit, different configurations are possible for special applications.

Tube material includes any that can be formed into a coil, but usually copper, copper alloys, and stainless steel are most common. The casing or shell material can be cast iron, cast steel, cast bronze, fabricated steel, stainless, and other high-alloy materials. Units are available with pressure vessel code conformance.

The data provided herein has been supplied by Graham Mfg. for their units called Heliflow.

**Applications** The common Heliflow applications are tank-vent condensers, sample coolers, pump-seal coolers, and steam-jet vacuum condensers. Instant water heaters, glycol/water services, and cryogenic vaporizers use the spiral tube's ability to reduce thermally induced stresses caused in these applications.

Many other applications are well suited for spiral tube units but many believe only small surface areas are possible with these units. Graham Mfg. states units are available to 60 m$^2$. Their ability to polish the surfaces, double-wall the coil, use finned coil, and insert static mixers, among others configurations in design, make them quite flexible. Tubeside design pressures can be up to 69000 kPa. A cross-flow design on the external surface of the coil is particularly useful in steam-jet ejector condensing service. These Heliflow units, can be made very cost-effective, especially in small units. The main differences, compared to spiral plate, is that the tubeside cannot be cleaned except chemically and that multiple flow passages make tubeside slurry applications (or fouling) impractical.

**Design** The fluid flow is similar to the spiral-plate exchangers, but through parallel tube passages. Graham Mfg. has a liquid-liquid sizing pamphlet available from their local distributor. An article by M.A. Noble, J.S. Kamlani, and J.J. McKetta "Heat Transfer in Spiral Coils", was published in *Petroleum Engineer,* April 1952 p. 723, discussing sizing techniques.

The tubeside fluid must be clean or at least chemically cleanable. With a large number of tubes in the coil, cleaning of inside surfaces is not totally reliable. Fluids that attack stressed materials such as chlorides should be reviewed as to proper coil-material selection. Fluids that contain solids can be a problem due to erosion of relatively thin coil materials unlike the thick plates in spiral-plate units and multiple, parallel, fluid passages compared to a single passage in spiral-plate units.

## GRAPHITE HEAT EXCHANGERS

Impervious graphite exchangers now come in a variety of geometries to suit the particular requirements of the service. They include cubic block form, drilled cylinder block, shell and tube, and plate and frame.

**Description** Graphite is one of three crystalline forms of carbon. The other two are diamond and charcoal. Graphite has a hexagonal crystal structure, diamond is cubic, and charcoal is amorphous. Graphite is inert to most chemicals and resists corrosion attack. It is however porous and to be used, it must be impregnated with a resin sealer. Two main resins used are phenolic and PTFE with furan (one currently being phased out of production). Selection of resins include chemical compatibility, operating temperatures, and type of unit to be used. For proper selection, consult with a graphite supplier.

Shell-and-tube units in graphite were started by Karbate in 1939. The European market started using block design in the 1940s. Both technologies utilize the high thermal conductivity of the graphite material to compensate for the poor mechanical strength. The thicker materials needed to sustain pressure do not adversely impede the heat transfer. Maximum design pressures range from 0.35 to 1.0 kPa depending on type and size of exchanger. Design temperature is dependent on the fluids and resin selection, the maximum is 230 °C.

In all situations, the graphite heat transfer surface is contained within a metal structure or a shell (graphite lined on process side) to maintain the design pressure. For shell and tube units, the design is a packed floating tubesheet at both ends within a shell and channel. For stacked block design, the standardize blocks are glued together with special adhesives and compressed within a framework that includes manifold connections for each fluid. The cylindrical block unit is a combination of the above two with blocks glued together and surrounded by a pressure retaining shell. Pressure vessel code conformance of the units is possible due to the metallic components of these designs. Since welding of graphite is not possible, the selection and application of the adhesives used are critical to the proper operating of these units. Tube to tubesheet joints are glued since rolling of tubes into tubesheet is not possible. The packed channels and gasketed manifold connections are two areas of additional concern when selecting sealants for these units.

**Applications and Design** The major applications for these units are in the acid-related industries. Sulfuric, phosphoric, and hydrochloric acids require either very costly metals or impervious graphite. Usually graphite is the more cost-effective material to be used. Applications are increasing in the herbicide and pharmaceutical industries as new products with chlorine and fluorine compounds expand. Services are coolers, condensers, and evaporators, basically all services requiring this material. Types of units are shell-and-tube, block-type (circular and rectangular), and plate-and-frame-type exchangers. The design of the shell-and-tube units are the same as any but the design characteristics of tubes, spacing, and thickness are unique to the graphite design. The block and plate and frame also can be evaluated using techniques previously addressed but again, the unique characteristics of the graphite materials require input from a reputable supplier. Most designs will need the supplier to provide the most cost-effective design for the immediate and future operation of the exchangers. Also, consider the entire system design as some condensers and/or evaporators can be integral with their associated column.

## CASCADE COOLERS

Cascade coolers are a series of standard pipes, usually manifolded in parallel, and connected in series by vertically or horizontally oriented U-bends. Process fluid flows inside the pipe entering at the bottom and water trickles from the top downward over the external pipe surface. The water is collected from a trough under the pipe sections, cooled, and recirculated over the pipe sections. The pipe material can be any of the metallic and also glass, impervious graphite, and ceramics. The tubeside coefficient and pressure drop is as in any circular duct. The water coefficient (with Re number less than 2100) is calculated from the following equation by W.H. McAdams, T.B. Drew, and G.S. Bays Jr., from the ASME trans. **62,** 627–631 (1940).

$$h = 218 * (G'/D_o)^{1/3} \text{ (W/m}^2\ ^\circ\text{C)} \quad (11\text{-}82)$$

$G' = m/(2L)$

$m$ = water rate (kg/hr)

$L$ = length of each pipe section (meter)

$D_o$ = outside diameter of pipe (meter)

LMTD corrections are per Fig. 11-4 *i* or *j* depending on U-bend orientation.

## BAYONET-TUBE EXCHANGERS

This type of exchanger gets its name from its design which is similar to a bayonet sword and its associated scabbard or sheath. The bayonet tube is a smaller-diameter tube inserted into a larger-diameter tube that has been capped at one end. The fluid flow is typically entering the inner tube, exiting, hitting the cap of the larger tube, and returning the opposite direction in the annular area. The design eliminates any thermal expansion problems. It also creates a unique nonfreeze-type tubeside for steam heating of cryogenic fluids, the inner tube steam keeps the annulus condensate from freezing against the cold shellside fluid. This design can be expensive on a surface-area basis due to the need of a double channel design and only the outer tube surface is used to transfer heat. LMTD calculations for nonisothermal fluid are quite extensive and those applications are far too few to attempt to define it. The heat transfer is like the annular calculation of a double-pipe unit. The shellside is a conventional-baffled shell-and-tube design.

## ATMOSPHERIC SECTIONS

These consist of a rectangular bundle of tubes in similar fashion to air cooler bundles, placed just under the cooled water distribution section of a cooling tower. It, in essence, combines the exchanger and cooling tower into a single piece of equipment. This design is only practical for single-service cooler/condenser applications, and expansion capabilities are not provided. The process fluid flows inside the tubes and the cooling tower provides cool water that flows over the outside of the tube bundle. Water quality is critical for these applications to prevent fouling or corrosive attack on the outside of the tube surfaces and to prevent blockage of the spray nozzles. The initial and operating costs are lower than separate cooling tower and exchanger. Principal applications now are in the HVAC, Refrigeration and Industrial systems. Sometimes these are called "Wet Surface Air Coolers."

$$h = 1729\ [(\text{m}^3/\text{hr})/\text{face area m}^2]^{1/3} \qquad (11\text{-}83)$$

## NONMETALLIC HEAT EXCHANGERS

Another growing field is that of nonmetallic heat exchanger designs which typically are of the shell and tube or coiled-tubing type. The graphite units were previously discussed but numerous other materials are available. The materials include Teflon, PVDF, glass, ceramic, and others as the need arises.

When using these types of products, consider the following topics and discuss the application openly with experienced suppliers.

1. The tube-to-tubesheet joint, how is it made? Many use "O" rings to add another material to the selection process. Preference should be given to a fusing technique of similar material.
2. What size tube or flow passage is available? Small tubes plug unless filtration is installed. Size of filtering is needed from the supplier.
3. These materials are very sensitive to temperature and pressure. Thermal or pressure shocks must be avoided.
4. Thermal conductivity of these materials is very low and affects the overall coefficient. When several materials are compatible, explore all of them, as final cost is not always the same as raw material costs.

## PVDF HEAT EXCHANGERS

These shell-and-tube-type exchangers are similar to the Teflon designs but have some mechanical advantages over Teflon units. First the tubes are available in 9.5 mm sizes which reduces the chances of plugging that are found in Teflon units with unfiltered fluids. Second, the material has higher strength even at lower temperatures almost double. Larger units are possible with PVDF materials.

Tube to tubesheet joints, a weakness of most nonmetallic units, are fused by special techniques that do not severely affect the chemical suitability of the unit. Some nonmetallics use Teflon or "O" rings that add an extra consideration to material selection.

The shell is usually a steel design and, like the graphite units before, can obtain pressure-vessel certification.

## CERAMIC HEAT EXCHANGERS

These include glass, silicon carbide, and similar variations. Even larger tubes are available in these materials, up to 19-mm diameter. They have high thermal conductivities and are usually very smooth surfaces to resist fouling. Very high material/fluid compatibility is seen for these products, not many fluids are excluded. Brittleness is a consideration of these materials and a complete discussion of the service with an experienced supplier is warranted. The major selection criteria to explore is the use of "O" rings and other associated joints at tubesheet. The shell is steel in most cases.

## TEFLON HEAT EXCHANGERS

Teflon tube shell-and-tube heat exchangers (Ametek) made with tubes of chemically inert Teflon fluorocarbon resin are available. The tubes are 0.25-in OD by 0.20-in ID, 0.175-in OD by 0.160-in ID, or 0.125-in OD by 0.100-in ID (in × 25.4 equal mm). The larger tubes are primarily used when pressure-drop limitations or particles reduce the effectiveness of smaller tubes. These heat exchangers generally operate at higher pressure drops than conventional units and are best suited for relatively clean fluids. Being chemically inert, the tubing has many applications in which other materials corrode. Fouling is negligible because of the antistick properties of Teflon.

The heat exchangers are of single-pass, countercurrent-flow design with removable tube bundles. Tube bundles are made of straight flexible tubes of Teflon joined together in integral honeycomb tube sheets. Baffles and O-ring gaskets are made of Teflon. Standard shell diameters are 102, 204, and 254 mm (4, 8, and 10 in). Tube counts range from 105 to 2000. Surface varies from 1.9 to 87 $\text{m}^2$ (20 to 940 $\text{ft}^2$). Tube lengths vary from 0.9 to 4.9 m (3 to 16 ft). At 37.8°C (100°F) maximum operating gauge pressures are 690 kPa (100 $\text{lbf/in}^2$) internal and 379 kPa (55 $\text{lbf/in}^2$) external. At 149°C (300°F) the maximum pressures are 207 kPa (30 $\text{lbf/in}^2$) internal and 124 kPa (18 $\text{lbf/in}^2$) external.

# HEAT EXCHANGERS FOR SOLIDS

This section describes equipment for heat transfer to or from solids by the indirect mode. Such equipment is so constructed that the solids load (burden) is separated from the heat-carrier medium by a wall; the two phases are never in direct contact. Heat transfer is by conduction based on diffusion laws. Equipment in which the phases are in direct contact is covered in other sections of this *Handbook,* principally in Sec. 20.

Some of the devices covered here handle the solids burden in a static or laminar-flowing bed. Other devices can be considered as continuously agitated kettles in their heat-transfer aspect. For the latter, unit-area performance rates are higher.

Computational and graphical methods for predicting performance are given for both major heat-transfer aspects in Sec. 10. In solids heat processing with indirect equipment, the engineer should remember that the heat-transfer capability of the wall is many times that of the solids burden. Hence the solids properties and bed geometry govern the rate of heat transfer. This is more fully explained earlier in this section. Only limited resultant (not predictive) and "experience" data are given here.

## EQUIPMENT FOR SOLIDIFICATION

A frequent operation in the chemical field is the removal of heat from a material in a molten state to effect its conversion to the solid state. When the operation is carried on batchwise, it is termed casting, but when done continuously, it is termed flaking. Because of rapid heat transfer and temperature variations, jacketed types are limited to an initial melt temperature of 232°C (450°F). Higher temperatures [to 316°C (600°F)] require extreme care in jacket design and cooling-liquid flow pattern. Best performance and greatest capacity are

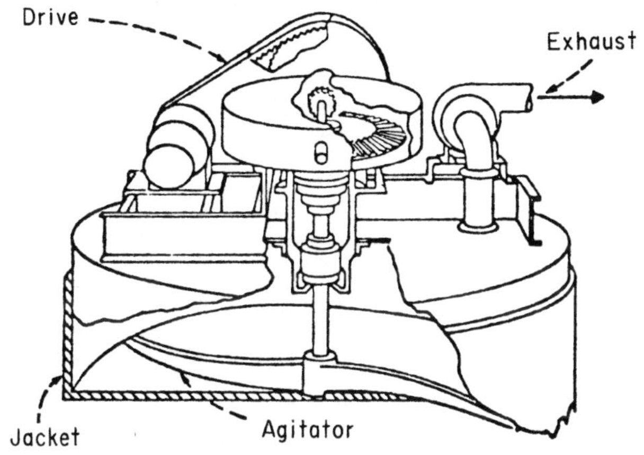


**FIG. 11-51** Heat-transfer equipment for solidification (with agitation); agitated-pan type.

obtained by (1) holding precooling to the minimum and (2) optimizing the cake thickness. The latter cannot always be done from the heat-transfer standpoint, as size specifications for the end product may dictate thickness.

**Table Type** This is a simple flat metal sheet with slightly upturned edges and jacketed on the underside for coolant flow. For many years this was the mainstay of food processors. Table types are still widely used when production is in small batches, when considerable batch-to-batch variation occurs, for pilot investigation, and when the cost of continuous devices is unjustifiable. Slab thicknesses are usually in the range of 13 to 25 mm (½ to 1 in). These units are homemade, with no standards available. Initial cost is low, but operating labor is high.

**Agitated-Pan Type** A natural evolution from the table type is a circular flat surface with jacketing on the underside for coolant flow and the added feature of a stirring means to sweep over the heat-transfer surface. This device is the agitated-pan type (Fig. 11-51). It is a batch-operation device. Because of its age and versatility it still serves a variety of heat-transfer operations for the chemical-process industries. While the most prevalent designation is agitated-pan dryer (in this mode, the burden is heated rather than cooled), considerable use is made of it for solidification applications. In this field, it is particularly suitable for processing burdens that change phase (1) slowly, by "thickening," (2) over a wide temperature range, (3) to an amorphous solid form, or (4) to a soft semigummy form (versus the usual hard crystalline structure).

The stirring produces the end product in the desired divided-solids form. Hence, it is frequently termed a "granulator" or a "crystallizer." A variety of factory-made sizes in various materials of construction are available. Initial cost is modest, while operating cost is rather high (as is true of all batch devices), but the ability to process "gummy" burdens and/or simultaneously effect two unit operations often yields an economical application.

**Vibratory Type** This construction (Fig. 11-52) takes advantage of the burden's special needs and the characteristic of vibratory actuation. A flammable burden requires the use of an inert atmosphere over it and a suitable nonhazardous fluid in the jacket. The vibratory action permits construction of rigid self-cleaning chambers with simple flexible connections. When solidification has been completed and vibrators started, the intense vibratory motion of the whole deck structure (as a rigid unit) breaks free the friable cake [up to 76 mm (3 in) thick], shatters it into lumps, and conveys it up over the dam to discharge. Heat-transfer performance is good, with overall coefficient $U$ of about 68 W/(m$^2$·°C) [12 Btu/(h·ft$^2$·°F)] and values of heat flux $q$ in the order of 11,670 W/m$^2$ [3700 Btu/(h·ft$^2$)]. Application of timing-cycle controls and a surge hopper for the discharge solids facilitates automatic operation of the caster and continuous operation of subsequent equipment.

**Belt Types** The patented metal-belt type (Fig. 11-53a), termed the "water-bed" conveyor, features a thin wall, a well-agitated fluid side for a thin water film (there are no rigid welded jackets to fail), a stainless-steel or Swedish-iron conveyor belt "floated" on the water with the aid of guides, no removal knife, and cleanability. It is mostly used for cake thicknesses of 3.2 to 15.9 mm (⅛ to ⅝ in) at speeds up to 15 m/min (50 ft/min), with 45.7-m (150-ft) pulley centers common. For 25- to 32-mm (1- to 1¼-in) cake, another belt on top to give two-sided cooling is frequently used. Applications are in food operations for cooling to harden candies, cheeses, gelatins, margarines, gums, etc.; and in chemical operations for solidification of sulfur, greases, resins, soaps, waxes, chloride salts, and some insecticides. Heat transfer is good, with sulfur solidification showing values of $q = 5800$ W/m$^2$ [1850 Btu/(h·ft$^2$)] and $U = 96$ W/(m$^2$·°C) [17 Btu/(h·ft$^2$·°F)] for a 7.9-mm (5⁄16-in) cake.

The submerged metal belt (Fig. 11-53b) is a special version of the metal belt to meet the peculiar handling properties of pitch in its solidification process. Although adhesive to a dry metal wall, pitch will not stick to the submerged wetted belt or rubber edge strips. Submergence helps to offset the very poor thermal conductivity through two-sided heat transfer.

A fairly recent application of the water-cooled metal belt to solidification duty is shown in Fig. 11-54. The operation is termed pastillizing

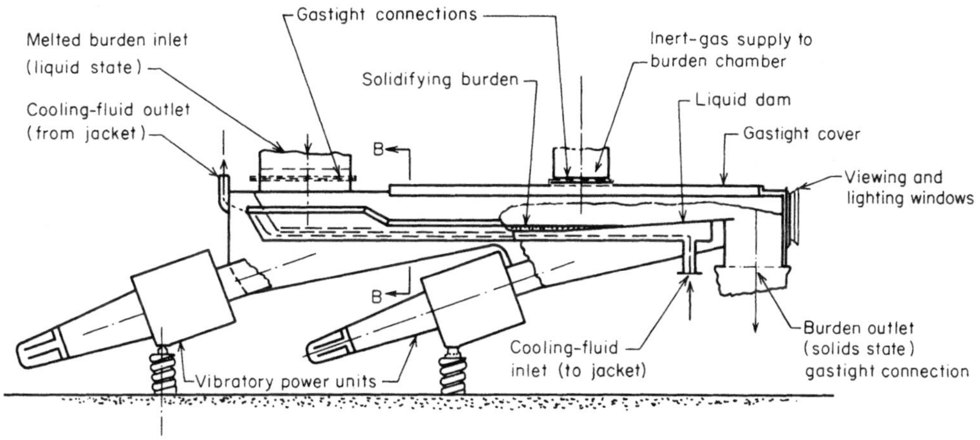


**FIG. 11-52** Heat-transfer equipment for batch solidification; vibrating-conveyor type. (*Courtesy of Jeffrey Mfg. Co.*)

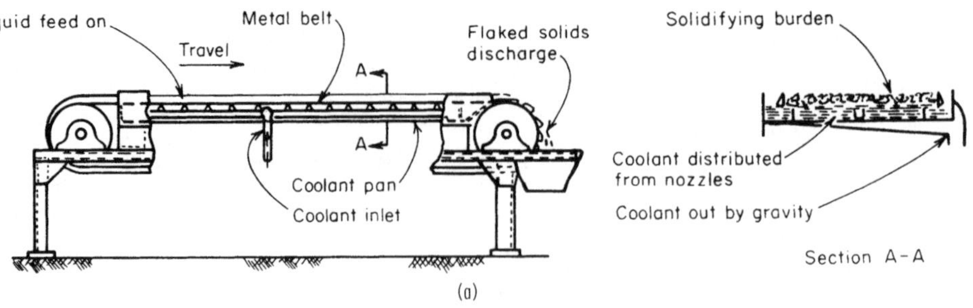


**FIG. 11-53** Heat-transfer equipment for continuous solidification. (*a*) Cooled metal belt. (*Courtesy of Sandvik, Inc.*) (*b*) Submerged metal belt. (*Courtesy of Sandvik, Inc.*)

from the form of the solidified end product, termed "pastilles." The novel feature is a one-step operation from the molten liquid to a fairly uniformly sized and shaped product without intermediate operations on the solid phase.

Another development features a nonmetallic belt [*Plast. Des. Process.*, **13** (July 1968)]. When rapid heat transfer is the objective, a glass-fiber, Teflon-coated construction in a thickness as little as 0.08 mm (0.003 in) is selected for use. No performance data are available, but presumably the thin belt permits rapid heat transfer while taking advantage of the nonsticking property of Teflon. Another development [*Food Process. Mark.*, **69** (March 1969)] is extending the capability of belt solidification by providing use of subzero temperatures.

**Rotating-Drum Type** This type (Fig. 11-55a and b) is not an adaptation of a material-handling device (though volumetric material throughput is a first consideration) but is designed specifically for heat-transfer service. It is well engineered, established, and widely used. The twin-drum type (Fig. 11-55b) is best suited to thin [0.4- to 6-mm (1/64 to 1/4-in)] cake production. For temperatures to 149°C (300°F) the coolant water is piped in and siphoned out. Spray application of coolant water to the inside is employed for high-temperature work, permitting feed temperatures to at least 538°C (1000°F), or double those for jacketed equipment. Vaporizing refrigerants are readily applicable for very low temperature work.

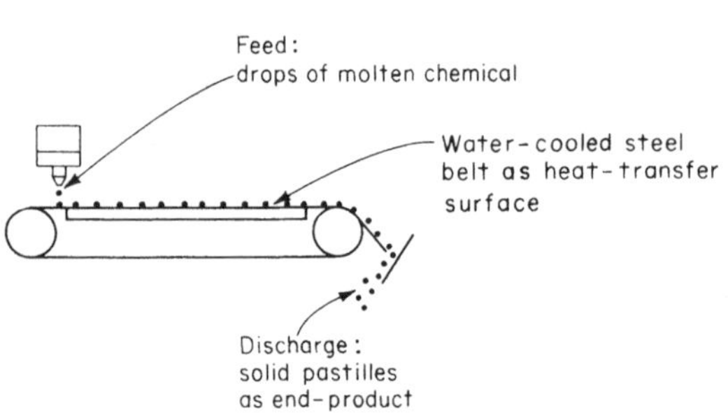


**FIG. 11-54** Heat-transfer equipment for solidification; belt type for the operation of pastillization. (*Courtesy of Sandvik, Inc.*)

The burden must have a definite solidification temperature to assure proper pickup from the feed pan. This limitation can be overcome by side feeding through an auxiliary rotating spreader roll. Application limits are further extended by special feed devices for burdens having oxidation-sensitive and/or supercooling characteristics. The standard double-drum model turns downward, with adjustable roll spacing to control sheet thickness. The newer twin-drum model (Fig. 11-55b) turns upward and, though subject to variable cake thickness, handles viscous and indefinite solidification-temperature-point burden materials well.

Drums have been successfully applied to a wide range of chemical products, both inorganic and organic, pharmaceutical compounds, waxes, soaps, insecticides, food products to a limited extent (including lard cooling), and even flake-ice production. A novel application is that of using a water-cooled roll to pick up from a molten-lead bath and turn out a 1.2-m- (4-ft-) wide continuous sheet, weighing 4.9 kg/m$^2$ (1 lb/ft$^2$), which is ideal for a sound barrier. This technique is more economical than other sheeting methods [*Mech. Eng.*, **631** (March 1968)].

**Heat-transfer performance** of drums, in terms of reported heat flux is: for an 80°C (176°F) melting-point wax, 7880 W/m$^2$ [2500 Btu/(h·ft$^2$)]; for a 130°C (266°F) melting-point organic chemical, 20,000 W/m$^2$ [6500 Btu/(h·ft$^2$)]; and for high- [318°C (604°F)] melting-point caustic soda (water-sprayed in drum), 95,000 to 125,000 W/m$^2$ [30,000 to 40,000 Btu/(h · ft$^2$)], with overall coefficients of 340 to 450 W/(m$^2$·°C) [60 to 80 Btu/(h·ft$^2$·°F)]. An innovation that is claimed often to increase these performance values by as much as 300 percent is the addition of hoods to apply impinging streams of heated air to the solidifying and drying solids surface as the drums carry it upward [*Chem. Eng.*, **74,** 152 (June 19, 1967)]. Similar rotating-drum indirect heat-transfer equipment is also extensively used for drying duty on liquids and thick slurries of solids (see Sec. 20).

**Rotating-Shelf Type** The patented **Roto-shelf** type (Fig. 11-55c) features (1) a large heat-transfer surface provided over a small floor space and in a small building volume, (2) easy floor cleaning, (3) nonhazardous machinery, (4) stainless-steel surfaces, (5) good control range, and (6) substantial capacity by providing as needed 1 to 10 shelves operated in parallel. It is best suited for thick-cake production and burden materials having an indefinite solidification temperature. Solidification of liquid sulfur into 13- to 19-mm- (1/2- to 3/4-in-) thick

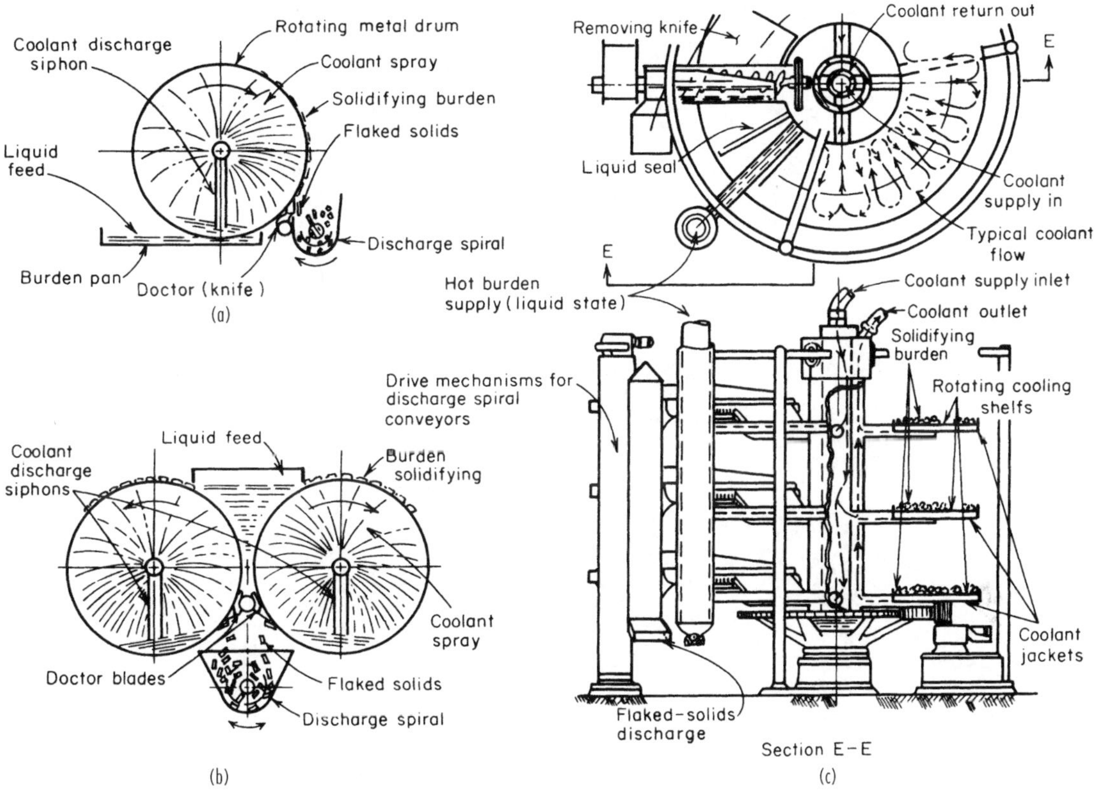


**FIG. 11-55** Heat-transfer equipment for continuous solidification. (*a*) Single drum. (*b*) Twin drum. (*c*) Roto-shelf. (*Courtesy of Buflovak Division, Blaw-Knox Food & Chemical Equipment, Inc.*)

lumps is a successful application. Heat transfer, by liquid-coolant circulation through jackets, limits feed temperatures to 204°C (400°F). Heat-transfer rate, controlled by the thick cake rather than by equipment construction, should be equivalent to the belt type. Thermal performance is aided by applying water sprayed directly to the burden top to obtain two-sided cooling.

## EQUIPMENT FOR FUSION OF SOLIDS

The thermal duty here is the opposite of solidification operations. The indirect heat-transfer equipment suitable for one operation is not suitable for the other because of the material-handling rather than the thermal aspects. Whether the temperature of transformation is a definite or a ranging one is of little importance in the selection of equipment for fusion. The burden is much agitated, but the beds are deep. Only fair overall coefficient values may be expected, although heat-flux values are good.

**Horizontal-Tank Type** This type (Fig. 11-56a) is used to transfer heat for melting or cooking dry powdered solids, rendering lard from meat-scrap solids, and drying divided solids. Heat-transfer coefficients are 17 to 85 W/(m$^2$·°C) [3 to 15 Btu/(h·ft$^2$·°F)] for drying and 28 to 140 W/(m$^2$·°C) [5 to 25 Btu/(h·ft$^2$·°F)] for vacuum and/or solvent recovery.

**Vertical Agitated-Kettle Type** Shown in Fig. 11-56b this type is used to cook, melt to the liquid state, and provide or remove reaction heat for solids that vary greatly in "body" during the process so that material handling is a real problem. The virtues are simplicity and 100 percent cleanability. These often outweigh the poor heat-transfer aspect. These devices are available from the small jacketed type illustrated to huge cast-iron direct-underfired bowls for calcining gypsum. Temperature limits vary with construction; the simpler jackets allow temperatures to 371°C (700°F) (as with Dowtherm), which is not true of all jacketed equipment.

**Mill Type** Figure 11-56c shows one model of roll construction used. Note the ruggedness, as it is a *power device* as well as one for *indirect* heat transfer, employed to knead and heat a mixture of dry powdered-solid ingredients with the objective of reacting and reforming via fusion to a consolidated product. In this compounding operation, frictional heat generated by the kneading may require heat-flow reversal (by cooling). Heat-flow control and temperature-level considerations often predominate over heat-transfer performance. Power and mixing considerations, rather than heat transfer, govern. The two-roll mill shown is employed in compounding raw plastic, rubber, and rubberlike elastomer stocks. Multiple-roll mills less knives (termed calenders) are used for continuous sheet or film production in widths up to 2.3 m (7.7 ft). Similar equipment is employed in the chemical compounding of inks, dyes, paint pigments, and the like.

## HEAT-TRANSFER EQUIPMENT FOR SHEETED SOLIDS

**Cylinder Heat-Transfer Units** Sometimes called "can" dryers or drying rolls, these devices are differentiated from drum dryers in that they are used for solids in flexible continuous-sheet form, whereas drum dryers are used for liquid or paste forms. The construction of the individual cylinders, or drums, is similar in most respects to that of drum dryers. Special designs are used to obtain uniform distribution of steam within large drums when uniform heating across the drum surface is critical.

A cylinder dryer may consist of one large cylindrical drum, such as the so-called Yankee dryer, but more often it comprises a number of

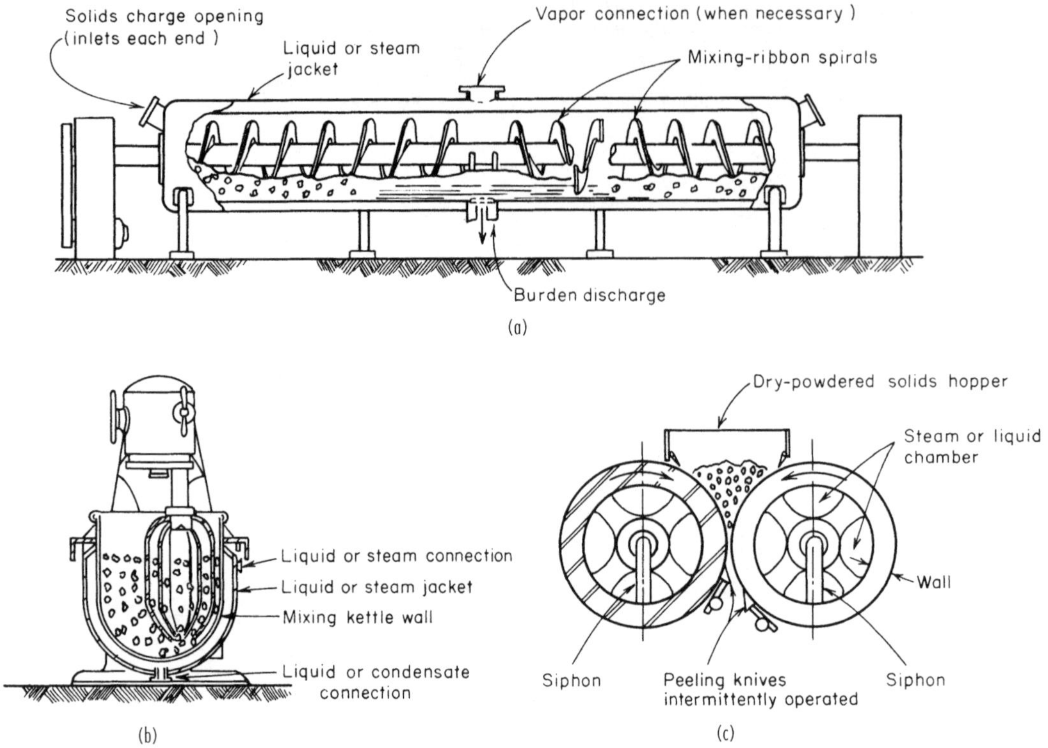


**FIG. 11-56** Heat-transfer equipment for fusion of solids. (*a*) Horizontal-tank type. (*Courtesy of Struthers Wells Corp.*) (*b*) Agitated kettle. (*Courtesy of Read-Standard Division, Capital Products Co.*) (*c*) Double-drum mill. (*Courtesy of Farrel-Birmingham Co.*)

drums arranged so that a continuous sheet of material may pass over them in series. Typical of this arrangement are Fourdrinier-paper-machine dryers, cellophane dryers, slashers for textile piece goods and fibers, etc. The multiple cylinders are arranged in various ways. Generally they are staggered in two horizontal rows. In any one row, the cylinders are placed close together. The sheet material contacts the undersurface of the lower rolls and passes over the upper rolls, contacting 60 to 70 percent of the cylinder surface. The cylinders may also be arranged in a single horizontal row, in more than two horizontal rows, or in one or more vertical rows. When it is desired to contact only one side of the sheet with the cylinder surface, unheated guide rolls are used to conduct the sheeting from one cylinder to the next. For sheet materials that shrink on processing, it is frequently necessary to drive the cylinders at progressively slower speeds through the dryer. This requires elaborate individual electric drives on each cylinder.

Cylinder dryers usually operate at atmospheric pressure. However, the Minton paper dryer is designed for operation under vacuum. The drying cylinders are usually heated by steam, but occasionally single cylinders may be gas-heated, as in the case of the Pease blueprinting machine. Upon contacting the cylinder surface, wet sheet material is first heated to an equilibrium temperature somewhere between the wet-bulb temperature of the surrounding air and the boiling point of the liquid under the prevailing total pressure. The heat-transfer resistance of the vapor layer between the sheet and the cylinder surface may be significant.

These cylinder units are applicable to almost any form of sheet material that is not injuriously affected by contact with steam-heated metal surfaces. They are used chiefly when the sheet possesses certain properties such as a tendency to shrink or lacks the mechanical strength necessary for most types of continuous-sheeting air dryers. Applications are to dry films of various sorts, paper pulp in sheet form, paper sheets, paperboard, textile piece goods and fibers, etc. In some cases, imparting a special finish to the surface of the sheet may be an objective.

The **heat-transfer performance capacity** of cylinder dryers is not easy to estimate without a knowledge of the sheet temperature, which, in turn, is difficult to predict. According to published data, steam temperature is the largest single factor affecting capacity. Overall evaporation rates based on the total surface area of the dryers cover a range of 3.4 to 23 kg water/(h·m$^2$) [0.7 to 4.8 lb water/(h·ft$^2$)].

The value of the **coefficient of heat transfer** from steam to sheet is determined by the conditions prevailing on the inside and on the surface of the dryers. Low coefficients may be caused by (1) poor removal of air or other noncondensables from the steam in the cylinders, (2) poor removal of condensate, (3) accumulation of oil or rust on the interior of the drums, and (4) accumulation of a fiber lint on the outer surface of the drums. In a test reported by Lewis et al. [*Pulp Pap. Mag. Can.*, **22** (February 1927)] on a sulfite-paper dryer, in which the actual sheet temperatures were measured, a value of 187 W/(m$^2$·°C) [33 Btu/(h·ft$^2$·°F)] was obtained for the coefficient of heat flow between the steam and the paper sheet.

Operating-cost data for these units are meager. Power costs may be estimated by assuming 1 hp per cylinder for diameters of 1.2 to 1.8 m (4 to 6 ft). Data on labor and maintenance costs are also lacking.

The size of commercial cylinder dryers covers a wide range. The individual rolls may vary in diameter from 0.6 to 1.8 m (2 to 6 ft) and up to 8.5 m (28 ft) in width. In some cases, the width of rolls decreases throughout the dryer in order to conform to the shrinkage of the sheet. A single-cylinder dryer, such as the Yankee dryer, generally has a diameter between 2.7 and 4.6 m (9 and 15 ft).

## HEAT-TRANSFER EQUIPMENT FOR DIVIDED SOLIDS

Most equipment for this service is some adaptation of a *material-handling* device whether or not the transport ability is desired. The old vertical tube and the vertical shell (fluidizer) are exceptions. Material-handling problems, plant transport needs, power, and maintenance are prime considerations in equipment selection and frequently overshadow heat-transfer and capital-cost considerations. Material handling is generally the most important aspect. Material-handling characteristics of the divided solids may vary during heat processing. The body changes are usually important in drying, occasionally significant for heating, and only on occasion important for cooling. The ability to minimize effects of changes is a major consideration in equipment selection. Dehydration operations are better performed on contactive apparatus (see Sec. 12) that provides air to carry off released water vapor before a semiliquid form develops.

Some types of equipment are convertible from heat removal to heat supply by simply changing the temperature level of the fluid or air. Other types require an auxiliary change. Others require constructional changes. Temperature limits for the equipment generally vary with the thermal operation. The kind of thermal operation has a major effect on heat-transfer values. For drying, overall coefficients are substantially higher in the presence of substantial moisture for the constant-rate period than in finishing. However, a stiff "body" occurrence due to moisture can prevent a normal "mixing" with an adverse effect on the coefficient.

**Fluidized-Bed Type** Known as the cylindrical fluidizer, this operates with a bed of **fluidized solids** (Fig. 11-57). It is an indirect heat-transfer version of the contactive type in Sec. 17. An application disadvantage is the need for batch operation unless some short circuiting can be tolerated. Solids-cooling applications are few, as they can be more effectively accomplished by the fluidizing gas via the contactive mechanism that is referred to in Sec. 11. Heating applications are many and varied. These are subject to one shortcoming, which is the dissipation of the heat input by carry-off in the fluidizing gas. Heat-transfer performance for the indirect mode to solids has been outstanding, with overall coefficients in the range of 570 to 850 W/($m^2 \cdot °C$) [100 to 150 Btu/($h \cdot ft^2 \cdot °F$)]. This device with its thin film does for solids what the falling-film and other thin-film techniques do for fluids, as shown by Holt (Pap. 11, 4th National Heat-Transfer Conference, August 1960). In a design innovation with high heat-transfer capability, heat is supplied indirectly to the fluidized solids through the walls of in-bed, horizontally placed, finned tubes [Petrie, Freeby, and Buckham, *Chem. Eng. Prog.*, **64**(7), 45 (1968)].

**Moving-Bed Type** This concept uses a single-pass tube bundle in a vertical shell with the divided solids flowing by gravity in the tubes. It is little used for solids. A major difficulty in divided-solids applications is the problem of charging and discharging with uniformity. A second is poor heat-transfer rates. Because of these limitations, this tube-bundle type is not the workhorse for solids that it is for liquid and gas-phase heat exchange.

However, there are applications in which the nature of a specific chemical reactor system requires indirect heating or cooling of a moving bed of divided solids. One of these is the segregation process which through a gaseous reaction frees chemically combined copper in an ore to a free copper form which permits easy, efficient subsequent recovery [Pinkey and Plint, *Miner. Process.*, 17–30 (June 1968)]. The apparatus construction and principle of operation are shown in Fig. 11-58. The functioning is abetted by a novel heat-exchange provision of a fluidized sand bed in the jacket. This provides a much higher unit heat-input rate (coefficient value) than would the usual low-density hot-combustion-gas flow.

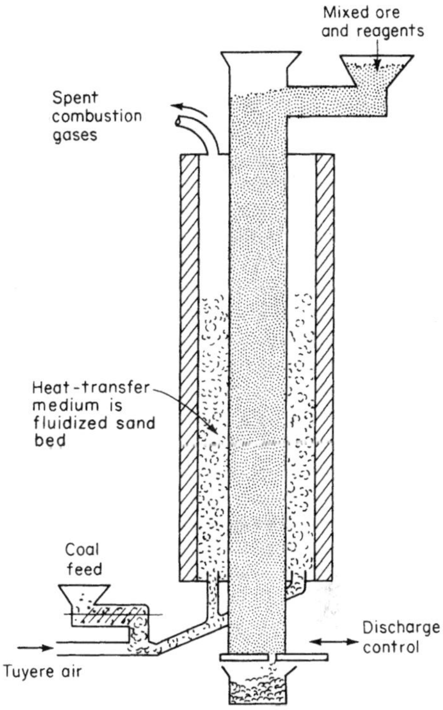


**FIG. 11-58** Stationary vertical-tube type of indirect heat-transfer equipment with divided solids inside tubes, laminar solids flow and steady-state heat conditions.

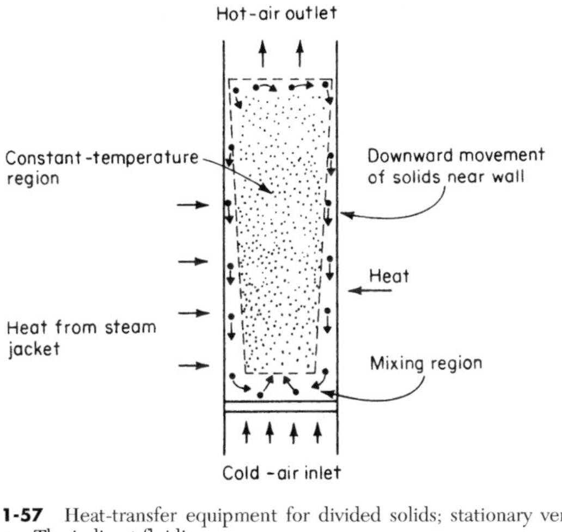


**FIG. 11-57** Heat-transfer equipment for divided solids; stationary vertical-shell type. The indirect fluidizer.

**Agitated-Pan Type** This device (Fig. 11-52) is not an adaptation of a material-handling device but was developed many years ago primarily for heat-transfer purposes. As such, it has found wide application. In spite of its batch operation with high attendant labor costs, it is still used for processing divided solids when no phase change is occurring. Simplicity and easy cleanout make the unit a wise selection for handling small, experimental, and even some production runs when quite a variety of burden materials are heat-processed. Both heating and cooling are feasible with it, but greatest use has been for drying [see Sec. 12 and Uhl and Root, *Chem. Eng. Prog.*, **63**(7), 8 (1967)]. This device, because it can be readily covered (as shown in the illustration) and a vacuum drawn or special atmosphere provided, features versatility to widen its use. For drying granular solids, the heat-transfer rate ranges from 28 to 227 W/($m^2 \cdot °C$) [5 to 40 Btu/($h \cdot ft^2 \cdot °F$)]. For atmospheric applications, thermal efficiency ranges from 65 to 75 percent. For vacuum applications, it is about 70 to 80 percent. These devices are available from several sources, fabricated of various metals used in chemical processes.

**Kneading Devices** These are closely related to the agitated pan but differ as being primarily mixing devices with heat transfer a secondary consideration. Heat transfer is provided by jacketed construction of the main body and is effected by a coolant, hot water, or steam. These devices are applicable for the compounding of divided solids by mechanical rather than chemical action. Application is largely in the pharmaceutical and food-processing industries. For a

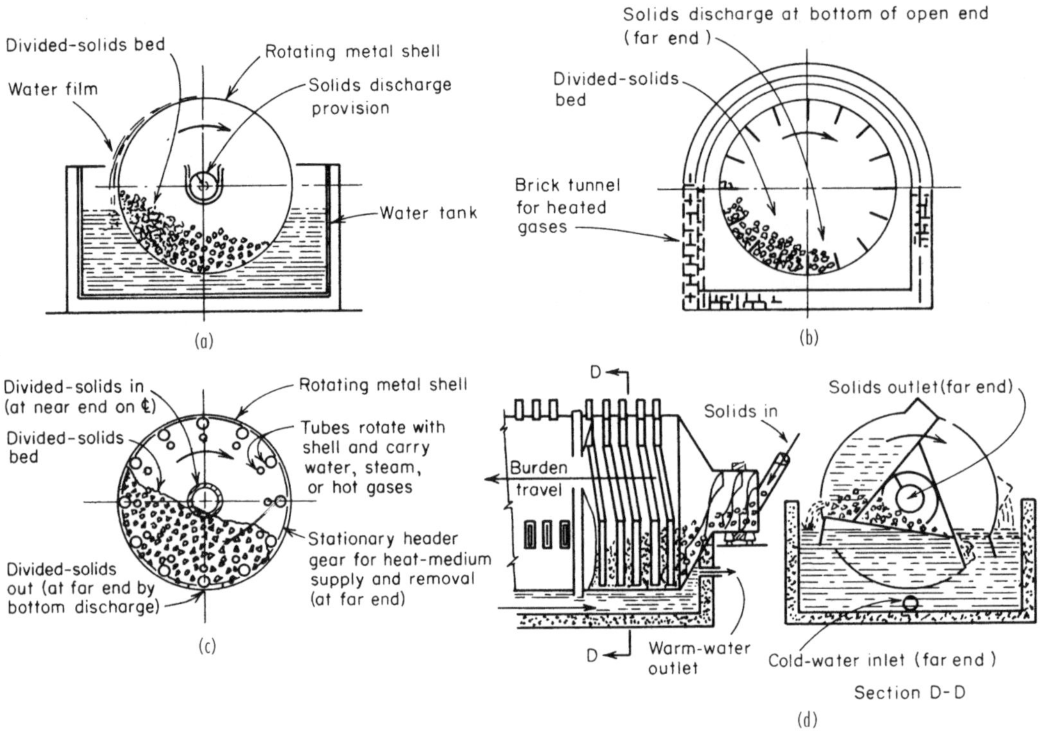


**FIG. 11-59** Rotating shells as indirect heat-transfer equipment. (*a*) Plain. (*Courtesy of BSP Corp.*) (*b*) Flighted. (*Courtesy of BSP Corp.*) (*c*) Tubed. (*d*) Deep-finned type. (*Courtesy of Link-Belt Co.*)

more complete description, illustrations, performance, and power requirements, refer to Sec. 19.

**Shelf Devices** Equipment having heated and/or cooled shelves is available but is little used for divided-solids heat processing. Most extensive use of stationary shelves is freezing of packaged solids for food industries and for freeze drying by sublimation (see Sec. 22).

**Rotating-Shell Devices** These (see Fig. 11-59) are installed horizontally, whereas stationary-shell installations are vertical. Material-handling aspects are of greater importance than thermal performance. Thermal results are customarily given in terms of overall coefficient on the basis of the total area provided, which varies greatly with the design. The effective use, chiefly percent fill factor, varies widely, affecting the reliability of stated coefficient values. For performance calculations see Sec. 10 on heat-processing theory for solids. These devices are variously used for cooling, heating, and drying and are the workhorses for heat-processing divided solids in the large-capacity range. Different modifications are used for each of the three operations.

The **plain** type (Fig. 11-59a) features simplicity and yet versatility through various end-construction modifications enabling wide and varied applications. Thermal performance is strongly affected by the "body" characteristics of the burden because of its dependency for material handling on frictional contact. Hence, performance ranges from well-agitated beds with good thin-film heat-transfer rates to poorly agitated beds with poor thick-film heat-transfer rates. Temperature limits in application are (1) low-range cooling with shell dipped in water, 400°C (750°F) and less; (2) intermediate cooling with forced circulation of tank water, to 760°C (1400°F); (3) primary cooling, above 760°C (1400°F), water copiously sprayed and loading kept light; (4) low-range heating, below steam temperature, hot-water dip; and (5) high-range heating by tempered combustion gases or ribbon radiant-gas burners.

The **flighted** type (Fig. 11-59b) is a first-step modification of the plain type. The simple flight addition improves heat-transfer performance. This type is most effective on semifluid burdens which slide readily. Flighted models are restricted from applications in which soft-cake sticking occurs, breakage must be minimized, and abrasion is severe. A special flighting is one having the cross section compartmented into four lesser areas with ducts between. Hot gases are drawn through the ducts en route from the outer oven to the stack to provide about 75 percent more heating surface, improving efficiency and capacity with a modest cost increase. Another similar unit has the flights made in a triangular-duct cross section with hot gases drawn through.

The **tubed-shell** type (Fig. 11-59c) is basically the same device more commonly known as a "steam-tube rotary dryer" (see Sec. 20). The rotation, combined with slight inclination from the horizontal, moves the shell-side solids through it continuously. This type features good mixing with the objective of increased heat-transfer performance. Tube-side fluid may be water, steam, or combustion gas. Bottom discharge slots in the shell are used so that heat-transfer-medium supply and removal can be made through the ends; these restrict wide-range loading and make the tubed type inapplicable for floody materials. These units are seldom applicable for sticky, soft-caking, scaling, or heat-sensitive burdens. They are not recommended for abrasive materials. This type has high thermal efficiency because heat loss is minimized. **Heat-transfer coefficient** values are: water, 34 W/($m^2 \cdot °C$) [6 Btu/($h \cdot ft^2 \cdot °F$)]; steam, same, with heat flux reliably constant at 3800 W/$m^2$ [1200 Btu/($h \cdot ft^2$)]; and gas, 17 W/($m^2 \cdot °C$) [3 Btu/($h \cdot ft^2 \cdot °F$)], with a high temperature difference. Although from the preceding discussion the device may seem rather limited, it is nevertheless widely used for drying, with condensing steam predominating as the heat-carrying fluid. But with water or refrigerants flowing in the tubes, it is also effective for cooling operations. The units are custom-built by several manufacturers in a wide range of sizes and materials.

A few fabricators that specialize in this type of equipment have accumulated a vast store of data for determining application sizing.

The patented **deep-finned** type in Fig. 11-59d is named the "Rotofin cooler." It features loading with a small layer thickness, excellent mixing to give a good effective diffusivity value, and a thin fluid-side film. Unlike other rotating-shell types, it is installed horizontally, and the burden is moved positively by the fins acting as an Archimedes spiral. Rotational speed and spiral pitch determine travel time. For cooling, this type is applicable to both secondary and intermediate cooling duties. Applications include solids in small lumps [9 mm (3⁄4 in)] and granular size [6 mm and less (1⁄4 to 0 in)] with no larger pieces to plug the fins, solids that have a free-flowing body characteristic with no sticking or caking tendencies, and drying of solids that have a low moisture and powder content unless special modifications are made for substantial vapor and dust handling. Thermal performance is very good, with overall coefficients to 110 W/(m$^2$·°C) [20 Btu/(h·ft$^2$·°F)], with one-half of these coefficients nominal for cooling based on the total area provided (nearly double those reported for other indirect rotaries).

**Conveyor-Belt Devices** The metal-belt type (Fig. 11-55) is the only device in this classification of material-handling equipment that has had serious effort expended on it to adapt it to indirect heat-transfer service with divided solids. It features a lightweight construction of a large area with a thin metal wall. Indirect-cooling applications have been made with poor thermal performance, as could be expected with a static layer. Auxiliary plowlike mixing devices, which are considered an absolute necessity to secure any worthwhile results for this service, restrict applications.

**Spiral-Conveyor Devices** Figure 11-60 illustrates the major adaptations of this widely used class of material-handling equipment to indirect heat-transfer purposes. These conveyors can be considered for heat-transfer purposes as continuously agitated kettles. The adaptation of Fig. 11-60*d* offers a batch-operated version for evaporation duty. For this service, all are package-priced and package-shipped items requiring few, if any, auxiliaries.

The **jacketed solid-flight** type (Fig. 11-60*a*) is the standard low-cost (parts-basis-priced) material-handling device, with a simple jacket added and employed for secondary-range heat transfer of an

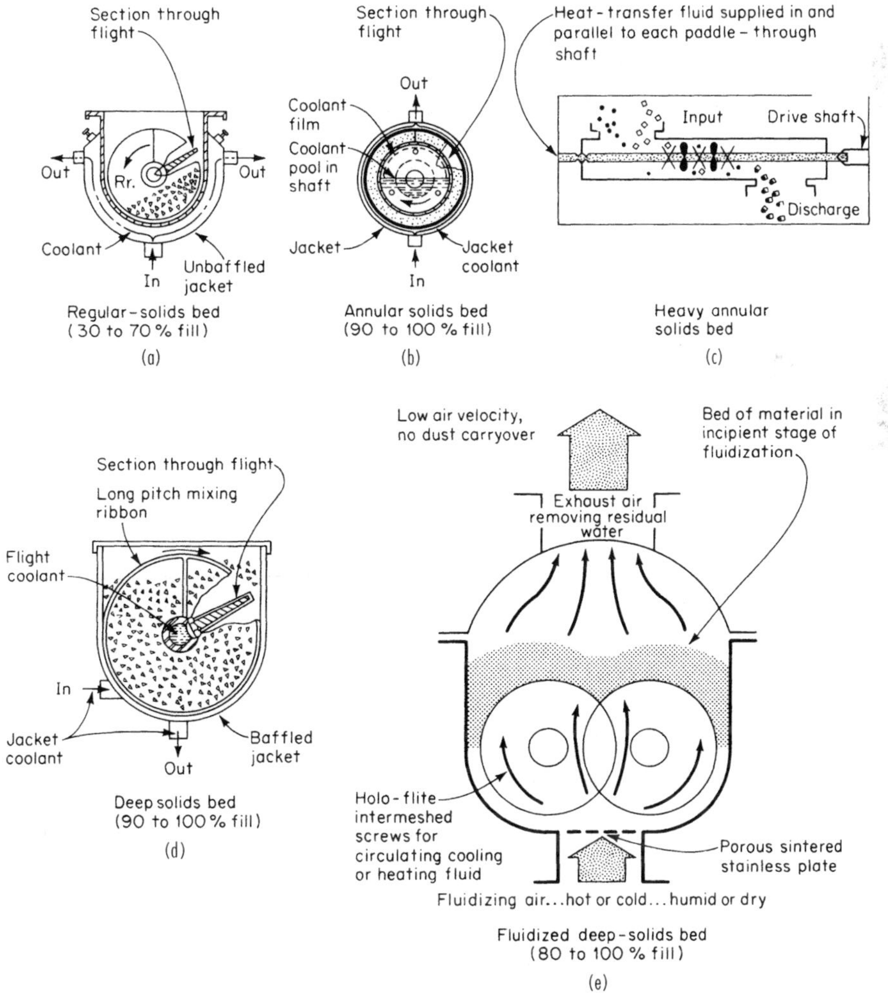


**FIG. 11-60** Spiral-conveyor adaptations as heat-transfer equipment. (*a*) Standard jacketed solid flight. (*Courtesy of Jeffrey Mfg. Co.*) (*b*) Small spiral, large shaft. (*Courtesy of Fuller Co.*) (*c*) "Porcupine" medium shaft. (*Courtesy of Bethlehem Corp.*) (*d*) Large spiral, hollow flight. (*Courtesy of Rietz Mfg. Co.*) (*e*) Fluidized-bed large spiral, helical flight. (*Courtesy of Western Precipitation Division, Joy Mfg. Co.*)

incidental nature. Heat-transfer coefficients are as low as 11 to 34 W/(m$^2$·°C) [2 to 6 Btu/(h·ft$^2$·°F)] on sensible heat transfer and 11 to 68 W/(m$^2$·°C) [2 to 12 Btu/(h·ft$^2$·°F)] on drying because of substantial static solids-side film.

The **small-spiral–large-shaft** type (Fig. 11-60*b*) is inserted in a solids-product line as pipe banks are in a fluid line, solely as a heat-transfer device. It features a thin burden ring carried at a high rotative speed and subjected to two-sided conductance to yield an estimated heat-transfer coefficient of 285 W/(m$^2$·°C) [50 Btu/(h·ft$^2$·°F)], thereby ranking thermally next to the shell-fluidizer type. This device for powdered solids is comparable with the Votator of the fluid field.

Figure 11-60*c* shows a fairly new spiral device with a medium-heavy annular solids bed and having the combination of a jacketed, stationary outer shell with moving paddles that carry the heat-transfer fluid. A unique feature of this device to increase volumetric throughput, by providing an overall greater temperature drop, is that the heat medium is supplied to and withdrawn from the rotor paddles by a parallel piping arrangement in the rotor shaft. This is a unique flow arrangement compared with the usual series flow. In addition, the rotor carries burden-agitating spikes which give it the trade name of Porcupine Heat-Processor (*Chem. Equip. News,* April 1966; and Uhl and Root, AIChE Prepr. **21,** 11th National Heat-Transfer Conference, August 1967).

The **large-spiral hollow-flight** type (Fig. 11-60*d*) is an adaptation, with external bearings, full fill, and salient construction points as shown, that is highly versatile in application. Heat-transfer coefficients are 34 to 57 W/(m$^2$·°C) [6 to 10 Btu/(h·ft$^2$·°F)] for poor, 45 to 85 W/(m$^2$·°C) [8 to 15 Btu/(h·ft$^2$·°F)] for fair, and 57 to 114 W/(m$^2$·°C) [10 to 20 Btu/(h·ft$^2$·°F)] for wet conductors. A popular version of this employs two such spirals in one material-handling chamber for a pug-mill agitation of the deep solids bed. The spirals are seldom heated. The shaft and shell are heated.

Another deep-bed spiral-activated solids-transport device is shown by Fig. 11-60*e*. The flights carry a heat-transfer medium as well as the jacket. A unique feature of this device which is purported to increase heat-transfer capability in a given equipment space and cost is the dense-phase fluidization of the deep bed that promotes agitation and moisture removal on drying operations.

**Double-Cone Blending Devices** The original purpose of these devices was mixing (see Sec. 19). Adaptations have been made; so many models now are primarily for indirect heat-transfer processing. A jacket on the shell carries the heat-transfer medium. The mixing action, which breaks up agglomerates (but also causes some degradation), provides very effective burden exposure to the heat-transfer surface. On drying operations, the vapor release (which in a static bed is a slow diffusional process) takes place relatively quickly. To provide vapor removal from the burden chamber, a hollow shaft is used. Many of these devices carry the hollow-shaft feature a step further by adding a rotating seal and drawing a vacuum. This increases thermal performance notably and makes the device a natural for solvent-recovery operations.

These devices are replacing the older tank and spiral-conveyor devices. Better provisions for speed and ease of fill and discharge (without powered rotation) minimize downtime to make this batch-operated device attractive. Heat-transfer coefficients ranging from 28 to 200 W/(m$^2$·°C) [5 to 35 Btu/(h·ft$^2$·°F)] are obtained. However, if caking on the heat-transfer walls is serious, then values may drop to 5.5 or 11 W/(m$^2$·°C) [1 or 2 Btu/(h·ft$^2$·°F)], constituting a misapplication. The double cone is available in a fairly wide range of sizes and construction materials. The users are the fine-chemical, pharmaceutical, and biological-preparation industries.

A novel variation is a cylindrical model equipped with a tube bundle to resemble a shell-and-tube heat exchanger with a bloated shell [*Chem. Process.,* **20** (Nov. 15, 1968)]. Conical ends provide for redistribution of burden between passes. The improved heat-transfer performance is shown by Fig. 11-61.

**Vibratory-Conveyor Devices** Figure 11-62 shows the various adaptations of vibratory material-handling equipment for indirect heat-transfer service on divided solids. The basic vibratory-equipment data are given in Sec. 21. These indirect heat-transfer adaptations feature simplicity, nonhazardous construction, nondegradation, nondusting, no wear, ready conveying-rate variation [1.5 to 4.5 m/min (5 to 15 ft/min)], and good heat-transfer coefficient—115 W/(m$^2$·°C)

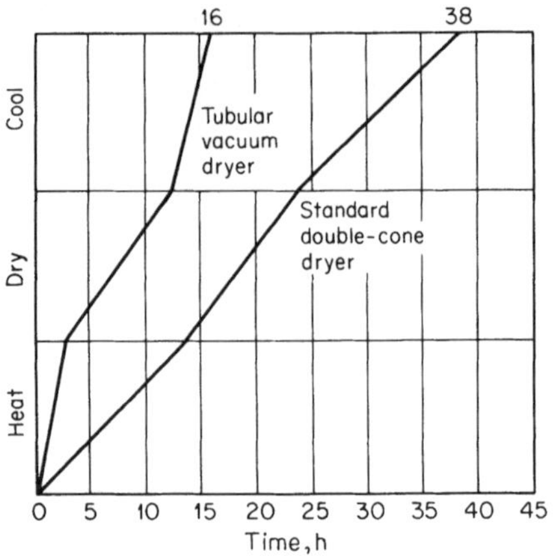


**FIG. 11-61** Performance of tubed, blender heat-transfer device.

[20 Btu/(h·ft$^2$·°F)] for sand. They usually require feed-rate and distribution auxiliaries. They are suited for heating and cooling of divided solids in powdered, granular, or moist forms but no sticky, liquefying, or floody ones. Terminal-temperature differences less than 11°C (20°F) on cooling and 17°C (30°F) on heating or drying operations are seldom practical. These devices are for medium and light capacities.

The **heavy-duty jacketed** type (Fig. 11-62a) is a special custom-built adaptation of a heavy-duty vibratory conveyor shown in Fig. 11-60. Its application is continuously to cool the crushed material [from about 177°C (350°F)] produced by the vibratory-type "caster" of Fig. 11-53. It does not have the liquid dam and is made in longer lengths that employ L, switchback, and S arrangements on one floor. The capacity rate is 27,200 to 31,700 kg/h (30 to 35 tons/h) with heat-transfer coefficients in the order of 142 to 170 W/(m$^2$·°C) [25 to 30 Btu/(h·ft$^2$·°F)]. For heating or drying applications, it employs steam to 414 kPa (60 lbf/in$^2$).

The **jacketed or coolant-spraying** type (Fig. 11-62b) is designed to assure a very thin, highly agitated liquid-side film and the same initial coolant temperature over the entire length. It is frequently employed for transporting substantial quantities of hot solids, with cooling as an incidental consideration. For heating or drying applications, hot water or steam at a gauge pressure of 7 kPa (1 lbf/in$^2$) may be employed. This type is widely used because of its versatility, simplicity, cleanability, and good thermal performance.

The **light-duty jacketed** type (Fig. 11-62c) is designed for use of air as a heat carrier. The flow through the jacket is highly turbulent and is usually counterflow. On long installations, the air flow is parallel to every two sections for more heat-carrying capacity and a fairly uniform surface temperature. The outstanding feature is that a wide range of temperature control is obtained by merely changing the heat-carrier temperature level from as low as atmospheric moisture condensation will allow to 204°C (400°F). On heating operations, a very good thermal efficiency can be obtained by insulating the machine and recycling the air. While heat-transfer rating is good, the heat-removal capacity is limited. Cooler units are often used in series with like units operated as dryers or when clean water is unavailable. Drying applications are for heat-sensitive [49 to 132°C (120° to 270°F)] products; when temperatures higher than steam at a gauge pressure of 7 kPa (1 lbf/in$^2$) can provide are wanted but heavy-duty equipment is too costly; when the jacket-corrosion hazard of steam is unwanted; when headroom space is at a premium; and for highly abrasive burden materials such as fritted or crushed glasses and porcelains.

The **tiered arrangement** (Fig. 11-62d) employs the units of Fig. 11-62 with either air or steam at a gauge pressure of 7 kPa (1 lbf/in$^2$) as a heat medium. These are custom-designed and built to provide a large amount of heat-transfer surface in a small space with the minimum of transport and to provide a complete processing system. These

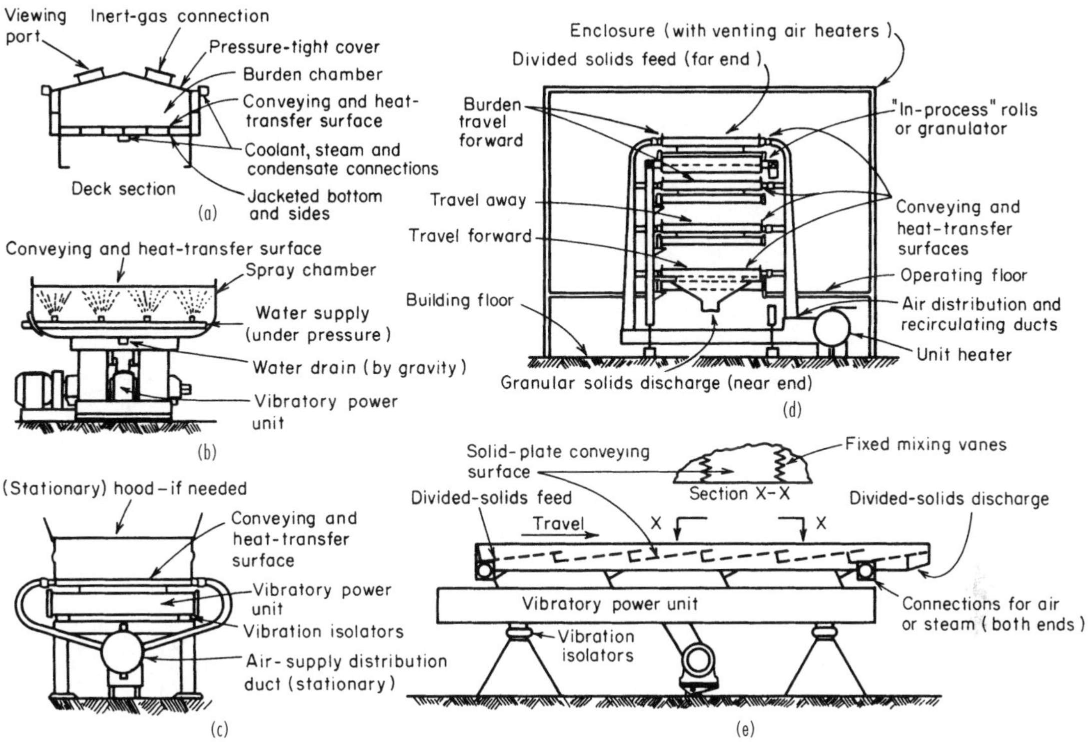


**FIG. 11-62** Vibratory-conveyor adaptations as indirect heat-transfer equipment. (*a*) Heavy-duty jacketed for liquid coolant or high-pressure steam. (*b*) Jacketed for coolant spraying. (*c*) Light-duty jacketed construction. (*d*) Jacketed for air or steam in tiered arrangement. (*e*) Jacketed for air or steam with Mix-R-Step surface. (*Courtesy of Jeffrey Mfg. Co.*)

receive a damp material, resize while in process by granulators or rolls, finish dry, cool, and deliver to packaging or tableting. The applications are primarily in the fine-chemical, food, and pharmaceutical manufacturing fields.

The **Mix-R-Step** type in Fig. 11-62e is an adaptation of a vibratory conveyor. It features better heat-transfer rates, practically doubling the coefficient values of the standard flat surface and trebling heat-flux values, as the layer depth can be increased from the normal 13 to 25 and 32 mm (½ to 1 and 1¼ in). It may be provided on decks jacketed for air, steam, or water spray. It is also often applicable when an infrared heat source is mounted overhead to supplement the indirect or as the sole heat source.

**Elevator Devices** The **vibratory elevating-spiral** type (Fig. 11-63) adapts divided-solids-elevating material-handling equipment to heat-transfer service. It features a large heat-transfer area over a small floor space and employs a reciprocating shaker motion to effect transport. Applications, layer depth, and capacities are restricted, as burdens must be of such "body" character as to convey uphill by the microhopping-transport principle. The type lacks self-emptying ability. Complete washdown and cleaning is a feature not inherent in any other elevating device. A typical application is the cooling of a low-density plastic powder at the rate of 544 kg/h (1200 lb/h).

Another elevator adaptation is that for a **spiral-type elevating** device developed for ground cement and thus limited to fine powdery burdens. The spiral operates inside a cylindrical shell, which is externally cooled by a falling film of water. The spiral not only elevates the material in a thin layer against the wall but keeps it agitated to achieve high heat-transfer rates. Specific operating data are not available [*Chem. Eng. Prog.*, **68**(7), 113 (1968)]. The falling-water film, besides being ideal thermally, by virtue of no jacket pressure very greatly reduces the hazard that the cooling water may contact the water-sensitive burden in process. Surfaces wet by water are accessible for cleaning. A fair range of sizes is available, with material-handling capacities to 60 tons/h.

**Pneumatic-Conveying Devices** See Sec. 21 for descriptions, ratings, and design factors on these devices. Use is primarily for transport purposes, and heat transfer is a very secondary consideration. Applications have largely been for plastics in powder and pellet forms.

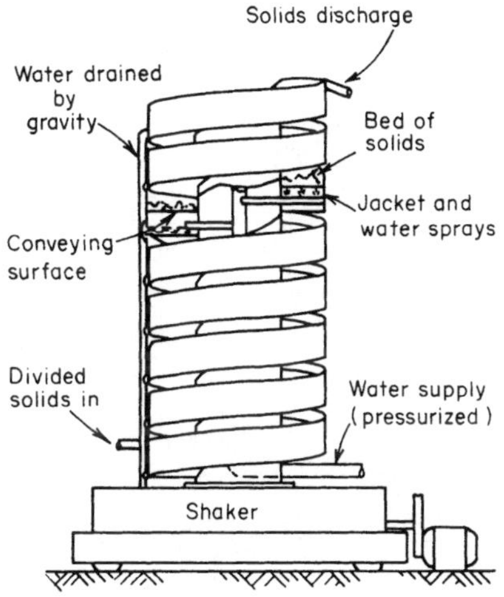


**FIG. 11-63** Elevator type as heat-transfer equipment. (*Courtesy of Carrier Conveyor Corp.*)

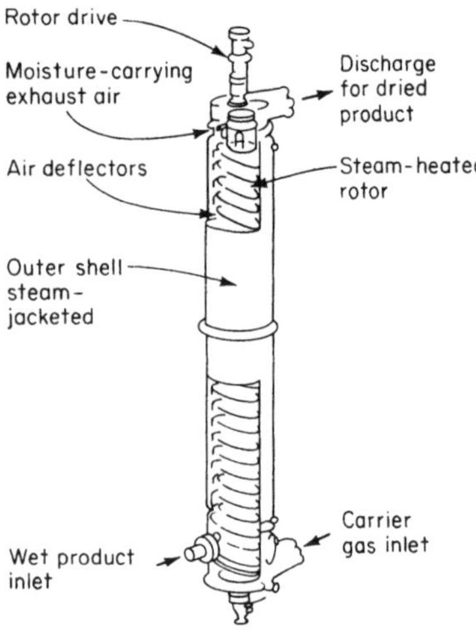


**FIG. 11-64** A pneumatic-transport adaptation for heat-transfer duty. (*Courtesy of Werner & Pfleiderer Corp.*)

By modifications, needed cooling operations have been simultaneously effected with transport to stock storage [*Plast. Des. Process.*, **28** (December 1968)].

Heat-transfer aspects and performance were studied and reported on by Depew and Farbar (ASME Pap. 62-HT-14, September 1962). Heat-transfer coefficient characteristics are similar to those shown in Sec. 11 for the indirectly heated fluid bed. Another frequent application on plastics is a small, rather incidental but necessary amount of drying required for plastic pellets and powders on receipt when shipped in bulk to the users. Pneumatic conveyors modified for heat transfer can handle this readily.

A pneumatic-transport device designed primarily for heat-sensitive products is shown in Fig. 11-64. This was introduced into the United States after 5 years' use in Europe [*Chem. Eng.*, **76,** 54 (June 16, 1969)].

Both the shell and the rotor carry steam as a heating medium to effect indirect transfer as the burden briefly contacts those surfaces rather than from the transport air, as is normally the case. The rotor turns slowly (1 to 10 r/min) to control, by deflectors, product distribution and prevent caking on walls. The carrier gas can be inert, as nitrogen, and also recycled through appropriate auxiliaries for solvent recovery. Application is limited to burdens that (1) are fine and uniformly grained for the pneumatic transport, (2) dry very fast, and (3) have very little, if any, sticking or decomposition characteristics. Feeds can carry 5 to 100 percent moisture (dry basis) and discharge at 0.1 to 2 percent. Wall temperatures range from 100 to 170°C (212 to 340°F) for steam and lower for a hot-water-heat source. Pressure drops are in order of 500 to 1500 mm$H_2O$ (20 to 60 in$H_2O$). Steam consumption approaches that of a contractive-mechanism dryer down to a low value of 2.9 kg steam/kg water (2.9 lb steam/lb water). Available burden capacities are 91 to 5900 kg/h (200 to 13,000 lb/h).

**Vacuum-Shelf Types** These are very old devices, being a version of the table type. Early-day use was for drying (see Sec. 12). Heat transfer is slow even when supplemented by vacuum, which is 90 percent or more of present-day use. The newer vacuum blender and cone devices are taking over many applications. The slow heat-transfer rate is quite satisfactory in a major application, freeze drying, which is a sublimation operation (see Sec. 22 for description) in which the water must be retained in the solid state during its removal. Then slow diffusional processes govern. Another extensive application is freezing packaged foods for preservation purposes.

Available sizes range from shelf areas of 0.4 to 67 $m^2$ (4 to 726 $ft^2$). These are available in several manufacturers' standards, either as system components or with auxiliary gear as packaged systems.

# THERMAL INSULATION

Materials or combinations of materials which have air- or gas-filled pockets or void spaces that retard the transfer of heat with reasonable effectiveness are thermal insulators. Such materials may be particulate and/or fibrous, with or without binders, or may be assembled, such as multiple heat-reflecting surfaces that incorporate air- or gas-filled void spaces.

The ability of a material to retard the flow of heat is expressed by its thermal **conductivity** (for unit thickness) or **conductance** (for a specific thickness). Low values for thermal conductivity or conductance (or high thermal resistivity or resistance value) are characteristics of thermal insulation.

Heat is transferred by radiation, conduction, and convection. Radiation is the primary mode and can occur even in a vacuum. The amount of heat transferred for a given area is relative to the temperature differential and emissivity from the radiating to the absorbing surface. Conduction is due to molecular motion and occurs within gases, liquids, and solids. The tighter the molecular structure, the higher the rate of transfer. As an example, steel conducts heat at a rate approximately 600 times that of typical thermal-insulation materials. Convection is due to mass motion and occurs only in fluids. The prime purpose of a thermal-insulation system is to minimize the amount of heat transferred.

## INSULATION MATERIALS

**Materials** Thermal insulations are produced from many materials or combinations of materials in various forms, sizes, shapes, and thickness. The most commonly available materials fall within the following categories:

*Fibrous or cellular—mineral.* Alumina, asbestos, glass, perlite, rock, silica, slag, or vermiculite.

*Fibrous or cellular—organic.* Cane, cotton, wood, and wood bark (cork).

*Cellular organic plastics.* Elastomer, polystyrene, polyisocyanate, polyisocyanurate, and polyvinyl acetate.

*Cements.* Insulating and/or finishing.

*Heat-reflecting metals (reflective).* Aluminum, nickel, stainless steel.

*Available forms.* Blanket (felt and batt), block, cements, loose fill, foil and sheet, formed or foamed in place, flexible, rigid, and semirigid.

The actual thicknesses of piping insulation differ from the nominal values. Dimensional data of ASTM Standard C585 appear in Table 11-21.

**Thermal Conductivity (*K* Factor)** Depending on the type of insulation, the thermal conductivity (*K* factor) can vary with age, manufacturer, moisture content, and temperature. Typical published values are shown in Fig. 11-65. Mean temperature is equal to the arithmetic average of the temperatures on both sides of the insulating material.

**Actual system heat loss** (or gain) will normally exceed calculated values because of projections, axial and longitudinal seams, expansion-contraction openings, moisture, workers' skill, and physical abuse.

**Finishes** Thermal insulations require an external covering (finish) to provide protection against entry of water or process fluids, mechanical damage, and ultraviolet degradation of foamed materials. In some cases the finish can reduce the flame-spread rating and/or provide fire protection.

The finish may be a coating (paint, asphaltic, resinous, or polymeric), a membrane (coated felt or paper, metal foil, or laminate of plastic, paper, foil or coatings), or sheet material (fabric, metal, or plastic).

**TABLE 11-21 Thicknesses of Piping Insulation**

| | Insulation, nominal thickness | | | | | | | | | | | | | | | |
|---|---|---|---|---|---|---|---|---|---|---|---|---|---|---|---|---|
| in<br>mm | Outer diameter | | 1<br>25 | | 1½<br>38 | | 2<br>51 | | 2½<br>64 | | 3<br>76 | | 3½<br>89 | | 4<br>102 | |
| | Approximate wall thickness | | | | | | | | | | | | | | | |
| Nominal iron-pipe size, in | in | mm | in | mm | in | mm | in | mm | in | mm | in | mm | in | mm | in | mm |
| ½ | 0.84 | 21 | 1.01 | 26 | 1.57 | 40 | 2.07 | 53 | 2.88 | 73 | 3.38 | 86 | 3.88 | 99 | 4.38 | 111 |
| ¾ | 1.05 | 27 | 0.90 | 23 | 1.46 | 37 | 1.96 | 50 | 2.78 | 71 | 3.28 | 83 | 3.78 | 96 | 4.28 | 109 |
| 1 | 1.32 | 33 | 1.08 | 27 | 1.58 | 40 | 2.12 | 54 | 2.64 | 67 | 3.14 | 80 | 3.64 | 92 | 4.14 | 105 |
| 1¼ | 1.66 | 42 | 0.91 | 23 | 1.66 | 42 | 1.94 | 49 | 2.47 | 63 | 2.97 | 75 | 3.47 | 88 | 3.97 | 101 |
| 1½ | 1.90 | 48 | 1.04 | 26 | 1.54 | 39 | 2.35 | 60 | 2.85 | 72 | 3.35 | 85 | 3.85 | 98 | 4.42 | 112 |
| 2 | 2.38 | 60 | 1.04 | 26 | 1.58 | 40 | 2.10 | 53 | 2.60 | 66 | 3.10 | 79 | 3.60 | 91 | 4.17 | 106 |
| 2½ | 2.88 | 73 | 1.04 | 26 | 1.86 | 47 | 2.36 | 60 | 2.86 | 73 | 3.36 | 85 | 3.92 | 100 | 4.42 | 112 |
| 3 | 3.50 | 89 | 1.02 | 26 | 1.54 | 39 | 2.04 | 52 | 2.54 | 65 | 3.04 | 77 | 3.61 | 92 | 4.11 | 104 |
| 3½ | 4.00 | 102 | 1.30 | 33 | 1.80 | 46 | 2.30 | 58 | 2.80 | 71 | 3.36 | 85 | 3.86 | 98 | 4.36 | 111 |
| 4 | 4.50 | 114 | 1.04 | 26 | 1.54 | 39 | 2.04 | 52 | 2.54 | 65 | 3.11 | 79 | 3.61 | 92 | 4.11 | 104 |
| 4½ | 5.00 | 127 | 1.30 | 33 | 1.80 | 46 | 2.30 | 58 | 2.86 | 73 | 3.36 | 85 | 3.86 | 98 | 4.48 | 114 |
| 5 | 5.56 | 141 | 0.99 | 25 | 1.49 | 38 | 1.99 | 51 | 2.56 | 65 | 3.06 | 78 | 3.56 | 90 | 4.18 | 106 |
| 6 | 6.62 | 168 | 0.96 | 24 | 1.46 | 37 | 2.02 | 51 | 2.52 | 64 | 3.02 | 77 | 3.65 | 93 | 4.15 | 105 |
| 7 | 7.62 | 194 | | | 1.52 | 39 | 2.02 | 51 | 2.52 | 64 | 3.15 | 80 | 3.65 | 93 | 4.15 | 105 |
| 8 | 8.62 | 219 | | | 1.52 | 39 | 2.02 | 51 | 2.65 | 67 | 3.15 | 80 | 3.65 | 93 | 4.15 | 105 |
| 9 | 9.62 | 244 | | | 1.52 | 39 | 2.15 | 55 | 2.65 | 67 | 3.15 | 80 | 3.65 | 93 | 4.15 | 105 |
| 10 | 10.75 | 273 | | | 1.58 | 40 | 2.08 | 53 | 2.58 | 66 | 3.08 | 78 | 3.58 | 91 | 4.08 | 104 |
| 11 | 11.75 | 298 | | | 1.58 | 40 | 2.08 | 53 | 2.58 | 66 | 3.08 | 78 | 3.58 | 91 | 4.08 | 104 |
| 12 | 12.75 | 324 | | | 1.58 | 40 | 2.08 | 53 | 2.58 | 66 | 3.08 | 78 | 3.58 | 91 | 4.08 | 104 |
| 14 | 14.00 | 356 | | | 1.46 | 37 | 1.96 | 50 | 2.46 | 62 | 2.96 | 75 | 3.46 | 88 | 3.96 | 101 |
| Over 14, up to and including 36 | | | | | 1.46 | 37 | 1.96 | 50 | 2.46 | 62 | 2.96 | 75 | 3.46 | 88 | 3.96 | 101 |

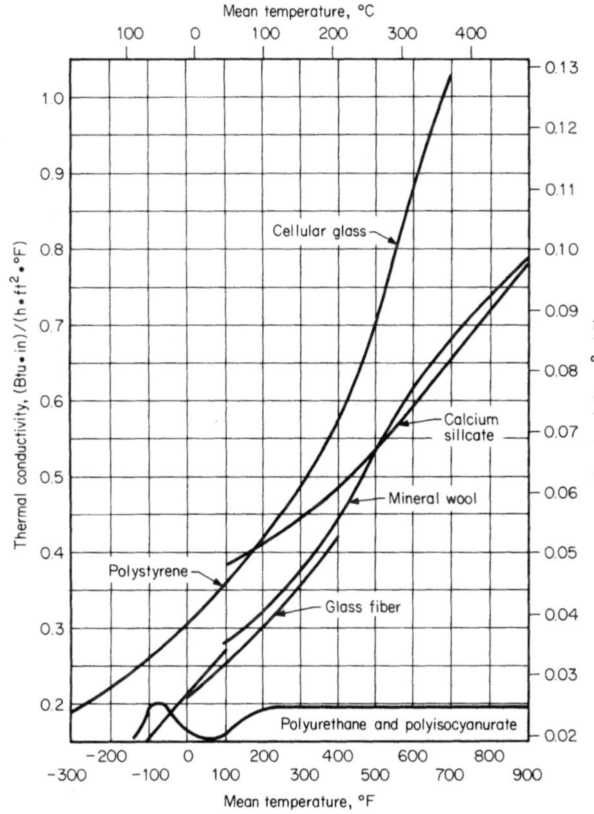


**FIG. 11-65** Thermal conductivity of insulating materials.

Finishes for systems operating below 2°C (35°F) must be sealed and retard vapor transmission. Those from 2°C (35°F) through 27°C (80°F) should retard vapor transmission (to prevent surface condensation), and those above 27°C (80°F) should prevent water entry and allow moisture to escape.

Metal finishes are more durable, require less maintenance, reduce heat loss, and, if uncoated, increase the surface temperature on hot systems.

## SYSTEM SELECTION

A combination of insulation and finish produces the thermal-insulation system. Selection of these components depends on the purpose for which the system is to be used. No single system performs satisfactorily from the cryogenic through the elevated-temperature range. Systems operating below freezing have a low vapor pressure, and atmospheric moisture is pushed into the insulation system, while the reverse is true for hot systems. Some general guidelines for system selection follow.

**Cryogenic [−273 to −101°C (−459 to −150°F)] High Vacuum** This technique is based on the Dewar flask, which is a double-walled vessel with reflective surfaces on the evacuated side to reduce radiation losses. Figure 11-66 shows a typical laboratory-size Dewar. Figure 11-67 shows a semiportable type. Radiation losses can be further reduced by filling the cavity with powders such as perlite or silica prior to pulling the vacuum.

***Multilayer*** Multilayer systems consist of series of radiation-reflective shields of low emittance separated by fillers or spacers of very low conductance and exposed to a high vacuum.

***Foamed or Cellular*** Cellular plastics such as polyurethane and polystyrene do not hold up or perform well in the cryogenic temperature range because of permeation of the cell structure by water vapor, which in turn increases the heat-transfer rate. Cellular glass holds up better and is less permeable.

**Low Temperature [−101 to −1°C (−150 to +30°F)]** Cellular glass, glass fiber, polyurethane foam, and polystyrene foam are frequently used for this service range. A vapor-retarder finish with a

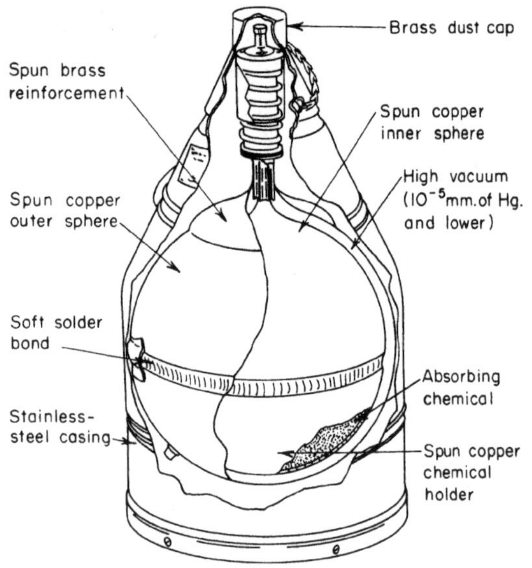


**FIG. 11-66** Dewar flask.

perm rating less than 0.02 is required. In addition, it is good practice to coat all contact surfaces of the insulation with a vapor-retardant mastic to prevent moisture migration when the finish is damaged or is not properly maintained. Closed-cell insulation should not be relied on as the vapor retarder. Hairline cracks can develop, cells can break

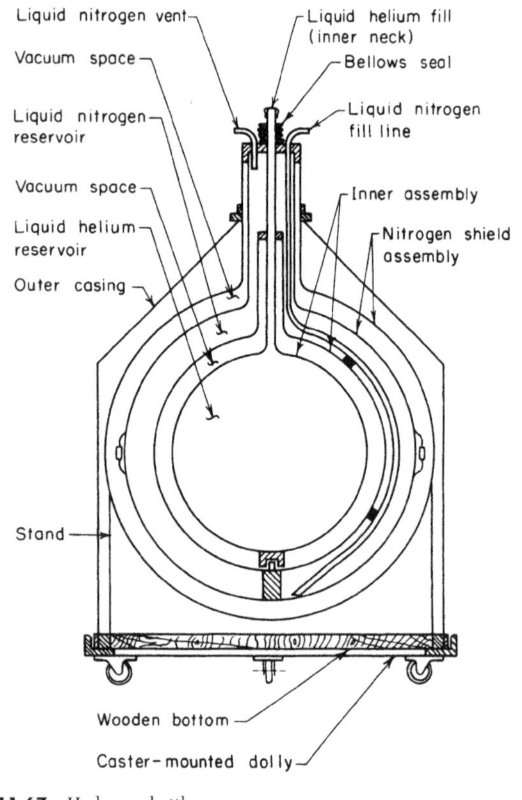


**FIG. 11-67** Hydrogen bottle.

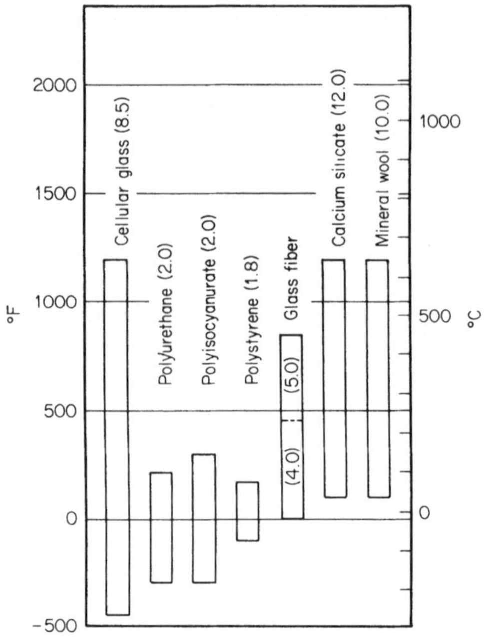


**FIG. 11-68** Insulating materials and applicable temperature ranges.

down, glass-fiber binders are absorbent, and moisture can enter at joints between all materials.

**Moderate and High Temperature [over 2°C (36°F)]** Cellular or fibrous materials are normally used. See Fig. 11-68 for nominal temperature range. Nonwicking insulation is desirable for systems operating below 100°C (212°F).

***Other Considerations*** **Autoignition** can occur if combustible fluids are absorbed by wicking-type insulations. **Chloride stress corrosion** of austenitic stainless steel can occur when chlorides are concentrated on metal surfaces at or above approximately 60°C (140°F). The chlorides can come from sources other than the insulation. Some calcium silicates are formulated to exceed the requirements of the MIL-I-24244A specification. **Fire resistance** of insulations varies widely. Calcium silicate, cellular glass, glass fiber, and mineral wool are fire-resistant but do not perform equally under actual fire conditions. A steel jacket provides protection, but aluminum does not.

Traced pipe performs better with a nonwicking insulation which has low thermal conductivity. Underground systems are very difficult to keep dry permanently. Methods of insulation include factory-preinsulated pouring types and conventionally applied types. Corrosion can occur under wet insulation. A protective coating, applied directly to the metal surface, may be required.

## ECONOMIC THICKNESS OF INSULATION

Optimal economic insulation thickness may be determined by various methods. Two of these are the minimum-total-cost method and the incremental-cost method (or marginal-cost method). The minimum-total-cost method involves the actual calculations of lost energy and insulation costs for each insulation thickness. The thickness producing the lowest total cost is the optimal economic solution. The optimum thickness is determined to be the point where the last dollar invested in insulation results in exactly $1 in energy-cost savings ("ETI—Economic Thickness for Industrial Insulation," Conservation Pap. **46,** Federal Energy Administration, August 1976). The incremental-cost method provides a simplified and direct solution for the least-cost thickness.

The total-cost method does *not* in general provide a satisfactory means for making most insulation investment decisions, since an economic return on investment is required by investors and the method does not properly consider this factor. Return on investment is considered by Rubin ("Piping Insulation—Economics and Profits," in *Practical Considerations in Piping Analysis,* ASME Symposium, vol. **69,** 1982, pp. 27–46). The incremental method used in this reference requires that each incremental ½ in of insulation provide the predetermined return on investment. The minimum thickness of installed insulation is used as a base for calculations. The incremental installed capital cost for each additional ½ in of insulation is determined. The energy saved for each increment is then determined. The value of this energy varies directly with the temperature level [e.g., steam at 538°C (1000°F) has a greater value than condensate at 100°C (212°F)]. The final increment selected for use is required either to provide a satisfactory return on investment or to have a suitable payback period.

**Recommended Thickness of Insulation** Indoor insulation thickness appears in Table 11-22, and outdoor thickness appears in Table 11-23. These selections were based upon calcium silicate insulation with a suitable aluminum jacket. However, the variation in thickness for fiberglass, cellular glass, and rock wool is minimal. Fiberglass is available for maximum temperatures of 260, 343, and 454°C (500, 650, and 850°F). Rock wool, cellular glass, and calcium silicate are used up to 649°C (1200°F).

**TABLE 11-22 Indoor Insulation Thickness, 80°F Still Ambient Air***

| | | Minimum pipe temperature, °F | | | | | | | |
|---|---|---|---|---|---|---|---|---|---|
| | | Energy cost, $/million Btu | | | | | | | |
| Pipe size, in | Insulation thickness, in | 1 | 2 | 3 | 4 | 5 | 6 | 7 | 8 |
| ¾ | 1½ | 950 | 600 | 550 | 400 | 350 | 300 | 250 | 250 |
| | 2 | | | | 1100 | 1000 | 900 | 800 | 750 |
| | 2½ | | | | 1750 | 1050 | 950 | 850 | 800 |
| | 3 | | | | | | | | 1200 |
| 1 | 1½ | 1200 | 800 | 600 | 500 | 450 | 400 | 350 | 300 |
| | 2 | | | 1200 | 1000 | 900 | 800 | 700 | 700 |
| | 2½ | | | | | 1200 | 1050 | 1000 | 900 |
| | 3 | | | | | | 1100 | 1150 | 950 |
| 1½ | 1½ | 1100 | 750 | 550 | 450 | 400 | 400 | 350 | 300 |
| | 2 | | | 1000 | 850 | 700 | 650 | 600 | 500 |
| | 2½ | | | | 1050 | 900 | 800 | 750 | 650 |
| | 3 | | | | | | 1150 | 1100 | 1000 |
| 2 | 1½ | 1050 | 700 | 500 | 450 | 400 | 350 | 300 | 300 |
| | 2 | | | 1050 | 850 | 750 | 700 | 600 | 600 |
| | 2½ | | | 1100 | 950 | 1000 | 750 | 700 | 650 |
| | 3 | | | | 1200 | 1050 | 950 | 850 | 800 |
| 3 | 1½ | 950 | 650 | 500 | 400 | 350 | 300 | 300 | 250 |
| | 2 | | 1100 | 900 | 700 | 600 | 550 | 500 | 450 |
| | 2½ | | | 1050 | 850 | 750 | 650 | 500 | 500 |
| | 3 | | | | 1050 | 950 | 800 | 750 | 700 |
| 4 | 1½ | 950 | 600 | 500 | 400 | 350 | 300 | 300 | 250 |
| | 2 | | 1100 | 850 | 700 | 600 | 550 | 500 | 450 |
| | 2½ | | | 1200 | 1000 | 850 | 750 | 700 | 650 |
| | 3 | | | | 1050 | 900 | 800 | 750 | 700 |
| | 3½ | | | | | | | 1150 | 1050 |
| 6 | 1½ | 600 | 350 | 300 | 250 | 250 | 200 | 200 | 200 |
| | 2 | | 1100 | 850 | 700 | 600 | 550 | 500 | 500 |
| | 2½ | | | 900 | 800 | 650 | 600 | 550 | 550 |
| | 3 | | | 1150 | 1000 | 850 | 750 | 700 | 600 |
| | 3½ | | | | | | 1100 | 1000 | 900 |
| | 4 | | | | | | | | 1200 |
| 8 | 2 | | 1000 | 800 | 650 | 550 | 500 | 450 | 400 |
| | 2½ | | 1050 | 850 | 700 | 600 | 550 | 500 | 450 |
| | 3 | | | | 1050 | 900 | 800 | 750 | 700 |
| | 3½ | | | | | 1200 | 1100 | 1000 | 900 |
| | 4 | | | | | | | 1150 | 1100 |
| 10 | 2 | | 1100 | 850 | 700 | 650 | 550 | 500 | 450 |
| | 2½ | | 1200 | 900 | 750 | 700 | 600 | 550 | 500 |
| | 3 | | | 1050 | 900 | 750 | 700 | 600 | 550 |
| | 3½ | | | | | 1200 | 1050 | 950 | 900 |
| | 4 | | | | | | | | 1200 |
| 12 | 2 | 1150 | 750 | 600 | 500 | 400 | 400 | 350 | 300 |
| | 2½ | | 1000 | 800 | 650 | 550 | 500 | 450 | 400 |
| | 3 | | | 1200 | 1000 | 900 | 800 | 700 | 650 |
| | 3½ | | | | | 1200 | 1100 | 1000 | 900 |
| | 4 | | | | | | 1150 | 1050 | 950 |
| | 4½ | | | | | | 1200 | 1100 | 1000 |
| 14 | 2 | 1050 | 650 | 550 | 450 | 400 | 350 | 300 | 300 |
| | 2½ | | 1000 | 800 | 650 | 550 | 500 | 450 | 400 |
| | 3 | | | 1100 | 950 | 800 | 700 | 650 | 600 |
| | 3½ | | | | | 1150 | 1000 | 950 | 850 |
| | 4 | | | | | 1200 | 1050 | 1000 | 900 |
| | 4½ | | | | | | 1200 | 1100 | 1000 |

**TABLE 11-22 Indoor Insulation Thickness, 80°F Still Ambient Air*** *(Concluded)*

| | | Minimum pipe temperature, °F | | | | | | | |
|---|---|---|---|---|---|---|---|---|---|
| | | Energy cost, $/million Btu | | | | | | | |
| Pipe size, in | Insulation thickness, in | 1 | 2 | 3 | 4 | 5 | 6 | 7 | 8 |
| 16 | 2 | 950 | 650 | 500 | 400 | 350 | 300 | 300 | 300 |
| | 2½ | | 1000 | 800 | 700 | 600 | 550 | 500 | 450 |
| | 3 | | 1200 | 950 | 800 | 700 | 600 | 550 | 500 |
| | 3½ | | | | | 1150 | 1050 | 950 | 850 |
| | 4 | | | | | 1200 | 1100 | 1000 | 900 |
| | 4½ | | | | | | 1150 | 1050 | 950 |
| 18 | 2 | 1000 | 650 | 500 | 400 | 350 | 350 | 300 | 300 |
| | 2½ | | 950 | 750 | 600 | 550 | 500 | 450 | 400 |
| | 3 | | 1150 | 900 | 750 | 650 | 550 | 500 | 500 |
| | 3½ | | | | | 1200 | 1100 | 1000 | 900 |
| | 4 | | | | | | 1150 | 1050 | 950 |
| | 4½ | | | | | | 1200 | 1100 | 1000 |
| 20 | 2 | 1050 | 700 | 550 | 450 | 400 | 350 | 350 | 300 |
| | 2½ | | 1000 | 800 | 600 | 550 | 500 | 450 | 400 |
| | 3 | | 1150 | 900 | 750 | 650 | 550 | 500 | 500 |
| | 3½ | | | | | | 1100 | 1000 | 950 |
| | 4 | | | | | | 1150 | 1050 | 1000 |
| | 4½ | | | | | | | 1200 | 1100 |
| 24 | 2 | 950 | 600 | 500 | 400 | 350 | 300 | 300 | 250 |
| | 2½ | | 1150 | 900 | 750 | 650 | 550 | 500 | 450 |
| | 3 | | | 1050 | 900 | 750 | 700 | 600 | 550 |
| | 3½ | | | | | 1100 | 1000 | 900 | 800 |
| | 4 | | | | | 1150 | 1050 | 950 | 850 |
| | 4½ | | | | | | 1150 | 1050 | 950 |

*Aluminum-jacketed calcium silicate insulation with an emissivity factor of 0.05. To convert inches to millimeters, multiply by 25.4, to convert dollars per 1 million British thermal units to dollars per 1 million kilojoules, multiply by 0.948, °C = 5/9 (°F − 32).

**TABLE 11-23 Outdoor Insulation Thickness, 7.5-mi/h Wind, 60°F Air***

| | | Minimum pipe temperature, °F | | | | | | | |
|---|---|---|---|---|---|---|---|---|---|
| | | Energy cost, $/million Btu | | | | | | | |
| Pipe size, in | Thickness, in | 1 | 2 | 3 | 4 | 5 | 6 | 7 | 8 |
| ¾ | 1 | 450 | 300 | 250 | 250 | 200 | 200 | 150 | 150 |
| | 1½ | 800 | 500 | 400 | 300 | 250 | 250 | 200 | 200 |
| | 2 | | | 1150 | 950 | 850 | 750 | 700 | 650 |
| | 2½ | | | 1100 | 1000 | 900 | 800 | 750 | 700 |
| 1 | 1 | 400 | 300 | 250 | 200 | 200 | 150 | 150 | 150 |
| | 1½ | 1000 | 650 | 500 | 400 | 350 | 300 | 300 | 250 |
| | 2 | | | 1100 | 900 | 800 | 700 | 600 | 600 |
| | 2½ | | | | 1200 | 1050 | 950 | 850 | 800 |
| | 3 | | | | | 1100 | 1000 | 900 | 850 |
| 1½ | 1 | 350 | 250 | 200 | 200 | 150 | 150 | 150 | 150 |
| | 1½ | 900 | 600 | 450 | 350 | 300 | 300 | 250 | 250 |
| | 2 | | 100 | 850 | 700 | 600 | 550 | 500 | 450 |
| | 2½ | | | 1150 | 950 | 800 | 750 | 700 | 600 |
| | 3 | | | | | 1200 | 1050 | 1000 | 900 |
| 2 | 1 | 350 | 250 | 200 | 150 | 150 | 150 | 150 | 150 |
| | 1½ | 900 | 550 | 450 | 400 | 300 | 300 | 250 | 250 |
| | 2 | | 1150 | 900 | 750 | 650 | 600 | 550 | 500 |
| | 2½ | | | 1000 | 850 | 750 | 650 | 600 | 550 |
| | 3 | | | | 1050 | 950 | 850 | 750 | 700 |
| 3 | 1 | 300 | 200 | 150 | 150 | 150 | 150 | 150 | 150 |
| | 1½ | 750 | 500 | 400 | 300 | 250 | 250 | 250 | 200 |
| | 2 | | 950 | 750 | 600 | 500 | 450 | 400 | 350 |
| | 2½ | | 1150 | 950 | 750 | 650 | 600 | 500 | 500 |
| | 3 | | | 1150 | 1000 | 850 | 750 | 650 | 600 |
| | 3½ | | | | | | | | 1150 |
| 4 | 1 | 250 | 200 | 150 | 150 | 150 | 150 | 150 | 150 |
| | 1½ | 750 | 500 | 350 | 300 | 250 | 250 | 200 | 200 |
| | 2 | | 950 | 750 | 600 | 500 | 450 | 400 | 350 |
| | 2½ | | | 1050 | 900 | 700 | 650 | 600 | 550 |
| | 3 | | | 1100 | 950 | 750 | 700 | 650 | 600 |
| | 3½ | | | | | | 1200 | 1100 | 1000 |

**TABLE 11-23 Outdoor Insulation Thickness, 7.5-mi/h Wind, 60°F Air*** *(Concluded)*

| | | Minimum pipe temperature, °F | | | | | | | |
|---|---|---|---|---|---|---|---|---|---|
| | | Energy cost, $/million Btu | | | | | | | |
| Pipe size, in | Thickness, in | 1 | 2 | 3 | 4 | 5 | 6 | 7 | 8 |
| 6 | 1 | 250 | 150 | 150 | 150 | 150 | 150 | 150 | 150 |
| | 1½ | 450 | 300 | 200 | 200 | 150 | 150 | 150 | 150 |
| | 2 | | 900 | 700 | 600 | 500 | 450 | 400 | 350 |
| | 2½ | | 1050 | 800 | 650 | 600 | 500 | 450 | 400 |
| | 3 | | | 1050 | 900 | 750 | 700 | 600 | 550 |
| | 3½ | | | | | 1150 | 1050 | 950 | 850 |
| | 4 | | | | | | | 1200 | 1150 |
| | 4½ | | | | | | | | 1200 |
| 8 | 1 | 250 | 200 | 150 | 150 | 150 | 150 | 150 | 150 |
| | 2 | | 850 | 650 | 550 | 450 | 400 | 350 | 350 |
| | 2½ | | 900 | 700 | 600 | 500 | 450 | 400 | 400 |
| | 3 | | | 1100 | 950 | 800 | 750 | 700 | 600 |
| | 3½ | | | | | 1150 | 1000 | 950 | 850 |
| | 4 | | | | | | | 1050 | 1000 |
| 10 | 2 | 200 | 150 | 150 | 150 | 150 | 150 | 150 | 150 |
| | 2½ | | 1000 | 800 | 650 | 550 | 500 | 450 | 400 |
| | 3 | | 1200 | 950 | 800 | 700 | 600 | 550 | 500 |
| | 3½ | | | | | 1100 | 1000 | 900 | 800 |
| | 4 | | | | | | | 1150 | 1050 |
| | 4½ | | | | | | | 1200 | 1100 |
| 12 | 1½ | 250 | 150 | 150 | 150 | 150 | 150 | 150 | 150 |
| | 2 | 950 | 600 | 500 | 400 | 350 | 300 | 250 | 250 |
| | 2½ | | 900 | 700 | 550 | 500 | 400 | 400 | 350 |
| | 3 | | | 1100 | 900 | 800 | 700 | 650 | 550 |
| | 3½ | | | | | 1100 | 1000 | 900 | 850 |
| | 4 | | | | | 1150 | 1050 | 950 | 900 |
| | 4½ | | | | | 1200 | 1100 | 1000 | 950 |
| 14 | 1½ | 250 | 150 | 150 | 150 | 150 | 150 | 150 | 150 |
| | 2 | 850 | 550 | 400 | 350 | 300 | 250 | 250 | 250 |
| | 2½ | | 850 | 650 | 550 | 500 | 400 | 400 | 400 |
| | 3 | | | 1000 | 850 | 700 | 650 | 550 | 500 |
| | 3½ | | | | 1200 | 1000 | 950 | 850 | 800 |
| | 4 | | | | | 1050 | 1000 | 900 | 850 |
| | 4½ | | | | | | 1100 | 1000 | 950 |
| | 1½ | 250 | 150 | 150 | 150 | 150 | 150 | 150 | 150 |
| 16 | 2 | 800 | 500 | 350 | 300 | 300 | 250 | 250 | 200 |
| | 2½ | | 900 | 700 | 550 | 500 | 450 | 400 | 350 |
| | 3 | | 1000 | 850 | 700 | 600 | 500 | 450 | 400 |
| | 3½ | | | | 1200 | 1000 | 950 | 850 | 800 |
| | 4 | | | | | 1100 | 1000 | 900 | 850 |
| | 4½ | | | | | 1150 | 1000 | 950 | 900 |
| | 1½ | 250 | 150 | 150 | 150 | 150 | 150 | 150 | 150 |
| 18 | 2 | 850 | 550 | 400 | 350 | 300 | 250 | 250 | 200 |
| | 2½ | | 800 | 650 | 500 | 450 | 400 | 350 | 350 |
| | 3 | | 1000 | 800 | 650 | 550 | 500 | 450 | 400 |
| | 3½ | | | | | 1100 | 1000 | 900 | 850 |
| | 1½ | 150 | 150 | 150 | 150 | 150 | 150 | 150 | 150 |
| 20 | 2 | 900 | 550 | 450 | 350 | 300 | 300 | 250 | 250 |
| | 2½ | | 850 | 650 | 550 | 450 | 400 | 350 | 350 |
| | 3 | | 1000 | 800 | 650 | 550 | 500 | 450 | 400 |
| | 3½ | | | | | 1150 | 1050 | 950 | 900 |
| | 4 | | | | | 1200 | 1100 | 1000 | 950 |
| | 4½ | | | | | | 1200 | 1100 | 1050 |
| | 1½ | 150 | 150 | 150 | 150 | 150 | 150 | 150 | 150 |
| 24 | 2 | 800 | 500 | 400 | 300 | 250 | 250 | 200 | 200 |
| | 2½ | | 950 | 750 | 650 | 550 | 500 | 450 | 400 |
| | 3 | | 1150 | 950 | 750 | 650 | 600 | 550 | 500 |
| | 3½ | | | | 1150 | 1000 | 900 | 800 | 750 |
| | 4 | | | | 1200 | 1050 | 950 | 850 | 800 |
| | 4½ | | | | | | 1050 | 950 | 850 |
| | 5 | | | | | | | | |

*Aluminum-jacketed calcium silicate insulation with an emissivity factor of 0.05. To convert inches to millimeters, multiply by 25.4; to convert miles per hour to kilometers per hour, multiply by 1.609; and to convert dollars per 1 million British thermal units to dollars per 1 million kilojoules, multiply by 0.948; °C = 5/9 (°F − 32).

The tables were based upon the cost of energy at the end of the first year, a 10 percent inflation rate on energy costs, a 15 percent interest cost, and a present-worth pretax profit of 40 percent per annum on the last increment of insulation thickness. Dual-layer insulation was used for 3½-in and greater thicknesses. The tables and a full explanation of their derivation appear in a paper by F. L. Rubin (op. cit.). Alternatively, the selected thicknesses have a payback period on the last nominal ½-in increment of 1.44 years as presented in a later paper by Rubin ["Can You Justify More Piping Insulation?" *Hydrocarbon Process.,* 152–155 (July 1982)].

***Example 1*** For 24-in pipe at 371°C (700°F) with an energy cost of $4/million Btu, select 2-in thickness for indoor and 2½-in for outdoor locations. [A 2½-in thickness would be chosen at 399°C (750°F) indoors and 3½-in outdoors.]

***Example 2*** For 16-in pipe at 343°C (650°F) with energy valued at $5/million Btu, select 2½-in insulation indoors [use 3-in thickness at 371°C (700°F)]. Outdoors choose 3-in insulation [use 3½-in dual-layer insulation at 538°C (1000°F)].

***Example 3*** For 12-in pipe at 593°C (1100°F) with an energy cost of $6/million Btu, select 3½-in thickness for an indoor installation and 4½-in thickness for an outdoor installation.

## INSTALLATION PRACTICE

**Pipe** Depending on diameter, pipe is insulated with cylindrical half, third, or quarter sections or with flat segmental insulation. Fittings and valves are insulated with preformed insulation covers or with individual pieces cut from sectional straight pipe insulation.

**Method of Securing** Insulation with factory-applied jacketing may be secured with adhesive on the overlap, staples, tape, or wire, depending on the type of jacket and the outside diameter. Insulation which has a separate jacket is wired or banded in place before the jacket (finish) is applied.

**Double Layer** Pipe expansion is a significant factor at temperatures above 600°F (316°C). Above this temperature, insulation should be applied in a double layer with all joints staggered to prevent excessive heat loss and high surface temperature at joints opened by pipe expansion. This procedure also minimizes thermal stresses in the insulation.

**Finish** Covering for cylindrical surfaces ranges from asphalt-saturated or saturated and coated organic and asbestos paper, through laminates of such papers and plastic films or aluminum foil, to medium-gauge aluminum, galvanized steel, or stainless steel. Fittings and irregular surfaces may be covered with fabric-reinforced mastics or preformed metal or plastic covers. Finish selection depends on function and location. Vapor-barrier finishes may be in sheet form or a mastic, which may or may not require reinforcing, depending on the method of application, and additional protection may be required to prevent mechanical abuse and/or provide fire resistance. Criteria for selecting other finishes should include protection of insulation against water entry, mechanical abuse, or chemical attack. Appearance, life-cycle cost, and fire resistance may also be determining factors. Finish may be secured with tape, adhesive, bands, or screws. Fasteners which will penetrate vapor-retarder finishes should not be used.

**Tanks, Vessels, and Equipment** Flat, curved, and irregular surfaces such as tanks, vessels, boilers, and breechings are normally insulated with flat blocks, beveled lags, curved segments, blankets, or spray-applied insulation. Since no general procedure can apply to all materials and conditions, it is important that manufacturers' specifications and instructions be followed for specific insulation applications.

**Method of Securing** On small-diameter cylindrical vessels, the insulation may be secured by banding around the circumference. On larger cylindrical vessels, banding may be supplemented with angle-iron ledges to support the insulation and prevent slipping. On large flat and cylindrical surfaces, banding or wiring may be supplemented with various types of welded studs or pins. Breather springs may be required with bands to accommodate expansion and contraction.

**Finish** The materials are the same as for pipe and should satisfy the same criteria. Breather springs may be required with bands.

ADDITIONAL REFERENCES: *ASHRAE Handbook and Product Directory: Fundamentals,* American Society of Heating, Refrigerating and Air Conditioning Engineers, Atlanta, 1981. Turner and Malloy, *Handbook of Thermal Insulation Design Economics for Pipes and Equipment,* Krieger, New York, 1980. Turner and Malloy, *Thermal Insulation Handbook,* McGraw-Hill, New York, 1981.

# AIR CONDITIONING

## INTRODUCTION

Air Conditioning is the process of treating air so as to control simultaneously its temperature, humidity, cleanliness, and distribution to meet the requirements of the conditioned space. The portions relating only to temperature and humidity control will be discussed here. For detailed discussions of air cleanliness and distribution, refer, for example, to the current edition of the *HVAC Applications* volume of the *A.S.H.R.A.E. Handbooks* (ASHRAE, 1791 Tullie Circle, N.E., Atlanta, Ga.). Applications of air conditioning include the promotion of human comfort and the maintenance of proper conditions for the manufacture, processing, and preserving of material and equipment. Also, in industrial environments where, for economical or other reasons, conditions cannot be made entirely comfortable, air conditioning may be used for maintaining the efficiency and health of workers at safe tolerance limits.

## COMFORT AIR CONDITIONING

Comfort is influenced by temperature, humidity, air velocity, radiant heat, clothing, and work intensity. Psychological factors may also influence comfort, but their discussion is beyond the scope of this handbook. The reader is referred to Chap. 42 of the *HVAC Applications* volume of the *A.S.H.R.A.E. Handbooks* for a full discussion of the control of noise, which must also be considered in air-conditioning design. Figure 5 in Chap. 8 of the *HVAC Fundamentals* volume of the *A.S.H.R.A.E. Handbooks* relates the variables of ambient temperature, dew point temperature (or humidity ratio) to comfort under clothing and activity conditions typical for office space occupancy. It also shows boundary values for ET°, the effective temperature index. This index combines temperature and moisture conditions into a single index representing the temperature of an environment at 50 percent relative humidity resulting in the same heat transfer from the skin as for the actual case. Hence, the ET° for 50 percent relative humidity is equal in value to the ambient dry-bulb temperature.

## INDUSTRIAL AIR CONDITIONING

Industrial buildings have to be designed according to their intended use. For instance, the manufacture of hygroscopic materials (paper, textiles, foods, etc.) will require relatively tight controls of relative humidity. On the other hand, the storage of furs will demand relatively low temperatures, while the ambient in a facility manufacturing refractories might be acceptable at notably higher temperatures. Chapter 12 of the *HVAC Applications* volume of the *A.S.H.R.A.E. Handbooks* provides extensive tables of suggested temperatures and humidities for industrial air conditioning.

## VENTILATION

In the design of comfort air-conditioning systems, odors arising from occupants, cooking, or other sources must be controlled. This is accomplished by introducing fresh air or purified recirculated air in sufficient quantities to reduce odor concentrations to an acceptable level by dilution. Recommended fresh-air requirements for different types of buildings are called for in A.S.H.R.A.E. standard 62-1989 "Ventilation for Acceptable Indoor Air Quality." These values range in the order of 15 to 30 cfm per person, according to application.

In industrial air-conditioning systems, harmful environmental gases, vapors, dusts, and fumes are often encountered. These contaminants can be controlled by exhaust systems at the source, by dilution ventilation, or by a combination of the two methods. When exhaust systems are used, it is necessary to introduce sufficient fresh air into the air-conditioned area to make up for that exhausted. Generally, low exhaust systems are used where the contaminant sources are concentrated and/or where the contaminant may be highly toxic. Where the contaminant comes from widely dispersed points, however, dilution ventilation is usually employed. Combinations of the two systems sometimes provide the least expensive installation. dilution ventilation is not appropriate for cases where large volumes of contaminant are released and cases where the employees must work near the contaminant source. The selection of dilution ventilation for cases with potential fire or smoke should be accompanied by careful study. Details for design of dilution ventilation systems are given in Chap. 25 of the *HVAC Applications* volume of the *A.S.H.R.A.E. Handbooks*. Chapter 27 of the same volume discusses industrial exhaust systems. Exhaust stacks should be high enough to adequately dispense the contaminated air and to prevent recirculation into fresh air intakes (Chap. 14 of the 1993 *HVAC Fundamentals* volume of the *A.S.H.R.A.E. Handbooks*). Depending on the contaminant and air pollution legislation, it may be necessary to reduce the contaminant emission rate by such methods as filtering, scrubbing, catalytic oxidation, or incineration.

## AIR-CONDITIONING EQUIPMENT

Basically, an air-conditioning system consists of a fan unit which forces a mixture of fresh outdoor air and room air through a series of devices which act upon the air to clean it, to increase or decrease its temperature, and to increase or decrease its water-vapor content or humidity. In general, air conditioning equipment can be classified into two broad types: central (sometimes called field erected) and unitary (or packaged).

## CENTRAL SYSTEMS

Figure 11-69 describes a typical central system. Either water or direct-expansion refrigerant coils or air washers may be used for cooling. Steam or hot-water coils are available for heating. Humidification may be provided by target-type water nozzles, pan humidifiers, air washers,

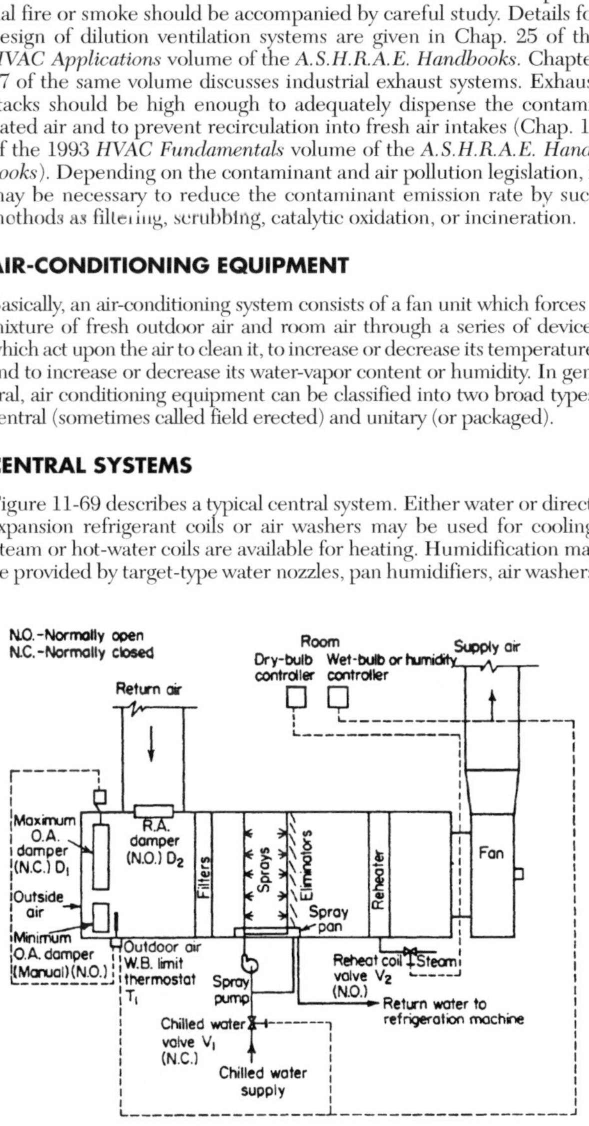


**FIG. 11-69** Typical central-station air-conditioning unit and control system. On a rising room wet-bulb temperature, the wet-bulb branch-line air pressure increases through the reverse-acting outdoor-air wet-bulb temperature-limit thermostat $T_1$ to open gradually the maximum outdoor-air damper $D_1$ and simultaneously closr return-air damper $D_2$, then gradually open chilled-water valve $V_1$. On a rising room dry-bulb temperature, the dry-bulb branch-line air pressure gradually increases to close reheat steam valve $V_2$. When outdoor wet-bulb temperature exceeds the set point of the outdoor-air wet-bulb-limit thermostat $T_1$, which is set at the return-air wet-bulb temperature, this thermostat decreases branch-line pressure to close gradually maximum outdoor damper $D_1$ and simultaneously open return-air damper $D_2$. The reverse sequences are followed during the heating season.

or sprayed coils. Air cleaning is usually provided by cleanable or throw-away filters. Central-station air-conditioning units in capacities up to about 50,000 cu ft/min are available in prefabricated units.

The principle types of refrigeration equipment used in large central systems are: Reciprocating (up to 300 hp); helical rotary (up to 750 tons); absorption (up to 2000 tons); and centrifugal (up to 10,000 tons). The drives for the reciprocating, rotary, and centrifugal compressors may be electric motors, gas or steam turbines, or gas or diesel engines. The heat rejected from the condensors usually calls for cooling towers or air-cooled condensors; in some cases evaporative cooling might be practical.

## UNITARY REFRIGERANT-BASED AIR-CONDITIONING SYSTEMS

These systems include window-mounted air conditioners and heat pumps, outdoor unitary equipment, indoor unitary equipment, unitary self-contained systems, and commercial self-contained systems. These are described in detail in the *HVAC Systems and Equipment* volume of the *A.S.H.R.A.E. Handbooks*. A detailed analysis of the proposed installation is usually necessary to select the air conditioning equipment which is best in overall performance. Each type of air conditioner has its own particular advantages and disadvantages. Important factors to be considered in the selection of air conditioning equipment are degree of temperature and humidity control required, investment, owning, and operating costs, and space requirements. Another important factor is the building itself, that is, whether it is new or existing construction. For example, for existing buildings where it may be inadvisable to install air-supply ducts, the self-contained or unit-type air conditioner may offer the greatest advantages in reduced installation costs. For large industrial processes where close temperature and humidity control are required, a central station system is usually employed.

## LOAD CALCULATION

First step in the solution of an air-conditioning problem is to determine the proper design temperature conditions. Since both outdoor and indoor temperatures greatly influence the size of the equipment, the designer must exercise good judgment in selecting the proper conditions for his/her particular case. Table 11-24 lists winter and summer outdoor temperature conditions in common use for comfort applications for various United States cities. For critical-process air conditioning, it may be desirable to use a different set of outdoor temperature conditions. However, it is seldom good practice to design for the extreme maximum or minimum outside conditions. (See the 1993 *HVAC Fundamentals* volume of the *A.S.H.R.A.E. Handbooks*).

After the proper inside and summer outside temperature conditions for comfort and temperature conditions for process air conditioning have been selected, the next step is to calculate the space cooling load, which is made up of sensible heat and latent heat loads. The sensible heat load consists of (1) transmission through walls, roofs, floors, ceilings, and window glass, (2) solar and sky radiation, (3) heat gains from infiltration of outside air, (4) heat gains from people, lights, appliances, and power equipment (including the supply-air fan motor), and (5) heat to be removed from materials or products brought in at higher than room temperature. The latent heat load includes loads due to moisture (1) given off from people, appliances, and products and (2) from infiltration of outside air. The space total heat load is the sum of the sensible heat load and latent heat load of the space. The total refrigeration load consists of the total space load plus the sensible and latent heat loads from the outside air introduced at the conditioning unit.

The procedure for load calculation in nonresidential buildings should account for thermal mass (storage) effects as well as occupancy and other uses affecting the load. The load can in turn be strongly dependent on the nature of the building utilization; as an example, lightning might be a major component in the thermal load for a high-rise office building causing a need for cooling even in winter days. There are various approaches to load calculation, some requiring elaborate computer models. Chapter 26 of the 1993 *HVAC Fundamentals* volume of the *A.S.H.R.A.E. Handbooks* presents a step-by-step outline of the current methods in practice for load calculation.

**TABLE 11-24 Outdoor Design Temperatures***

| City-state | Winter Dry bulb, °F. | Summer Dry bulb, °F. | Summer Wet bulb, °F. | City-state | Winter Dry bulb, °F. | Summer Dry bulb, °F. | Summer Wet bulb, °F. |
|---|---|---|---|---|---|---|---|
| Akron, Ohio | −5 | 95 | 75 | Milwaukee, Wis. | −15 | 95 | 75 |
| Albany, N.Y. | −10 | 93 | 75 | Minneapolis, Minn. | −20 | 95 | 75 |
| Albuquerque, N.M. | 0 | 95 | 70 | Nashville, Tenn. | 0 | 95 | 78 |
| Atlanta, Ga. | 10 | 95 | 76 | New Haven, Conn. | 0 | 95 | 75 |
| Baltimore, Md. | 0 | 95 | 78 | New Orleans, La. | 20 | 95 | 80 |
| Billings, Mont. | −25 | 90 | 66 | New York, N.Y. | 0 | 95 | 75 |
| Birmingham, Ala. | 10 | 95 | 78 | Newark, N.J. | 0 | 95 | 75 |
| Bloomfield, N.J. | 0 | 95 | 75 | Norfolk, Va. | 15 | 95 | 78 |
| Boise, Idaho | −10 | 95 | 65 | Oakland, Calif. | 30 | 85 | 65 |
| Boston, Mass. | 0 | 92 | 75 | Oklahoma City, Okla. | 0 | 101 | 77 |
| Bridgeport, Conn. | 0 | 95 | 75 | Omaha, Nebr. | −10 | 95 | 78 |
| Buffalo, N.Y. | −5 | 93 | 73 | Peoria, Ill. | −10 | 96 | 76 |
| Charleston, S.C. | 15 | 95 | 78 | Philadelphia, Pa. | 0 | 95 | 78 |
| Chattanooga, Tenn. | 10 | 95 | 76 | Phoenix, Ariz. | 0 | 105 | 76 |
| Chicago, Ill. | −10 | 95 | 75 | Pittsburgh, Pa. | −5 | 95 | 75 |
| Cincinnati, Ohio | 0 | 95 | 78 | Portland, Me. | −5 | 90 | 73 |
| Cleveland, Ohio | 0 | 95 | 75 | Portland, Ore. | 10 | 90 | 68 |
| Columbus, Ohio | −10 | 95 | 76 | Providence, R.I. | 0 | 93 | 75 |
| Dallas, Tex. | 0 | 100 | 78 | Reno, Nev. | −5 | 95 | 65 |
| Dayton, Ohio | 0 | 95 | 78 | Richmond, Va. | 15 | 95 | 78 |
| Denver, Colo. | −10 | 95 | 64 | Roanoke, Va. | 0 | 95 | 76 |
| Des Moines, Iowa | −15 | 95 | 78 | Rochester, N.Y. | −5 | 95 | 75 |
| Detroit, Mich. | −10 | 95 | 75 | St. Louis, Mo. | 0 | 95 | 78 |
| Duluth, Minn. | −25 | 93 | 73 | St. Paul, Minn. | −20 | 95 | 75 |
| East Orange, N.J. | 0 | 95 | 75 | Salt Lake City, Utah | −10 | 95 | 65 |
| El Paso, Tex. | 10 | 100 | 69 | San Antonio, Tex. | 20 | 100 | 78 |
| Erie, Pa. | −5 | 93 | 75 | San Francisco, Calif. | 35 | 85 | 65 |
| Fitchburg, Mass. | −10 | 93 | 75 | Schenectady, N.Y. | −10 | 93 | 75 |
| Flint, Mich. | −10 | 95 | 75 | Scranton, Pa. | −5 | 95 | 75 |
| Fort Wayne, Ind. | −10 | 95 | 75 | Seattle, Wash. | 15 | 85 | 65 |
| Fort Worth, Tex. | 10 | 100 | 78 | Shreveport, La. | 20 | 100 | 78 |
| Grand Rapids, Mich. | −10 | 95 | 75 | Sioux City, Iowa | −20 | 95 | 78 |
| Hartford, Conn. | 0 | 93 | 75 | Spokane, Wash. | −15 | 93 | 65 |
| Houston, Tex. | 20 | 95 | 80 | Springfield, Mass. | −10 | 93 | 75 |
| Indianapolis, Ind. | −10 | 95 | 76 | Syracuse, N.Y. | −10 | 93 | 75 |
| Jacksonville, Fla. | 25 | 95 | 78 | Tampa, Fla. | 30 | 95 | 78 |
| Jersey City, N.J. | 0 | 95 | 75 | Toledo, Ohio | −10 | 95 | 75 |
| Kansas City, Mo. | −10 | 100 | 76 | Tucson, Ariz. | 25 | 105 | 72 |
| Lincoln, Nebr. | −10 | 95 | 78 | Tulsa, Okla. | 0 | 101 | 77 |
| Little Rock, Ark. | 5 | 95 | 78 | Washington, D.C. | 0 | 95 | 78 |
| Long Beach, Calif. | 35 | 90 | 70 | Wichita, Kans. | −10 | 100 | 75 |
| Los Angeles, Calif. | 35 | 90 | 70 | Wilmington, Del. | 0 | 95 | 78 |
| Louisville, Ky. | 0 | 95 | 78 | Worcester, Mass. | 0 | 93 | 75 |
| Memphis, Tenn. | 0 | 95 | 78 | Youngstown, Ohio | −5 | 95 | 75 |
| Miami, Fla. | 35 | 91 | 79 | | | | |

*Carrier, Cherne, Grant, and Roberts, *Modern Air Conditioning, Heating, and Ventilating*, 3d ed., p. 531, Pitman, New York, 1959.

# REFRIGERATION

## INTRODUCTION

Refrigeration is a process where heat is transferred from a lower- to a higher-temperature level by doing work on a system. In some systems heat transfer is used to provide the energy to drive the refrigeration cycle. All refrigeration systems are heat pumps ("pumps energy from a lower to a higher potential"). The term heat pump is mostly used to describe refrigeration system applications where heat rejected to the condenser is of primary interest.

There are many means to obtain refrigerating effect, but here three will be discussed: mechanical vapor refrigeration cycles, absorption and steam jet cycles due to their significance for industry.

**Basic Principles** Since refrigeration is the practical application of the thermodynamics, comprehending the basic principles of thermodynamics is crucial for full understanding of refrigeration. Section 4 includes a through approach to the theory of thermodynamics. Since our goal is to understand refrigeration processes, cycles are of the crucial interest.

The Carnot refrigeration cycle is reversible and consists of adiabatic (isentropic due to reversible character) compression (1-2), isothermal rejection of heat (2-3), adiabatic expansion (3-4) and isothermal addition of heat (4-1). The temperature-entropy diagram is shown in Fig. 11-70. The Carnot cycle is an unattainable ideal which serves as a standard of comparison and it provides a convenient guide to the temperatures that should be maintained to achieve maximum effectiveness.

The measure of the system performance is *coefficient of performance* (COP). For refrigeration applications COP is the ratio of heat removed from the low-temperature level ($Q_{low}$) to the energy input ($W$):

$$COP_R = \frac{Q_{low}}{W} \tag{11-84}$$

For the heat pump (HP) operation, heat rejected at the high temperature ($Q_{high}$) is the objective, thus:

$$COP_{HP} = \frac{Q_{high}}{W} = \frac{Q + W}{W} = COP_R + 1 \tag{11-85}$$

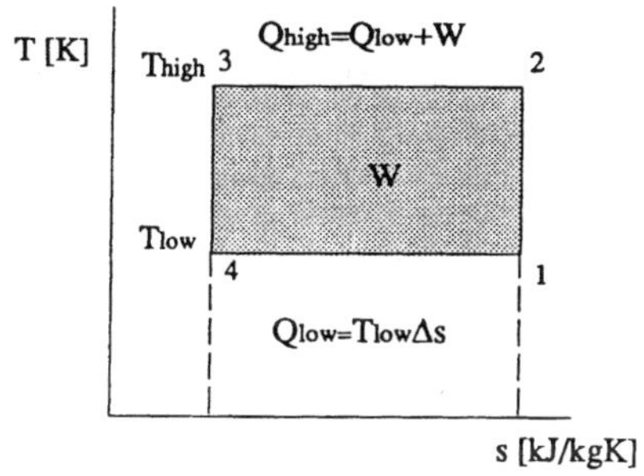


**FIG. 11-70** Temperature-entropy diagram of the Carnot cycle.

For a Carnot cycle (where $\Delta Q = T\Delta s$), the COP for the refrigeration application becomes (note than $T$ is absolute temperature [K]):

$$COP_R = \frac{T_{low}}{T_{high} - T_{low}} \tag{11-86}$$

and for heat pump application:

$$COP_{HP} = \frac{T_{high}}{T_{high} - T_{low}} \tag{11-87}$$

The COP in real refrigeration cycles is always less than for the ideal (Carnot) cycle and there is constant effort to achieve this ideal value.

**Basic Refrigeration Methods** Three basic methods of refrigeration (mentioned above) use similar processes for obtaining refrigeration effect: evaporation in the evaporator, condensation in the condenser where heat is rejected to the environment, and expansion in a flow restrictor. The main difference is in the way compression is being done (Fig. 11-71): using mechanical work (in compressor), thermal energy (for absorption and desorption), or pressure difference (in ejector).

In the next figure (Fig. 11-72) basic refrigeration systems are displayed more detailed. More elaborated approach is presented in the text.

## MECHANICAL REFRIGERATION (VAPOR-COMPRESSION SYSTEMS)

**Vapor-Compression Cycles** The most widely used refrigeration principle is vapor compression. Isothermal processes are realized through isobaric evaporation and condensation in the tubes. Standard vapor compression refrigeration cycle (counterclockwise Rankine cycle) is marked in Fig. 11-72*a*) by 1, 2, 3, 4.

Work that could be obtained in turbine is small, and iturbine is substituted for an expansion valve. For the reasons of proper compressor function, wet compression is substituted for an compression of dry vapor.

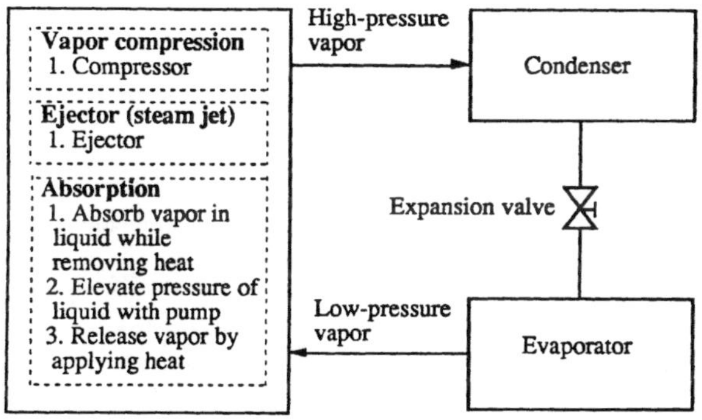


**FIG. 11-71** Methods of transforming low-pressure vapor into high-pressure vapor in refrigeration systems (Stoecker, Refrigeration and Air Conditioning).

Although the *T-s* diagram is very useful for thermodynamic analysis, the pressure enthalpy diagram is used much more in refrigeration practice due to the fact that both evaporation and condensation are isobaric processes so that heat exchanged is equal to enthalpy difference $\Delta Q = \Delta h$. For the ideal, isentropic compression, the work could be also presented as enthalpy difference $\Delta W = \Delta h$. The vapor compression cycle (Rankine) is presented in Fig. 11-73 in *p-h* coordinates.

Figure 11-74 presents actual versus standard vapor-compression cycle. In reality, flow through the condenser and evaporator must be accompanied by pressure drop. There is always some subcooling in the condenser and superheating of the vapor entering the compressor-suction line, both due to continuing process in the heat exchangers and the influence of the environment. Subcooling and superheating are usually desirable to ensure only liquid enters the expansion device. Superheating is recommended as a precaution against droplets of liquid being carried over into the compressor.

There are many ways to increase cycle efficiency (COP). Some of them are better suited to one, but not for the other refrigerant. Sometimes, for the same refrigerant, the impact on COP could be different for various temperatures. One typical example is the use of a liquid-to-suction heat exchanger (Fig. 11-75).

The suction vapor coming from the evaporator could be used to subcool the liquid from the condenser. Graphic interpretation in *T-s* diagram for such a process is shown in Fig. 11-76. The result of the use of suction line heat exchanger is to increase the refrigeration effect $\Delta Q$ and to increase the work by $\Delta W$. The change in COP is then:

$$\Delta COP = COP' - COP = \frac{Q + \Delta Q}{(P + \Delta P) - Q/P} \tag{11-88}$$

When dry, or superheated, vapor is used to subcool the liquid, the COP in R12 systems will increase, and decrease the COP in $NH_3$ systems. For R22 systems it could have both effects, depending on the operating regime. Generally, this measure is advantageous (COP is improved) for fluids with high, specific heat of liquid (less-inclined saturated-liquid line on the *p-h* diagram), small heat of evaporation $h_{fg}$, when vapor-specific heat is low (isobars in superheated regions are steep), and when the difference between evaporation and condensation temperature is high. Measures to increase COP should be studied for every refrigerant. Sometimes the purpose of the suction-line heat exchanger is not only to improve the COP, but to ensure that only the vapor reaches the compressor, particularly in the case of a malfunctioning expansion valve.

The system shown in Fig. 11-75 is direct expansion where dry or slightly superheated vapor leaves the evaporator. Such systems are predominantly used in small applications because of their simplicity and light weight. For the systems where efficiency is crucial (large industrial systems), recirculating systems (Fig. 11-77) are more appropriate.

Ammonia refrigeration plants are almost exclusively built as recirculating systems. The main advantage of recirculating versus direct expansion systems is better utilization of evaporator surface area. The diagram showing influence of quality on the local heat-transfer coefficients is shown in figure 11-90. It is clear that heat-transfer characteristics will be better if the outlet quality is lower than 1. Circulation could be achieved either by pumping (mechanical or gas) or using gravity (thermosyphon effect: density of pure liquid at the evaporator entrance is higher than density of the vapor-liquid mixture leaving the evaporator). The circulation ratio (ratio of actual mass flow rate to the evaporated mass flow rate) is higher than 1 and up to 5. Higher values are not recommended due to a small increase in heat-transfer rate for a significant increase in pumping costs.

**Multistage Systems** When the evaporation and condensing pressure (or temperature) difference is large, it is prudent to separate compression in two stages. The use of multistage systems opens up the opportunity to use flash-gas removal and intercooling as measures to improve performance of the system. One typical two-stage system with two evaporating temperatures and both flash-gas removal and intercooling is shown in figure 11-78. The purpose of the flash-tank intercooler is to: (1) separate vapor created in the expansion process,

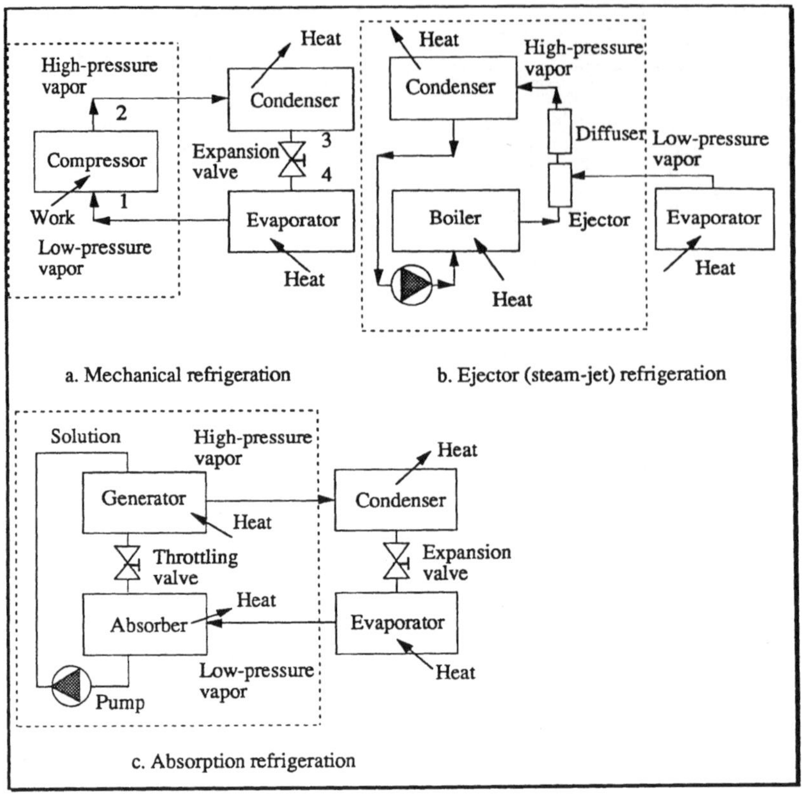


**FIG. 11-72** Basic refrigeration systems.

(2) cool superheated vapor from compressor discharge, and (3) to eventually separate existing droplets at the exit of the medium-temperature evaporator. The first measure will decrease the size of the low-stage compressor because it will not wastefully compress the portion of the flow which cannot perform the refrigeration and second will decrease the size of the high-stage compressor due to lowering the specific volume of the vapor from the low-stage compressor discharge, positively affecting operating temperatures of the high-stage compressor due to cooling effect.

If the refrigerating requirement at a low-evaporating temperature is $Q_l$ and at the medium level is $Q_m$, then mass flow rates ($m_l$ and $m_m$

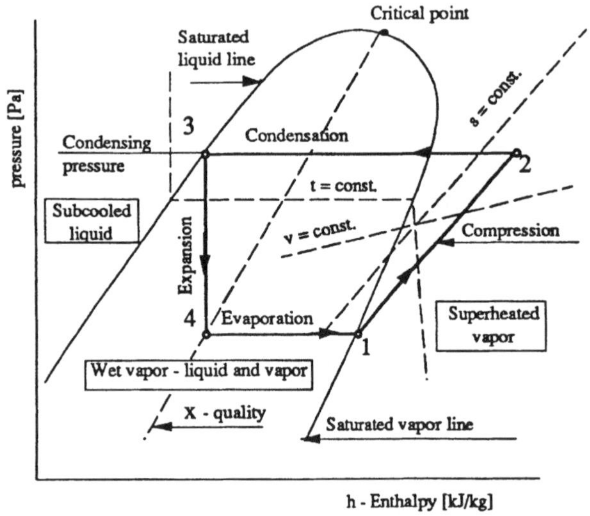


**FIG. 11-73** *p-h* diagram for vapor-compression cycle.

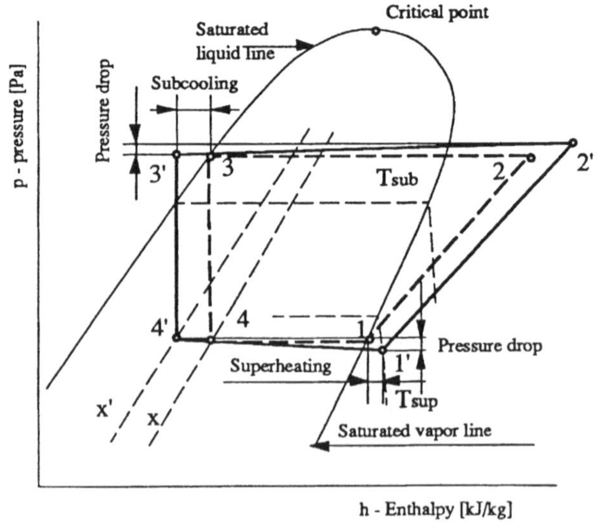


**FIG. 11-74** Actual vapor-compression cycle compared with standard cycle.

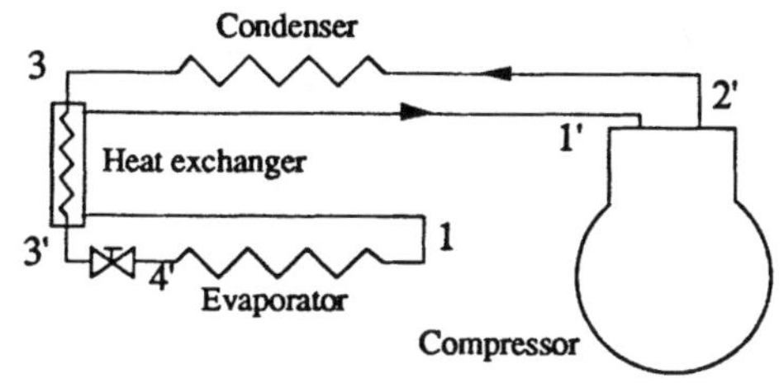


**FIG. 11-75** Refrigeration system with a heat exchanger to subcool the liquid from the condenser.

respectively) needed are:

$$m_l = \frac{Q_l}{h_1 - h_8} = \frac{Q_l}{h_1 - h_7} \tag{11-89}$$

$$m_m = \frac{Q_m}{h_3 - h_6} \tag{11-90}$$

The mass flow rate at the flash-tank inlet $m_i$ consists of three components ($m_i = m_l + m_{sup} + m_{flash}$):

$m_l$ = liquid at $p_m$ feeding low temperature evaporator,

$m_{sup}$ = liquid at $p_m$ to evaporate in flash tank to cool superheated discharge,

$m_{flash}$ = flashed refrigerant, used to cool remaining liquid.

Vapor component is:

$$m_{flash} = x_m * m_i \tag{11-91}$$

and liquid component is:

$$(1 - x_m) * m_i = m_l + m_{sup} \tag{11-92}$$

Liquid part of flow to cool superheated compressor discharge is determined by:

$$m_{sup} = \frac{Q_l}{h_1 - h_8} * \frac{h_2 - h_3}{h_3 - h_7} = m_l * \frac{h_2 - h_3}{h_{fgm}} \tag{11-93}$$

Since the quality $x_m$ is:

$$x_m = \frac{h_6 - h_7}{h_3 - h_7} \tag{11-94}$$

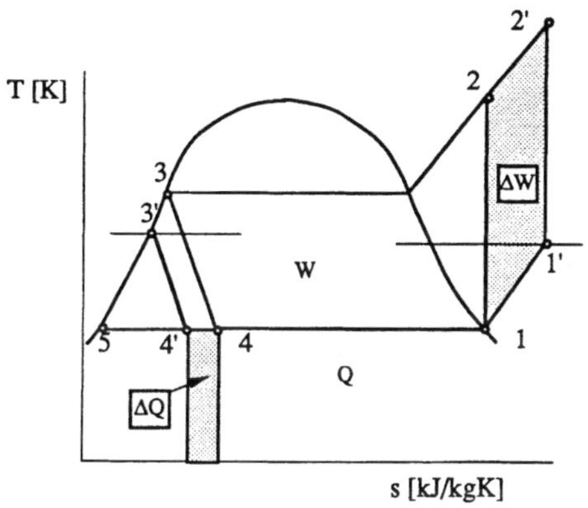


**FIG. 11-76** Refrigeration system with a heat exchanger to subcool the liquid from the condenser.

mass flow rate through condenser and high-stage compressor $m_h$ is finally:

$$m_h = m_m + m_i \tag{11-95}$$

The optimum intermediate pressure for the two-stage refrigeration cycles is determined as the geometric mean between evaporation

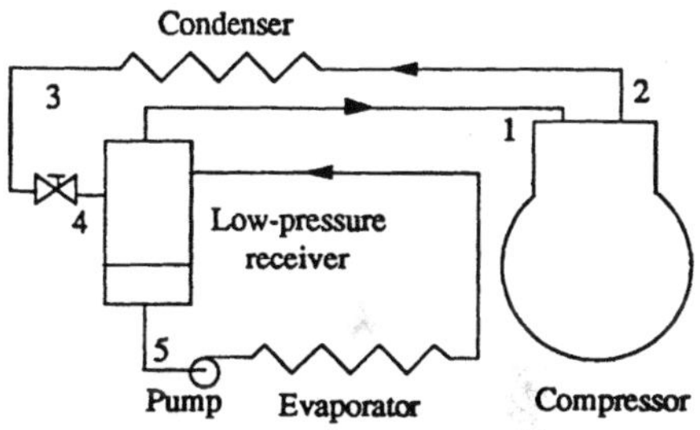


**FIG. 11-77** Recirculation system.

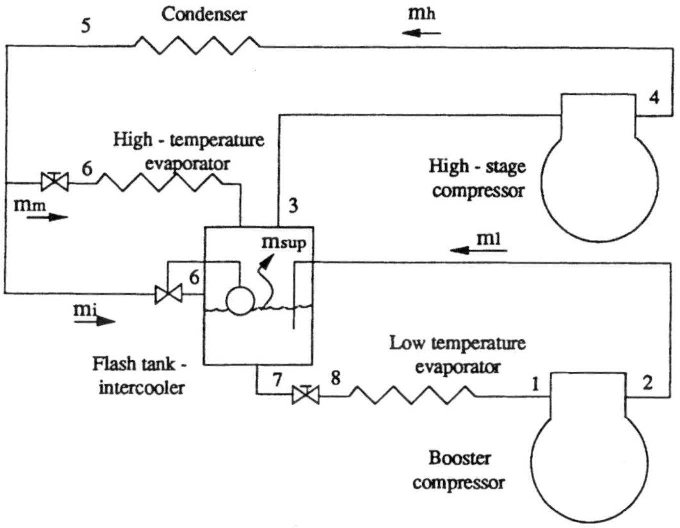


**FIG. 11-78** Typical two-stage system with two evaporating temperatures, flash-gas removal, and intercooling.

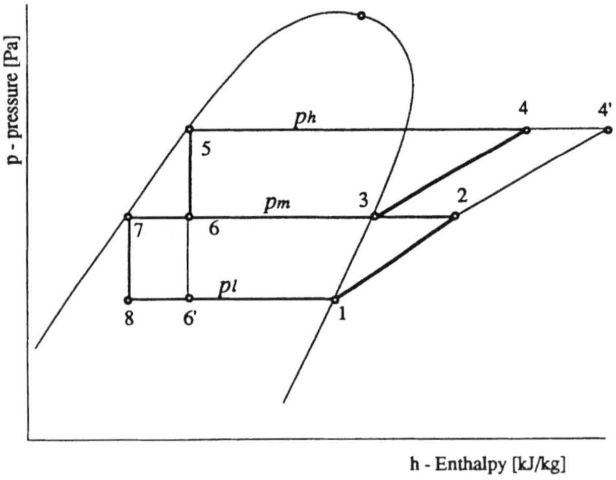


**FIG. 11-79** Pressure enthalpy diagram for typical two-stage system with two evaporating temperatures, flash-gas removal, and intercooling.

pressure ($p_l$) and condensing pressure ($p_h$, Fig. 11-79):

$$p_m = (\text{sqrt})\left(\frac{p_h}{p_l}\right)$$

based on equal pressure ratios for low- and high-stage compressors. Optimum interstage pressure is slightly higher than the geometric mean of the suction and the discharge pressures, but, due to very flat optimum of power versus interstage pressure relation geometric mean, it is widely accepted for determining the intermediate pressure. Required pressure of intermediate-level evaporator may dictate interstage pressure other than determined as optimal.

Two-stage systems should be seriously considered when the evaporating temperature is below −20°C. Such designs will save on power and reduce compressor discharge temperatures, but will increase initial cost.

**Cascade System** This is a reasonable choice in cases when the evaporating temperature is very low (below −60°C). When condensing pressures are to be in the rational limits, the same refrigerant has a high, specific volume at very low temperatures, requiring a large compressor. The evaporating pressure may be below atmospheric, which could cause moisture and air infiltration into the system if there is a leak. In other words, when the temperature difference between the medium that must be cooled and the environment is too high to be served with one refrigerant, it is wise to use different refrigerants in the high and low stages. Figure 11-80 shows a cascade system schematic diagram. There are basically two independent systems linked via a heat exchanger: the evaporator of the high-stage and the condenser of the low-stage system.

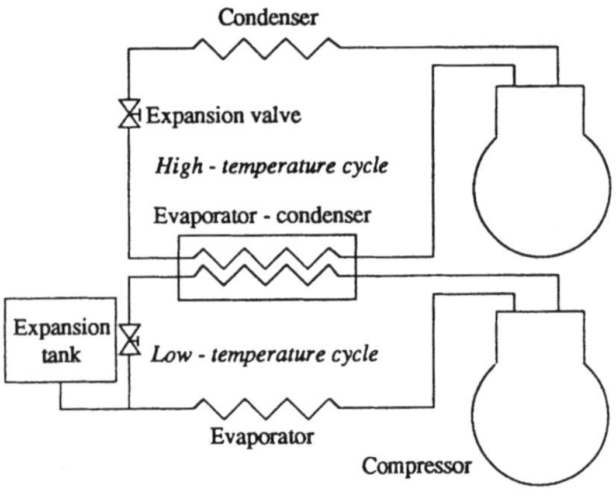


**FIG. 11-80** Cascade system.

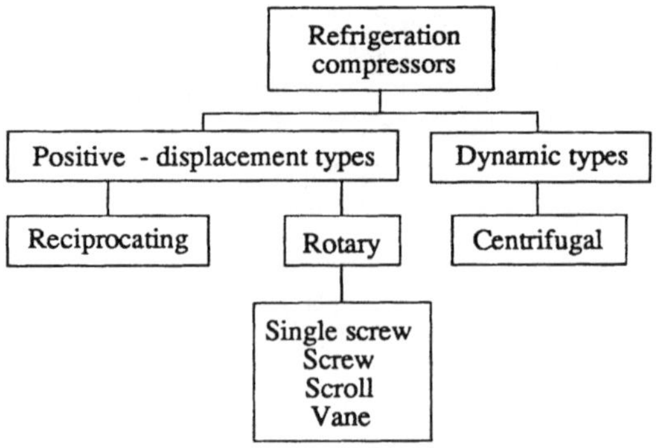


**FIG. 11-81** Types of refrigeration compressors.

## EQUIPMENT

**Compressors** These could be classified by one criteria (the way the increase in pressure is obtained) as positive-displacement and dynamic types as shown in Fig. 11-81 (see Sec. 10 for drawings and mechanical description of the various types of compressors). *Positive-displacement compressors* (PDC) are the machines that increase the pressure of the vapor by reducing the volume of the chamber. Typical PDC are reciprocating (in a variety of types) or rotary as screw (with one and two rotors), vane, scroll, and so on. Centrifugal or turbocompressors are machines where the pressure is raised converting some of kinetic energy obtained by a rotating mechanical element which continuously adds angular momentum to a steadily flowing fluid, similar to a fan or pump.

Generally, reciprocating compressors dominate in the range up to 300 kW refrigeration capacity. Centrifugal compressors are more accepted for the range over 500 kW, while screw compressors are in between with a tendency to go toward smaller capacities. The vane and the scroll compressors are finding their places primarily in very low capacity range (domestic refrigerators and the air conditioners), although vane compressors could be found in industrial compressors. Frequently, screw compressors operate as boosters, for the base load, while reciprocating compressors accommodate the variation of capacity, in the high stage. The major reason is for such design is the advantageous operation of screw compressors near full load and in design conditions, while reciprocating compressors seem to have better efficiencies at part-load operation than screw.

Using other criteria, compressors are classified as *open, semihermetic (accessible)*, or *hermetic*. Open type is characterized by shaft extension out of compressor where it is coupled to the driving motor. When the electric motor is in the same housing with the compressor mechanism, it could be either hermetic or accessible (semihermetic). Hermetic compressors have welded enclosures, not designed to be repaired, and are generally manufactured for smaller capacities (seldom over 30 kW), while semihermetic or an accessible type is located in the housing which is tightened by screws. Semihermetic compressors have all the advantages of hermetic (no sealing of moving parts, e.g., no refrigerant leakage at the seal shaft, no external motor mounting, no coupling alignment) and could be serviced, but it is more expensive.

Compared to other applications, refrigeration capacities in the chemical industry are usually high. That leads to wide usage of either centrifugal, screw, or high-capacity rotary compressors. Most centrifugal and screw compressors use economizers to minimize power and suction volume requirements. Generally, there is far greater use of open-drive type compressors in the chemical plants than in air-conditioning, commercial, or food refrigeration. Very frequently, compressor lube oil systems are provided with auxiliary oil pumps, filters, coolers, and other equipment to permit maintenance and repair without shut down.

**Positive-Displacement Compressors** *Reciprocating compressors* are built in different sizes (up to about one megawatt refrigeration capacity per unit). Modern compressors are high-speed, mostly direct-coupled, single-acting, from one to mostly eight, and occasionally up to sixteen cylinders.

Two characteristics of compressors for refrigeration are the most important: refrigerating capacity and power. Typical characteristics are as presented in the Fig. 11-82.

Refrigerating capacity $Q_e$ is the product of mass flow rate of refrigerant $m$ and refrigerating effect $R$ which is (for isobaric evaporation) $R = h_{\text{evaporator outlet}} - h_{\text{evaporator inlet}}$. Power $P$ required for the compression, necessary for the motor selection, is the product of mass flow rate $m$ and work of compression $W$. The latter is, for the isentropic compression, $W = h_{\text{discharge}} - h_{\text{suction}}$. Both of these characteristics could be calculated for the ideal (without losses) and for the actual compressor. Ideally, the mass flow rate is equal to the product of the compressor displacement $V_i$ per unit time and the gas density ρ: $m = V_i * \rho$. The compressor displacement rate is volume swept through by the pistons (product of the cylinder number $n$, and volume of cylinder $V = \text{stroke} * d^2\pi/4$) per second. In reality, the actual compressor delivers less refrigerant.

Ratio of the actual flow rate (entering compressor) to the displacement rate is the volumetric efficiency $\eta_{va}$. The volumetric efficiency is less than unity due to: reexpansion of the compressed vapor in clearance volume, pressure drop (through suction and discharge valves, strainers, manifolds, etc.), internal gas leakage (through the clearance between piston rings and cylinder walls, etc.), valve inefficiencies, and due to expansion of the vapor in the suction cycle caused by the heat exchanged (hot cylinder walls, oil, motor, etc.).

Similar to volumetric efficiency, isentropic (adiabatic) efficiency $\eta_a$ is the ratio of the work required for isentropic compression of the gas to work input to the compressor shaft. The adiabatic efficiency is less than one mainly due to pressure drop through the valve ports and other restricted passages and the heating of the gas during compression.

Figure 11-83 presents the compression on a pressure-volume diagram for an ideal compressor with clearance volume (thin lines) and

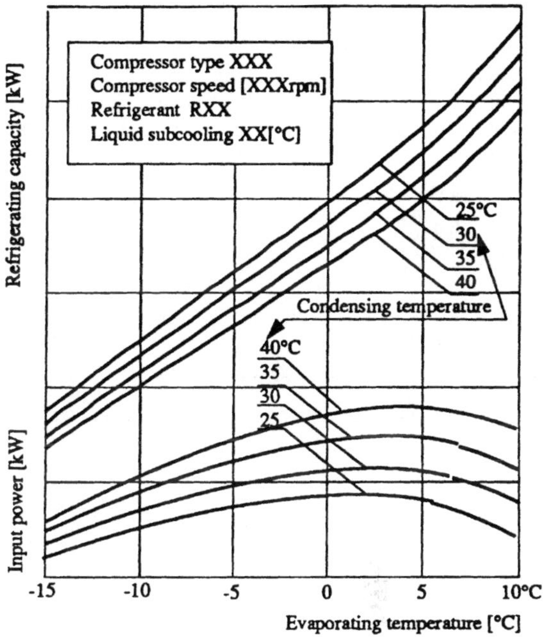


**FIG. 11-82** Typical capacity and power-input curves for reciprocating compressor.

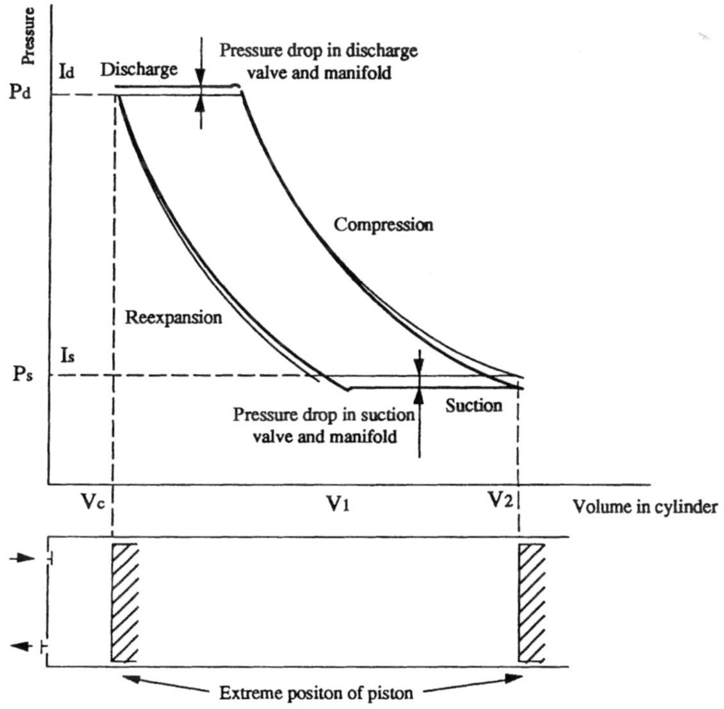


**FIG. 11-83** Pressure-volume diagram of an ideal (thin line) and actual (thick line) reciprocating compressor.

actual (thick lines). Compression in an ideal compressor without clearance is extended using dashed lines to the points $I_d$ (end of discharge), line $I_d - I_s$ (suction), and $I_s$ (beginning of suction). The area surrounded by the lines of compression, discharge, reexpansion and intake presents the work needed for compression. Actual compressor only appears to demand less work for compression due to smaller area in the $p$-$V$ diagram. Mass flow rate for an ideal compressor is higher, which cannot be seen in the diagram. In reality, an actual compressor will have diabatic compression and reexpansion and higher-discharge and lower-suction pressures due to pressure drops in valves and lines. The slight increase in the pressure at the beginning of the discharge and suction is due to forces needed to initially open valves.

When the suction pressure is lowered, the influence of the clearance will increase, causing in the extreme cases the entire volume to be used for reexpansion, which drives the volumetric efficiency to zero.

There are various options for capacity control of reciprocating refrigeration compressors:

1. Opening the suction valves by some external force (oil from the lubricating system, discharge gas, electromagnets . . . ).
2. Gas bypassing—returning discharge gas to suction (within the compressor or outside the compressor).
3. Controlling suction pressure by throttling in the suction line.
4. Controlling discharge pressure.
5. Adding reexpansion volume.
6. Changing the stroke.
7. Changing the compressor speed.

The first method is used most frequently. The next preference is for the last method, mostly used in small compressors due to problems with speed control of electrical motors. Other means of capacity control are very seldom utilized due to thermodynamic inefficiencies and design difficulties. Energy losses in a compressor, when capacity regulation is provided by lifting the suction valves, are due to friction of gas flowing in and out the unloaded cylinder. This is shown in Fig. 11-84 where the comparison is made for ideal partial load operation, reciprocating, and screw compressors.

Rotary compressors are also PDC types, but where refrigerant flow rotates during compression. Unlike the reciprocating type, rotary compressors have a built-in volume ratio which is defined as volume in cavity when the suction port is closed ($V_s = m * v_s$) over the volume in the cavity when the discharge port is uncovered ($V_d = m * v_d$). Built-in volume ratio determines for a given refrigerant and conditions the pressure ratio which is:

$$\frac{p_d}{p_s} = \left(\frac{v_s}{v_d}\right)^n \qquad (11\text{-}96)$$

where $n$ represents the politropic exponent of compression.

In other words, in a reciprocating compressor the discharge valve opens when the pressure in the cylinder is slightly higher than the pressure in the high-pressure side of the system, while in rotary compressors the discharge pressure will be established only by inlet conditions and built-in volume ratio regardless of the system discharge pressure. Very seldom are the discharge and system (condensing) pressure equal, causing the situation shown in Fig. 11-85. When condensing pressure ($p$) is lower than discharge ($p_2$), shown as case ($a$), "over compression" will cause energy losses presented by the horn on the diagram. If the condensing pressure is higher, in the moment when the discharge port uncovers there will be flow of refrigerant backwards into the compressor, causing losses shown in Fig. 11-85$b$ and the last stage will be only discharge without compression. The case when the compressor discharge pressure is equal to the condensing pressure is shown in the Fig. 11-85$c$.

*Double helical rotary (twin) screw compressors* consist of two mating helically grooved rotors (male and female) with asymmetric profile, in a housing formed by two overlapped cylinders, with inlet and outlet ports. Developed relatively recently (in 1930s) the first twin screw compressors were used for air, and later (1950s) became popular for refrigeration. Screw compressors have some advantages over reciprocating compressors (fewer moving parts and more compact) but also some drawbacks (lower efficiency at off-design conditions, as discussed above, higher manufacturing cost due to complicated screw geometry, large separators and coolers for oil which is important as a sealant). Figure 11-86 shows the oil circuit of a screw compressor. Oil cooling could be provided by water, glycol, or refrigerant either in the heat-exchanger-utilizing-thermosyphon effect or the using-direct-expansion concept.

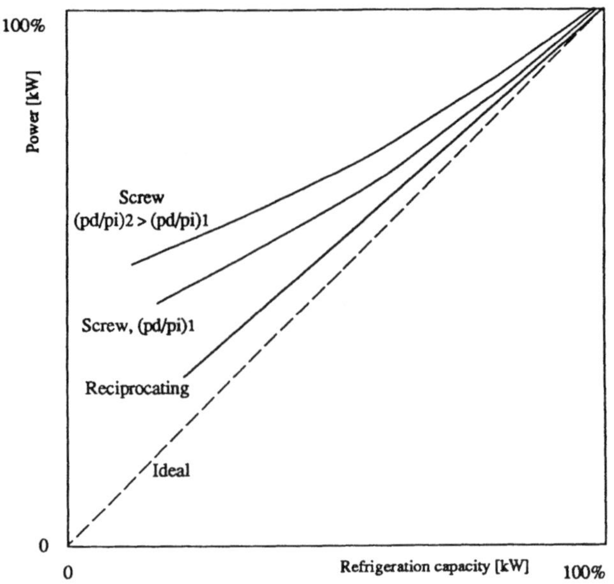


**FIG. 11-84** Typical power-refrigeration capacity data for different types of compressors during partial, unloaded operation.

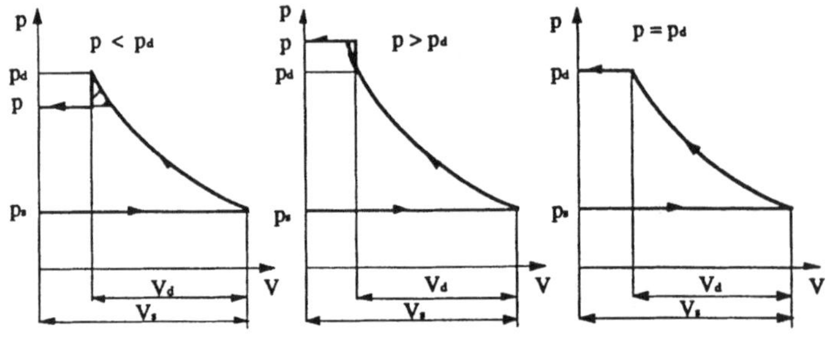


**FIG. 11-85** Matching compressor built-in pressure ratio with actual pressure difference.

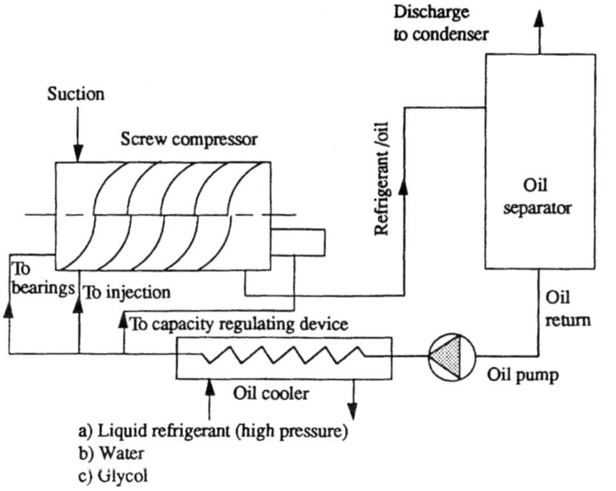


**FIG. 11-86** Oil cooling in a screw compressor.

In order to overcome some inherent disadvantages, screw compressors have been initially used predominantly as booster (low-stage) compressors, and following development in capacity control and decreasing prices, they are widely used for high-stage applications. There are several methods for capacity regulation of screw compressors. One is variable speed drive, but a more economical first-cost concept is a slide valve that is used in some form by practically all screw compressors.

The slide is located in the compressor casting below the rotors, allowing internal gas recirculation without compression. Slide valve is operated by a piston located in a hydraulic cylinder and actuated by high-pressure oil from both sides. When the compressor is started, the slide valve is fully open and the compressor is unloaded. To increase capacity, a solenoid valve on the hydraulic line opens, moving the piston in the direction of increasing capacity. In order to increase part-load efficiency, the slide valve is designed to consist of two parts, one traditional slide valve for capacity regulation and other for built-in volume adjustment.

*Single screw compressors* are a newer design (early 1960s) compared to twin screw compressors, and are manufactured in the range of capacity from 100 kW to 4 MW. The compressor screw is cylindrical with helical grooves mated with two star wheels (gaterotors) rotating in opposite direction from one another. Each tooth acts as the piston in the rotating "cylinder" formed by screw flute and cylindrical main-rotor casting.

As compression occurs concurrently in both halves of the compressor, radial forces are oppositely directed, resulting in negligible net-radial loads on the rotor bearings (unlike twin screw compressors), but there are some loads on the star wheel shafts.

*Scroll compressors* are currently used in relatively small-sized installations, predominantly for residential air-conditioning (up to 50 kW). They are recognized for low-noise operation. Two scrolls (free-standing, involute spirals bounded on one side by a flat plate) facing each other form a closed volume while one moves in a controlled orbit around a fixed point on the other, fixed scroll.

The suction gas which enters from the periphery is trapped by the scrolls. The closed volumes move radially inward until the discharge port is reached, when vapor is pressed out. The orbiting scroll is driven by a short-throw crank mechanism. Similar to screw compressors, internal leakage should be kept low, and is occurring in gaps between cylindrical surfaces and between the tips of the involute and the opposing scroll base plate.

Similar to the screw compressor, the scroll compressor is a constant-volume-ratio machine. Losses occur when operating conditions of the compressor do not match the built-in volume ratio (see Fig. 11-85).

*Vane compressors* are used in small, hermetic units, but sometimes as booster compressors in industrial applications. Two basic types are *fixed* (roller) or single-vane type and the *rotating* or multiple-vane type. In the single-vane type the rotor (called roller) is eccentrically placed in the cylinder so these two are always in contact. The contact line make the first separation between the suction and discharge chambers while the vane (spring-loaded divider) makes the second. In the multiple-vane compressors the rotor and the cylinder are not in the contact. The rotor has two or more sliding vanes which are held against the cylinder by centrifugal force. In the vane rotary compressors, no suction valves are needed. Since the gas enters the compressor continuously, gas pulsations are at minimum. Vane compressors have a high volumetric efficiency because of the small clearance volume and consequent low reexpansion losses. Rotary vane compressors have a low weight-to-displacement ratio, which makes them suitable for transport applications.

**Centrifugal Compressors** These are sometimes called turbocompressors and mostly serve refrigeration systems in the capacity range 200 to 10,000 kW. The main component is a spinning impeller wheel, backwards curved, which imparts energy to the gas being compressed. Some of the kinetic energy converts into pressure in a volute. Refrigerating centrifugal compressors are predominantly multistage, compared to other turbocompressors, that produce high-pressure ratios.

The torque $T$ (Nm) the impeller ideally imparts to the gas is:

$$T = m\ (u_{\text{tang.out}}\ r_{\text{out}} - u_{\text{tang.in}}\ r_{\text{in}}) \tag{11-97}$$

where: $m$ (kg/s) = mass flow rate
$r_{\text{out}}$ (m) = radius of exit of impeller
$r_{\text{in}}$ (m) = radius of exit of impeller
$u_{\text{tang.out}}$ (m/s) = tangential velocity of refrigerant leaving impeller
$u_{\text{tang.in}}$ (m/s) = tangential velocity of refrigerant entering impeller

When refrigerant enters essentially radially, $u_{\text{tang.in}} = 0$ and torque becomes:

$$T = m * u_{\text{tang.out}} * r_{\text{out}} \tag{11-98}$$

The power $P$ (W), is the product of torque and rotative speed $\omega$ [l/s] so is

$$P = T * \omega = m * u_{\text{tang.out}} * r_{\text{out}} * \omega \tag{11-99}$$

which for $u_{\text{tang.out}} = r_{\text{out}} * \omega$ becomes

$$P = m * u^2_{\text{tang.out}} \tag{11-100}$$

or for isentropic compression

$$P = m * \Delta h \tag{11-101}$$

The performance of a centrifugal compressor (discharge to suction-pressure ratio vs. the flow rate) for different speeds is shown in Fig. 11-87. Lines of constant efficiencies show the maximum efficiency. Unstable operation sequence, called surging, occurs when compressors fails to operate in the range left of the surge envelope. It is characterized by noise and wide fluctuations of load on the compressor and the motor. The period of the cycle is usually 2 to 5 s, depending upon the size of the installation.

The capacity could be controlled by: (1) adjusting the prerotation vanes at the impeller inlet, (2) varying the speed, (3) varying the condenser pressure, and (4) bypassing discharge gas. The first two methods are predominantly used.

**Condensers** These are heat exchangers that convert refrigerant vapor to a liquid. Heat is transferred in three main phases: (1) desuperheating, (2) condensing, and (3) subcooling. In reality condensation occurs even in the superheated region and subcooling occurs in the condensation region. Three main types of refrigeration condensers are: air cooled, water cooled, and evaporative.

Air-cooled condensers are used mostly in air-conditioning and for smaller-refrigeration capacities. The main advantage is availability of cooling medium (air) but heat-transfer rates for the air side are far below values when water is used as a cooling medium. Condensation always occurs inside tubes, while the air side uses extended surface (fins).

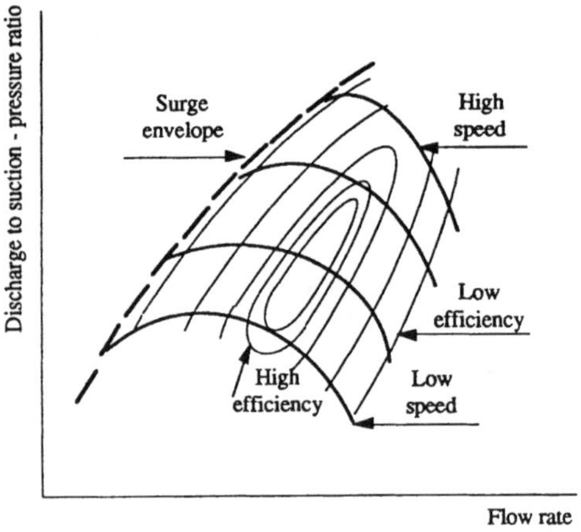


**FIG. 11-87** Performance of the centrifugal compressor.

The most common types of water-cooled refrigerant condensers are: (1) shell-and-tube, (2) shell-and-coil, (3) tube-in-tube, and (4) brazed-plate. *Shell-and-tube* condensers are built up to 30 MW capacity. Cooling water flows through the tubes in a single or multipass circuit. Fixed-tube sheet and straight-tube construction are common. Horizontal layout is typical, but sometimes vertical is used. Heat-transfer coefficients for the vertical types are lower due to poor condensate drainage, but less water of lower purity can be utilized. Condensation always occurs on the tubes, and often the lower portion of the shell is used as a receiver. In *shell-and-coil* condensers water circulates through one or more continuous or assembled coils contained within the shell while refrigerant condenses outside. The tubes cannot be mechanically cleaned nor replaced. *Tube-in-tube* condensers could be found in versions where condensation occurs either in the inner tube or in the annulus. Condensing coefficients are more difficult to predict, especially in the cases where tubes are formed in spiral. Mechanical cleaning is more complicated, sometimes impossible, and tubes are not replaceable. *Brazed-plate* condensers are constructed of plates brazed together to make up an assembly of separate channels. The plates are typically stainless steel, wave-style corrugated, enabling high heat-transfer rates. Performance calculation is difficult, with very few correlations available. The main advantage is the highest performance/volume (mass) ratio and the lowest refrigerant charge. The last mentioned advantage seems to be the most important feature for many applications where minimization of charge inventory is crucial.

*Evaporative condensers* (Fig. 11-88) are widely used due to lower condensing temperatures than in the air-cooled condensers and also lower than the water-cooled condenser combined with the cooling tower. Water demands are far lower than for water-cooled condensers. The chemical industry uses shell-and-tube condensers widely, although the use of air-cooled condensing equipment and evaporative condensers is on the increase.

Generally, cooling water is of a lower quality then normal, having also higher mud and silt content. Sometimes even replaceable copper tubes in shell-and-tube heat exchangers are required. It is advisable to use cupronickel instead of copper tubes (when water is high in chlorides) and to use conservative water side velocities (less then 2 m/s for copper tubes).

Evaporative condensers are used quite extensively. In most cases commercial evaporative condensers are not totally suitable for chemical plants due to the hostile atmosphere which usually abounds in vapor and dusts which can cause either chemical (corrosion) or mechanical problems (plugging of spray nozzles).

Air-cooled condensers are similar to evaporative in that the service dictates either the use of more expensive alloys in the tube construction or conventional materials of greater wall thickness.

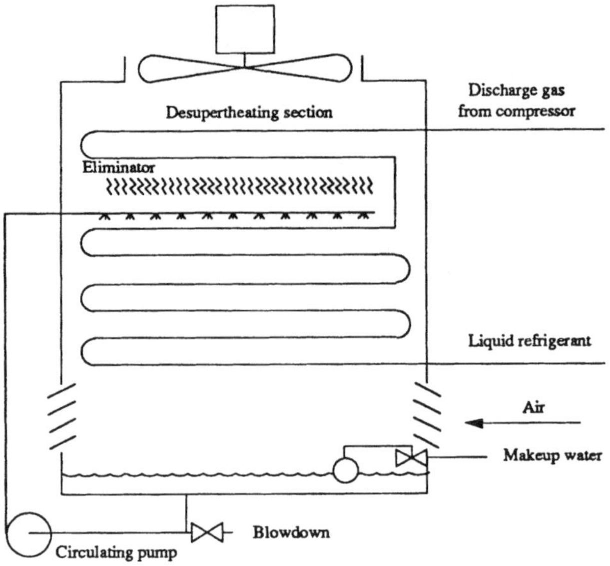


**FIG. 11-88** Evaporative condenser with desuperheating coil.

Heat rejected in the condenser $Q_{Cd}$ consists of heat absorbed in the evaporator $Q_{Eevap}$ and energy W supplied by the compressor:

$$Q_{Cd} = Q_{Eevap} + W \tag{11-102}$$

For the actual systems, compressor work will be higher than for ideal for the isentropic efficiency and other losses. In the case of hermetic or accessible compressors where an electrical motor is cooled by the refrigerant, condenser capacity should be:

$$Q_{Cd} = Q_{Eevap} + P_{EM} \tag{11-103}$$

It is common that compressor manufacturers provide data for the ratio of the heat rejected at the condenser to the refrigeration capacity as shown in Fig. 11-89. The solid line represents data for the open compressors while the dotted line represents the hermetic and accessible compresors. The difference between solid and dotted line is due

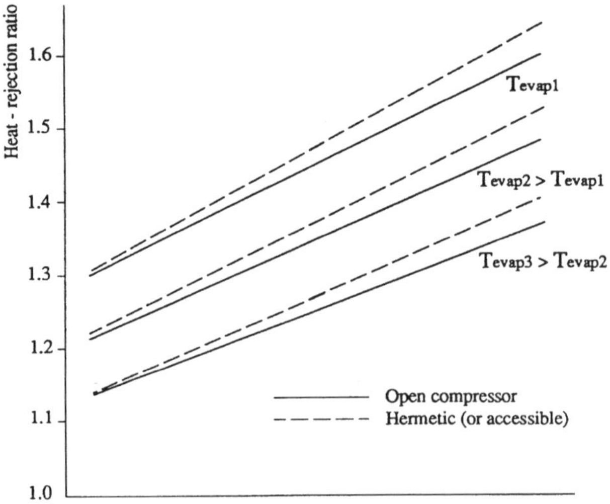


**FIG. 11-89** Typical values of the heat-rejection ratio of the heat rejected at the condenser to the refrigerating capacity.

to all losses (mechanical and electrical in the electrical motor). Condenser design is based on the value:

$$Q_{Cd} = Q_{Eevap} * \text{heat-rejection ratio} \qquad (11\text{-}104)$$

Thermal and mechanical design of heat exchangers (condensers and evaporators) is presented earlier in this section.

**Evaporators** These are heat exchangers where refrigerant is evaporated while cooling the product, fluid, or body. Refrigerant could be in direct contact with the body that is being cooled, or some other medium could be used as secondary fluid. Mostly air, water, or antifreeze are fluids that are cooled. Design is strongly influenced by the application. Evaporators for air cooling will have in-tube evaporation of the refrigerant, while liquid chillers could have refrigerant evaporation inside or outside the tube. The heat-transfer coefficient for evaporation inside the tube (vs. length or quality) is shown in the Fig. 11-90. Fundamentals of the heat transfer in evaporators, as well as design aspects, are presented in Sec. 11. We will point out only some specific aspects of refrigeration applications.

Refrigeration evaporators could be classified according to the method of feed as either direct (dry) expansion or flooded (liquid overfeed). In dry-expansion the evaporator's outlet is dry or slightly superheated vapor. This limits the liquid feed to the amount that can be completely vaporized by the time it reaches the end of evaporator. In the liquid overfeed evaporator, the amount of liquid refrigerant circulating exceeds the amount evaporated by the circulation number. Decision on the type of the system to be used is one of the first in the design process. Direct-expansion evaporator is generally applied in smaller systems where compact design and the low first costs are crucial. Control of the refrigerant mass flow is then obtained by either a thermoexpansion valve or a capillary tube. Figure 11-90 suggests that the evaporator surface is the most effective in the regions with quality which is neither low nor high. In dry-expansion evaporators, inlet qualities are 10–20 percent, but when controlled by the thermoexpansion valve, vapor at the outlet is not only dry, but even superheated.

In recirculating systems saturated liquid ($x = 0$) is entering the evaporator. Either the pump or gravity will deliver more refrigerant liquid then will evaporate, so outlet quality could be lower than one. The ratio of refrigerant flow rate supplied to the evaporator overflow rate of refrigerant vaporized is the circulation ratio, $n$. When $n$ increases, the coefficient of heat transfer will increase due to the wetted outlet of the evaporator and the increased velocity at the inlet (Fig. 11-91). In the range of $n$ = 2 to 4, the overall $U$ value for air cooler increases roughly by 20 to 30 percent compared to the direct-expansion case. Circulation rates higher than four are not efficient.

The price for an increase in heat-transfer characteristics is a more complex system with more auxiliary equipment: low-pressure receivers, refrigerant pumps, valves, and controls. Liquid refrigerant is predominantly pumped by mechanical pumps, however, sometimes gas at condensing pressure is used for pumping, in the variety of concepts.

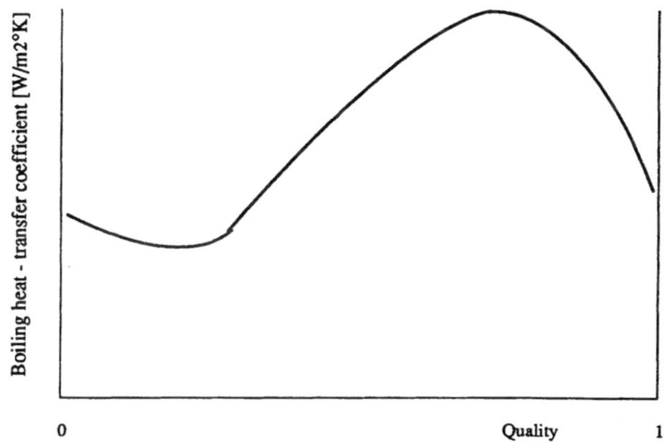


**FIG. 11-90** Heat-transfer coefficient for boiling inside the tube.

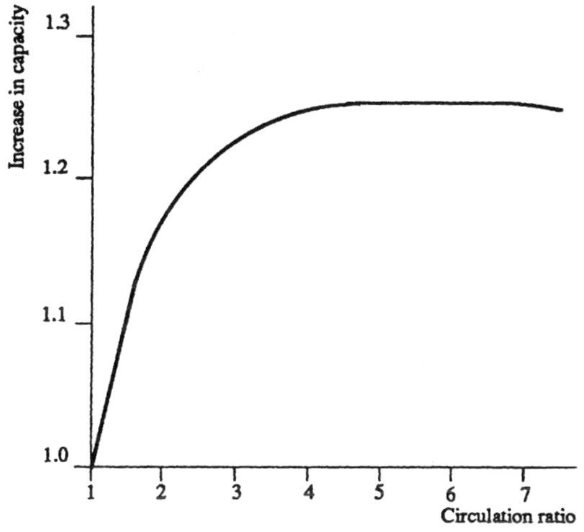


**FIG. 11-91** Effect of circulation ratio on the overall heat-transfer coefficient of an air-cooling coil.

The important characteristics of the refrigeration evaporators is the presence of the oil. The system is contaminated with oil in the compressor, in spite of reasonably efficient oil separators. Some systems will recirculate oil, when miscible with refrigerant, returning it to the compressor crankcase. This is found mostly in the systems using halocarbon refrigerants. Although oils that are miscible with ammonia exist, immiscibles are predominantly used. This inhibits the ammonia systems from recirculating the oil. In systems with oil recirculation when halocarbons are used special consideration should be given to proper sizing and layout of the pipes. Proper pipeline configuration, slopes, and velocities (to ensure oil circulation under all operating loads) are essential for good system operation. When refrigerant is lighter than the oil in systems with no oil recirculation, oil will be at the bottom of every volume with a top outlet. Then oil must be drained periodically to avoid decreasing the performance of the equipment.

It is essential for proper design to have the data for refrigerant-oil miscibility under all operating conditions. Some refrigerant-oil combinations will always be miscible, some always immiscible, but some will have both characteristics, depending on temperatures and pressures applied. Defrosting is the important issue for evaporators which are cooling air below freezing. Defrosting is done periodically, actuated predominantly by time relays, but other frost indicators are used (temperature, visual, or pressure-drop sensors). Defrost technique is determined mostly by fluids available and tolerable complexity of the system. Defrosting is done by the following mechanisms when the system is off:

- Hot (or cool) refrigerant gas (the predominant method in industrial applications)
- Water (defrosting from the outside, unlike hot gas defrost)
- Air (only when room temperature is above freezing)
- Electricity (for small systems where hot-gas defrost will be to complex and water is not available)
- Combinations of above.

**System Analysis** Design calculations are made on the basis of the close to the highest refrigeration load, however the system operates at the design conditions very seldom. The purpose of regulating devices is to adjust the system performance to cooling demands by decreasing the effect or performance of some component. Refrigeration systems have inherent self-regulating control which the engineer could rely on to a certain extent. When the refrigeration load starts to decrease, less refrigerant will evaporate. This causes a drop in evaporation temperature (as long as compressor capacity is unchanged) due to the imbalance in vapor being taken by the compressor and produced by evaporation in evaporator. With a drop in evaporation pressure, the compressor capacity will decrease due to: (1) lower vapor density (lower mass flow for the same volumetric flow rate) and (2) decrease in volumetric efficiency. On the other hand, when the evaporation temperature drops, for the unchanged temperature of the medium being cooled, the evaporator capacity will increase due to increase in the mean-temperature difference between refrigerant and cooled medium, causing a positive effect (increase) on the cooling load. With a decrease in the evaporation temperature the heat-rejection factor will increase causing an increase of heat rejected to the condenser, but refrigerant mass flow rate will decrease due to compressor characteristics. These will have an opposite effect on condenser load. Even a simplified analysis demonstrates the necessity for better understanding of system performance under different operating conditions. Two methods could be used for more accurate analysis. The traditional method of refrigeration-system analysis is through determination of balance points, while in recent years, system analysis is performed by system simulation or mathematical modeling, using mathematical (equation solving) rather than graphical (intersection of two curves) procedures. Systems with a small number of components such as the vapor-compression refrigeration system could be analyzed both ways. Graphical presentation, better suited for understanding trends is not appropriate for more complex systems, more detailed component description, and frequent change of parameters. There is a variety of different mathematical models tailored to fit specific systems, refrigerants, resources available, demands, and complexity. Although limited in its applications, graphical representation is valuable as the starting tool and for clear understanding of the system performance.

Refrigeration capacity $q_e$ and power $P$ curves for the reciprocating compressor are shown in Fig. 11-92. They are functions of temperatures of evaporation and condensation:

$$q_e = q_e(t_{\text{evap}}, t_{\text{cd}}) \qquad (11\text{-}105a)$$

and

$$P = P(t_{\text{evap}}, t_{\text{cd}}) \qquad (11\text{-}105b)$$

where $q_e$ (kW) = refrigerating capacity
$P$ (kW) = power required by the compressor
$t_{\text{evap}}$ (°C) = evaporating temperature
$t_{\text{cd}}$ (°C) = condensing temperature.

A more detailed description of compressor performance is shown in the section on the refrigeration compressors.

Condenser performance, shown in figure 11-93, could be simplified as:

$$q_{\text{cd}} = F(t_{\text{cd}} - t_{\text{amb}}) \qquad (11\text{-}105c)$$

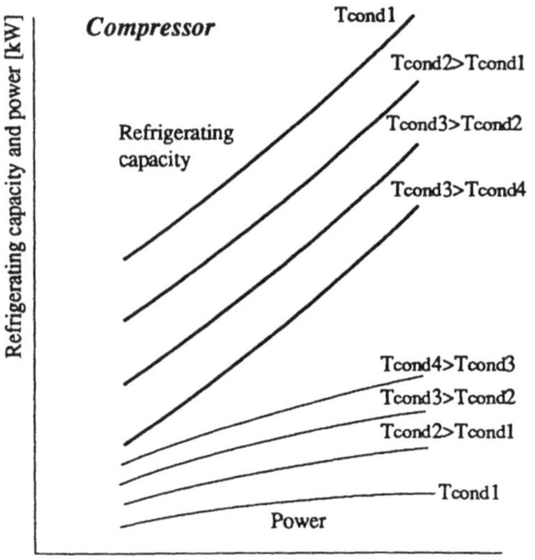


**FIG. 11-92** Refrigerating capacity and power requirement for the reciprocating compressor.

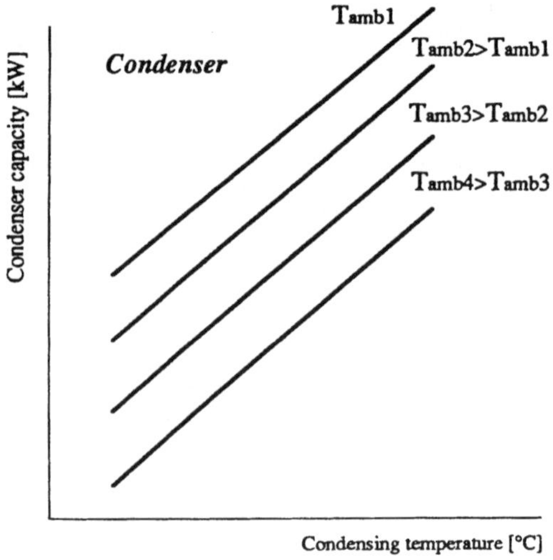


**FIG. 11-93** Condenser performance.

where $q_{\text{cd}}$ (kW) = capacity of condenser;
$F$ (kW/°C) = capacity of condenser per unit inlet temperature difference ($F = U * A$);
$t_{\text{amb}}$ (°C) = ambient temperature (or temperature of condenser cooling medium).

In this analysis $F$ will be constant but it could be described more accurately as a function of parameters influencing heat transfer in the condenser (temperature, pressure, flow rate, fluid thermodynamical, and thermophysical characteristics . . . ).

Condenser performance should be expressed as "evaporating effect" to enable matching with compressor and evaporator performance. Condenser "evaporating effect" is the refrigeration capacity of an evaporator served by a particular condenser. It is the function of the cycle, evaporating temperature, and the compressor. The "evaporating effect" could be calculated from the heat-rejection ratio $q_{\text{Cd}}/q_e$:

$$q_e = \frac{q_{\text{Cd}}}{\text{heat-rejection ratio}} \qquad (11\text{-}105d)$$

The heat-rejection rate is presented in Fig. 11-94 (or Fig. 11-89).

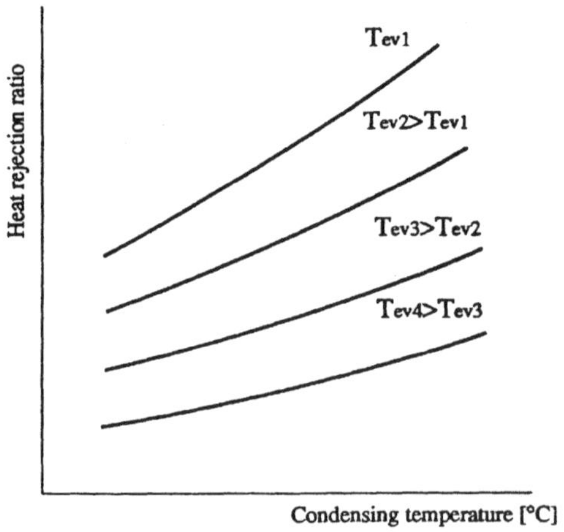


**FIG. 11-94** Heat-rejection ratio.

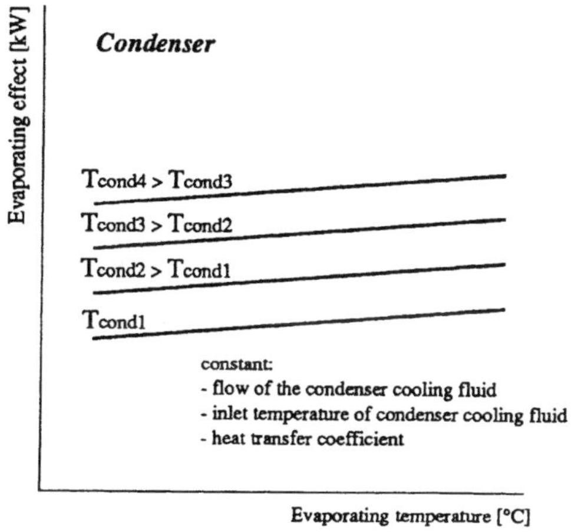


**FIG. 11-95** Condenser evaporating effect.

Finally, the evaporating effect of the condenser is shown in Fig. 11-95.

The performance of the condensing-unit (compressor and condenser) subsystem could be developed as shown in Fig. 11-96 by superimposing two graphs, one for compressor performance and the other for condenser evaporating effect.

Evaporator performance could be simplified as:

$$q_e = F_{\text{evap}}(t_{\text{amb}} - t_{\text{evap}}) \tag{11-106}$$

where $q_e$(kW) = evaporator capacity
$F_e$ (kW/°C) = evaporator capacity per unit inlet temperature difference
$t_{\text{amb}}$ (°C) = ambient temperature (or temperature of cooled body or fluid).

The diagram of the evaporator performance is shown in the Fig. 11-97. The character of the curvature of the lines (variable heat-transfer rate) indicates that the evaporator is cooling air. Influences of the flow rate of cooled fluid are also shown in this diagram; i.e., higher flow rate will increase heat transfer. The same effect could be shown in the condenser-performance curve. It is omitted only for the reasons of simplicity.

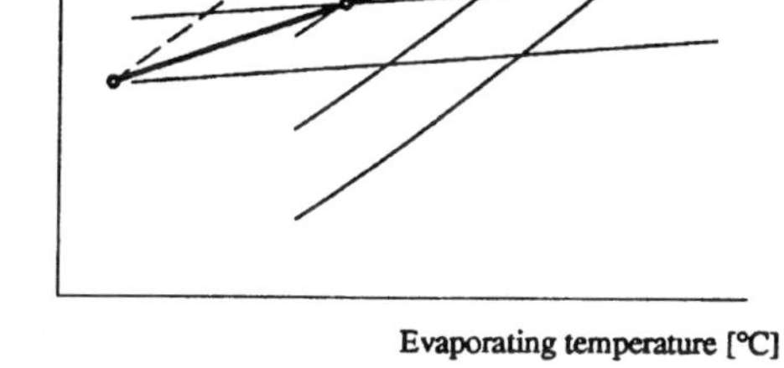


**FIG. 11-96** Balance points of compressor and condenser determines performance of condensing unit for fixed temperature of condenser cooling fluid (flow rate and heat-transfer coefficient are constant).

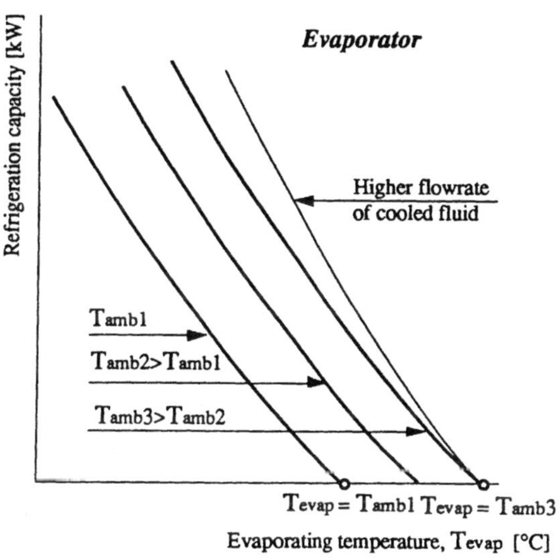


**FIG. 11-97** Refrigerating capacity of evaporator.

The performance of the complete system could be predicted by superimposing the diagrams for the condensing unit and the evaporator, as shown in Fig. 11-98. Point 1 reveals a balance for constant flows and inlet temperatures of chilled fluid and fluid for condenser cooling. When this point is transferred in the diagram for the condensing unit in the Figs. 11-95 or 11-96, the condensing temperature could be determined. When the temperature of entering fluid in the evaporator $t_{\text{amb1}}$ is lowered to $t_{\text{amb2}}$ the new operating conditions will be determined by the state at point 2. Evaporation temperature drops from $t_{\text{evap1}}$ to $t_{\text{evap2}}$. If the evaporation temperature should be unchanged, the same reduction of inlet temperature could be achieved by reducing the capacity of the condensing unit from Cp to Cp°. The new operating point 3 shows reduction in capacity for Δ due to the reduction in the compressor or the condenser capacity.

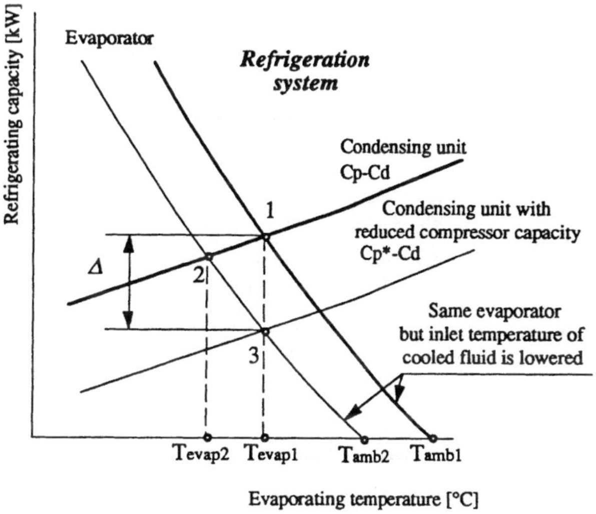


**FIG. 11-98** Performance of complete refrigeration system (1), when there is reduction in heat load (2), and when for the same ambient (or inlet in evaporator) evaporation temperature is maintained constant by reducing capacity of compressor/condenser part (3).

Mathematical modeling is essentially the same process, but the description of the component performance is generally much more complex and detailed. This approach enables a user to vary more parameters easier, look into various possibilities for intervention, and predict the response of the system on different influences. Equation-solving does not necessarily have to be done by successive substitution or iteration as this procedure could suggest.

**System, Equipment, and Refrigerant Selection** There is no universal rule which can be used to decide which system, equipment type, or refrigerant is the most appropriate for a given application. A number of variables influence the final-design decision:

- Refrigeration load
- Type of installation
- Temperature level of medium to be cooled
- Condensing media characteristics: type (water, air, . . . ), temperature level, available quantities
- Energy source for driving the refrigeration unit (electricity, natural gas, steam, waste heat)
- Location and space available (urban areas, sensitive equipment around, limited space . . . )
- Funds available (i.e. initial vs. run-cost ratio)
- Safety requirements (explosive environment, aggressive fluids, . . . )
- Other demands (compatibility with existing systems, type of load, compactness, level of automatization, operating life, possibility to use process fluid as refrigerant)

Generally, vapor *compression systems* are considered first. They can be used for almost every task. Whenever it is possible, prefabricated elements or complete units are recommended. Reciprocating compressors are widely used for lower rates, more uneven heat loads (when frequent and wider range of capacity reduction is required). They ask for more space and have higher maintenance costs then centrifugal compressors, but are often the most economical in first costs. Centrifugal compressors are considered for huge capacities, when the evaporating temperature is not too low. Screw compressors are considered first when space in the machine room is limited, when system is operating long hours, and when periods between service should be longer.

Direct-expansions are more appropriate for smaller systems which should be compact, and where there are just one or few evaporators. Overfeed (recirculation) systems should be considered for all applications where first cost for additional equipment (surge drums, low-pressure receivers, refrigerant pumps, and accessories) is lower than the savings for the evaporator surface.

Choice of refrigerant is complex and not straightforward. For industrial applications, advantages of ammonia (thermodynamical and economical) overcome drawbacks which are mostly related to possible low-toxic and panics created by accidental leaks when used in urban areas. Halocarbons have many advantages (not toxic, not explosive, odorless . . . ), but environmental issues and slightly inferior ther-modynamical and thermophysical properties compared to ammonia or hydrocarbons as well as rising prices are giving the chance to other options. When this text was written the ozone-depletion issue was not resolved, R22 was still used but facing phase-out, and R134a was considered to be the best alternative for CFCs and HCFCs, having similar characteristics to the already banned R12. Very often, fluid to be cooled is used as a refrigerant in the chemical industry. Use of secondary refrigerants in combination with the ammonia central-refrigeration unit is becoming a viable alternative in many applications.

Absorption systems will be considered when there is low-cost low-pressure steam or waste heat available and evaporation temperature and refrigeration load are relatively high. Typical application range is for water chilling at 7–10°C, and capacities from 300 kW to 5 MW in a single unit. The main drawback is the difficulty in maintaining a tight system with the highly corrosive lithium bromide, and an operating pressure in the evaporator and the absorber below atmospheric.

Ejector (steam-jet) refrigeration systems are used for similar applications, when chilled water-outlet temperature is relatively high, when relatively cool condensing water and cheap steam at 7 bar are available, and for similar high duties (0.3–5 MW). Even though these systems usually have low first and maintenance costs, there are not many steam-jet systems running.

## OTHER REFRIGERATION SYSTEMS APPLIED IN THE INDUSTRY

**Absorption Refrigeration Systems** Two main absorption systems are used in industrial application: lithium bromide-water and ammonia-water. Lithium bromide-water systems are limited to evaporation temperatures above freezing because water is used as the refrigerant, while the refrigerant in an ammonia-water system is ammonia and consequently it can be applied for the lower-temperature requirements.

Single-effect indirect-fired lithium bromide cycle is shown in Fig. 11-99. The machine consists of five major components:

*Evaporator* is the heat exchanger where refrigerant (water) evaporates (being sprayed over the tubes) due to low pressure in the vessel. Evaporation chills water flow inside the tubes that bring heat from the external system to be cooled.

*Absorber* is a component where strong absorber solution is used to absorb the water vapor flashed in the evaporator. A solution pump sprays the lithium bromide over the absorber tube section. Cool water is passing through the tubes taking refrigeration load, heat of dilution, heat to cool condensed water, and sensible heat for solution cooling.

*Heat exchanger* is used to improve efficiency of the cycle, reducing consumption of steam and condenser water.

*Generator* is a component where heat brought to a system in a tube section is used to restore the solution concentration by boiling off the water vapor absorbed in the absorber.

*Condenser* is an element where water vapor, boiled in the generator, is condensed, preparing pure water (refrigerant) for discharge to an evaporator.

Heat supplied to the generator is boiling weak (dilute) absorbent solution on the outside of the tubes. Evaporated water is condensed on the outside of the condenser tubes. Water utilized to cool the condenser is usually cooled in the cooling tower. Both condenser and generator are located in the same vessel, being at the absolute pressure of about 6 kPa. The water condensate passes through a liquid trap and enters the evaporator. Refrigerant (water) boils on the evaporator tubes and cools the water flow that brings the refrigeration load. Refrigerant that is not evaporated flows to the recirculation pump to be sprayed over the evaporator tubes. Solution with high water concentration that enters the generator increases in concentration as water evaporates. The resulting strong, absorbent solution (solution with low water concentration) leaves the generator on its way to the heat exchanger. There the stream of high water concentration that flows to the generator cools the stream of solution with low water concentration that flows to the second vessel. The solution with low water concentration is distributed over the absorber tubes. Absorber and evaporator are located in the same vessel, so the refrigerant evaporated on the evaporator tubes is readily absorbed into the absorbent solution. The pressure in the second vessel during the operation is 7 kPa (absolute). Heat of absorption and dilution are removed by cooling water (usually from the cooling tower). The resulting solution with high water concentration is pumped through the heat exchanger to the generator, completing the cycle. Heat exchanger increases the efficiency of the system by preheating, that is, reducing the amount of heat that must be added to the high water solution before it begins to evaporate in the generator.

The absorption machine operation is analyzed by the use of a lithium bromide-water equilibrium diagram, as shown in Fig. 11-100. Vapor pressure is plotted against the mass concentration of lithium bromide in the solution. The corresponding saturation temperature for a given vapor pressure is shown on the left-hand side of the diagram. The line in the lower right corner of the diagram is the crystallization line. It indicates the point at which the solution will begin to change from liquid to solid, and this is the limit of the cycle. If the solution becomes overconcentrated, the absorption cycle will be interrupted owing to solidification, and capacity will not be restored until the unit is desolidified. This normally requires the addition of heat to the outside of the solution heat exchanger and the solution pump.

The diagram in Fig. 11-101 presents enthalpy data for LiBr-water solutions. It is needed for the thermal calculation of the cycle. Enthalpies for water and water vapor can be determined from the table of properties of water. The data in Fig. 11-101 are applicable to

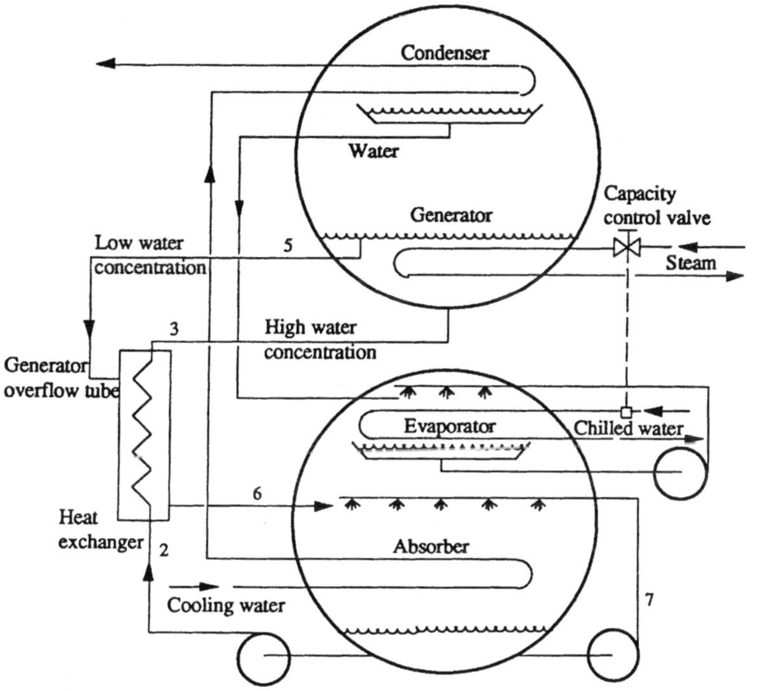


**FIG. 11-99** Two-shell lithium bromide-water cycle chiller.

saturated or subcooled solutions and are based on a zero enthalpy of liquid water at 0°C and a zero enthalpy of solid LiBr at 25°C. Since the zero enthalpy for the water in the solution is the same as that in conventional tables of properties of water, the water property tables can be used in conjunction with diagram in Fig. 11-100.

Coefficient of performance of the absorption cycle is defined on the same principle as for the mechanical refrigeration:

$$COP_{abs} = \frac{\text{useful effect}}{\text{heat input}} = \frac{\text{refrigeration rate}}{\text{heat input at generator}}$$

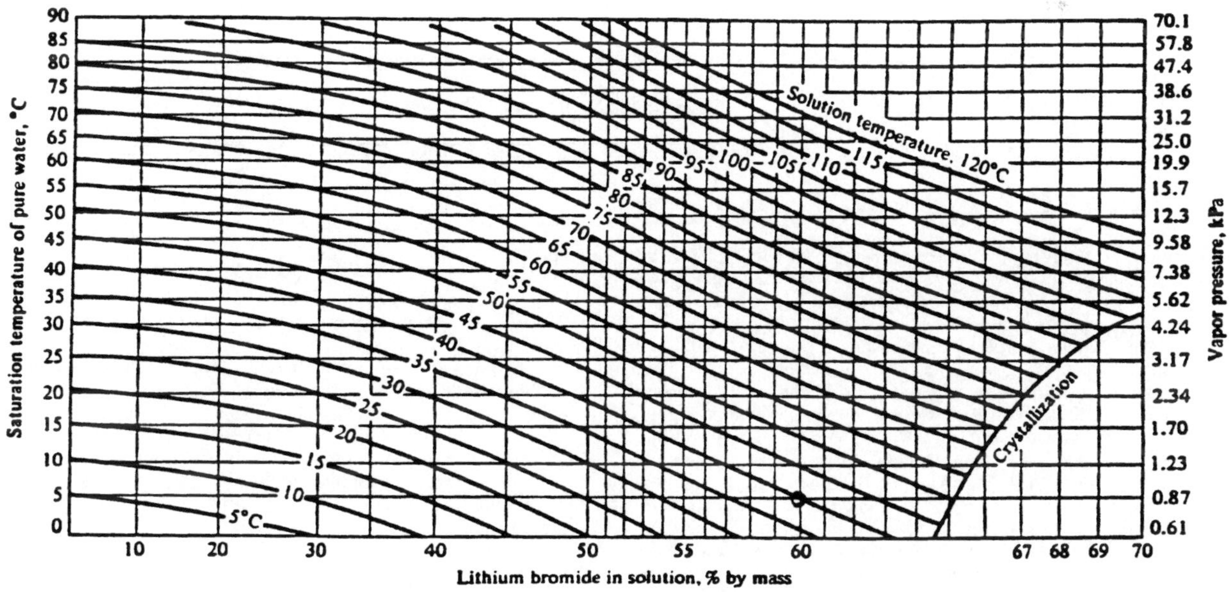


**FIG. 11-100** Temperature-pressure-concentration diagram of saturated LiBr-water solutions (*W. F. Stoecker and J. W. Jones:* Refrigeration and Air-Conditioning)

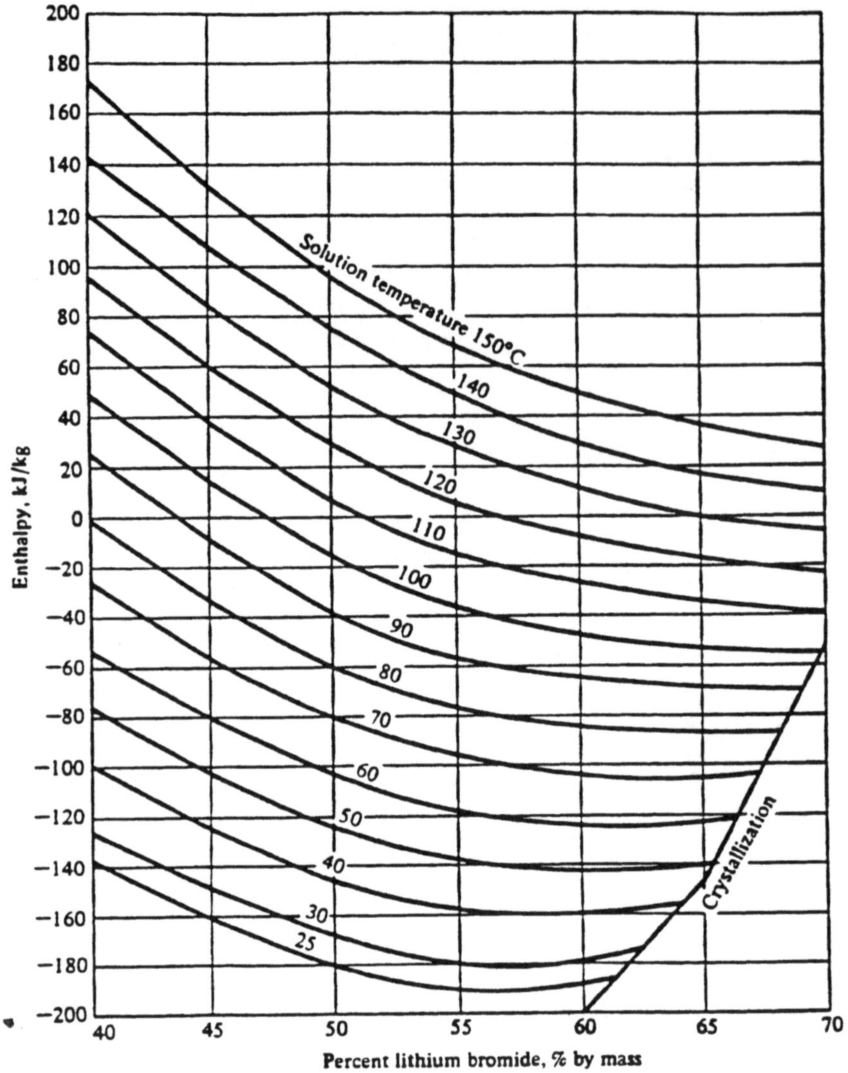


**FIG. 11-101** Enthalpy of LiBr-water solutions (*W. F. Stoecker and J. W. Jones:* Refrigeration and Air-Conditioning).

but it should be noted that here denominator for the $COP_{abs}$ is heat while for the mechanical refrigeration cycle it is work. Since these two forms of energy are not equal, $COP_{abs}$ is not as low (0.6–0.8) as it appears compared to COP for mechanical system (2.5–3.5).

The double-effect absorption unit is shown in Fig. 11-102. All major components and operation of the double-effect absorption machine is similar to that for the single-effect machine. The primary generator, located in the vessel 1, is using an external heat source to evaporate water from dilute-absorbent (high water concentration) solution. Water vapor readily flows to the generator II where it is condensed on the tubes. The absorbent (LiBr) intermediate solution from generator I will pass through the heat exchanger on the way to the generator II where it is heated by the condensing water vapor. The throttling valve reduces pressure from vessel 1 (about 103 kPa absolute) to that of vessel 2. Following the reduction of pressure some water in the solution flashes to vapor, which is liquefied at the condenser. In the high temperature heat exchanger intermediate solution heats the weak (high water concentration) solution stream coming from the low temperature heat exchanger. In the low temperature heat exchanger strong solution is being cooled before entering the absorber. The absorber is on the same pressure as the evaporator. The double-effect absorption units achieve higher COPs than the single stage.

The ammonia-water absorption system was extensively used until the fifties when the LiBr-water combination became popular. Figure 11-103 shows a simplified ammonia-water absorption cycle. The refrigerant is ammonia, and the absorbent is dilute aqueous solution of ammonia. Ammonia-water systems differ from water-lithium bromide equipment to accommodate major differences: Water (here absorbent) is also volatile, so the regeneration of weak water solution to strong water solution is a fractional distillation. Different refrigerant (ammonia) causes different, much higher pressures: about 1100–2100 kPa absolute in condenser.

Ammonia vapor from the evaporator and the weak water solution from the generator are producing strong water solution in the absorber. Strong water solution is then separated in the rectifier producing (1)

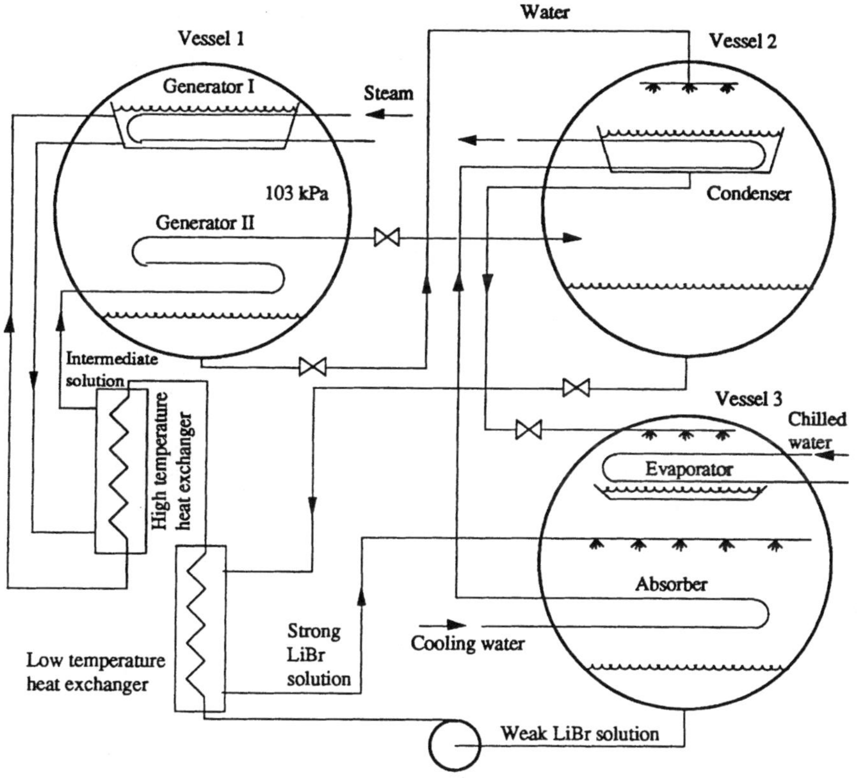


**FIG. 11-102** Double-effect absorption unit.

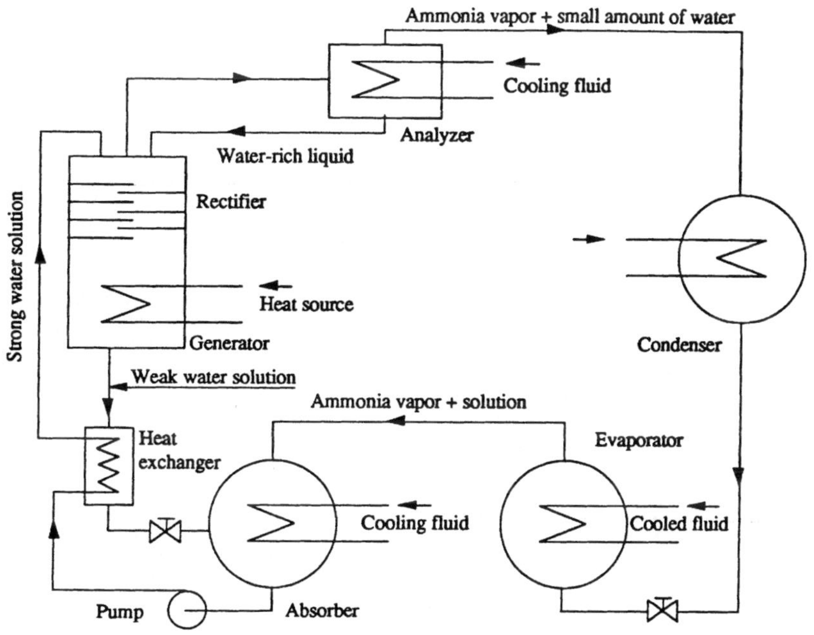


**FIG. 11-103** Simplified ammonia-water absorption cycle.

ammonia with some water vapor content and (2) very strong water solution at the bottom, in the generator. Heat in the generator vaporizes ammonia and the weak solution returns to absorber. On its way to the absorber the weak solution stream passes through the heat exchanger, heating strong solution from the absorber on the way to the rectifier. The other stream, mostly ammonia vapor but with some water-vapor content flows to the condenser. To remove water as much as possible, the vapor from the rectifier passes through the analyzer where it is additionally cooled. The remaining water escaped from the analyzer pass as liquid through the condenser and the evaporator to the absorber.

Ammonia-water units can be arranged for single-stage or cascaded two-stage operation. The advantage of two-staging is it creates the possibility of utilizing only part of the heat on the higher and the rest on the lower temperature level but the price is increase in first cost and heat required.

Ammonia-water and lithium bromide-water systems operate under comparable COP. The ammonia-water system is capable achieving evaporating temperatures below 0°C because the refrigerant is ammonia. Water as the refrigerant limits evaporating temperatures to no lower than freezing, better to 3°C. Advantage of the lithium bromide-water system is that it requires less equipment and operates at lower pressures. But this is also a drawback, because pressures are below atmospheric, causing air infiltration in the system which must be purged periodically. Due to corrosion problems, special inhibitors must be used in the lithium-bromide-water system. The infiltration of air in the ammonia-water system is also possible, but when evaporating temperature is below −33°C. This can result in formation of corrosive ammonium carbonate.

***Further readings:*** *ASHRAE Handbook* 1994 Refrigeration Systems and Applications; Bogart, M., 1981: *Ammonia Absorption Refrigeration in Industrial Processes,* Gulf Publishing Co. Houston; Stoecker, W. F., and Jones, J. W., 1982: *Refrigeration and Air-Conditioning,* McGraw-Hill Book Company, New York.

**Steam-Jet (Ejector) Systems** These systems substitute an ejector for a mechanical compressor in a vapor compression system. Since refigerant is water, maintaining temperatures lower than the environment requires that the pressure of water in the evaporator must be below atmospheric. A typical arrangement for the steam-jet refrigeration cycle is shown in Fig. 11-104.

***Main Components*** The main components of steam-jet refrigeration systems are:

1. *Primary steam ejector.* A kinetic device that utilizes the momentum of a high-velocity jet to entertain and accelerate a slower-moving medium into which it is directed. High-pressure steam is delivered to the nozzle of the ejector. The steam expands while flowing through the nozzle where the velocity increases rapidly. The velocity of steam leaving the nozzle is around 1200 m/s. Because of this high velocity, flash vapor from the tank is continually aspired into the moving steam. The mixture of steam and flash vapor then enters the diffuser section where the velocity is gradually reduced because of increasing cross-sectional area. The energy of the high-velocity steam compresses the vapor during its passage through the diffuser, raising it's temperature above the temperature of the condenser cooling water.

2. *Condenser.* The component of the system where the vapor condenses and where the heat is rejected. The rate of heat rejected is:

$$Q_{cond} = (W_s + W_w)\, hfg \qquad (11\text{-}107)$$

where: $Q_{cond}$ = heat rejection (kW)
$W_s$ = primary booster steam rate (kg/s)
$W_w$ = flash vapor rate (kg/s)
$hfg$ = latent heat (kJ/kg)

The condenser design, surface area, and condenser cooling water quantity should be based on the highest cooling water temperature likely to be encountered. If the inlet cooling water temperature becomes hotter then the design, the primary booster (ejector) may cease functioning because of the increase in condenser pressure.

Two types of condensers could be used: the surface condenser (shown in Fig. 11-104) and the barometric or jet condenser (Fig. 11-105). The surface condenser is of shell-and-tube design with water flowing through the tubes and steam condensed on the outside surface. In the jet condenser, condensing water and the steam being condensed are mixed directly, and no tubes are provided. The jet condenser can be barometric or a low-level type. The barometric condenser requires a height of ~10 m above the level of the water in the hot well. A tailpipe of this length is needed so that condenser water and condensate can drain by gravity. In the low-level jet type, the tailpipe is eliminated, and it becomes necessary to remove the condenser water and condensate

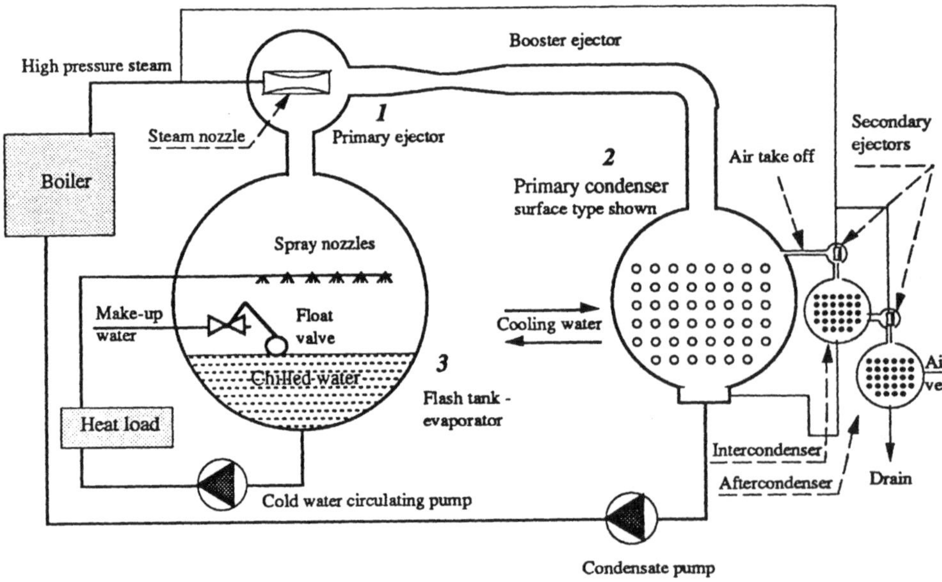


**FIG. 11-104** Ejector (steam-jet) refrigeration cycle (with surface-type condenser).

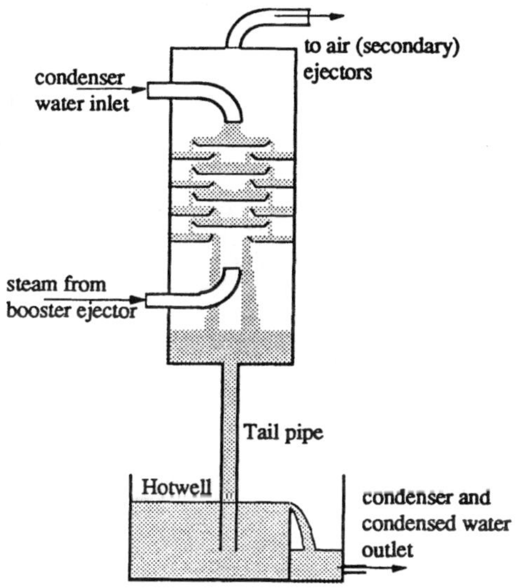


**FIG. 11-105** Barometric condenser for steam-jet system.

by pumping from the condenser to the hot well. The main advantages of the jet condenser are low maintenance with the absence of tubes and the fact that condenser water of varying degrees of cleanliness may be used.

3. *Flash tank.* This is the evaporator of the ejector system and is usually a large-volume vessel where large water-surface area is needed for efficient evaporative cooling action. Warm water returning from the process is sprayed into the flash chamber through nozzles (sometimes cascades are used for maximizing the contact surface, being less susceptible to carryover problems) and the cooled effluent is pumped from the bottom of the flash tank.

When the steam supply to one ejector of a group is closed, some means must be provided to for preventing the pressure in the condenser and flash tank from equalizing through that ejector. A compartmental flash tank is frequently used for such purposes. With this arrangement, partitions are provided so that each booster is operating on its own flash tank. When the steam is shut off to any booster, the valve to the inlet spray water to that compartment also is closed.

A float valve is provided to control the supply of makeup water to replace the water vapor that has flashed off. The flash tank should be insulated.

***Applications*** The steam-jet refrigeration is suited for:

1. Processes where direct vaporization is used for concentration or drying of heat-sensitive foods and chemicals, where, besides elimination of the heat exchanger, preservation of the product quality is an important advantage.

2. Enabling the use of hard or even sea water for heat rejection e.g. for absorption of gases ($CO_2$, $SO_2$, $ClO_2$ . . . ) in chilled water (desorption is provided simultaniously with chilling) when a direct contact barometric condenser is used.

Despite being simple, rugged, reliable, requiring low maintenance, low cost, and vibration free, steam-jet systems are not widely accepted in water chilling for air-conditioning due to characteristics of the cycle.

***Factors Affecting Capacity*** Ejector (steam-jet) units become attractive when cooling relatively high-temperature chilled water with a source of about 7 bar gauge waste steam and relatively cool condensing water. The factors involved with steam-jet capacity include the following:

1. *Steam pressure.* The main boosters can operate on steam pressures from as low as 0.15 bar up to 7 bar gauge. The quantity of steam required increases rapidly as the steam pressure drops (Fig. 11-106). The best steam rates are obtained with about 7 bar. Above this pressure the change in quantity of steam required is practically

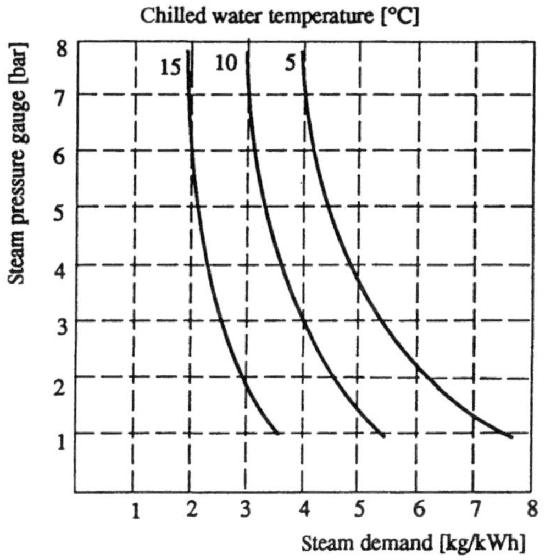


**FIG. 11-106** Effect of steam pressure on steam demand at 38°C condenser temperature (ASHRAE 1983 Equipment Handbook).

negligible. Ejectors must be designed for the highest available steam pressure, to take advantage of the lower steam consumption for various steam-inlet pressures.

The secondary ejector systems used for removing air require steam pressures of 2.5 bar or greater. When the available steam pressure is lower than this, an electrically driven vacuum pump is used for either the final secondary ejector or for the entire secondary group. The secondary ejectors normally require 0.2–0.3 kg/h of steam per kW of refrigeration capacity.

2. *Condenser water temperature.* In comparison with other vapor-compression systems, steam-jet machines require relatively large water quantities for condensation. The higher the inlet-water temperature, the higher are the water requirements (Fig. 11-107). The condensing water temperature has an important effect on steam rate per refrigeration effect, rapidly decreasing with colder condenser cooling water. Figure 11-108 presents data on steam rate versus condenser water inlet for given chilled-water outlet temperatures and steam pressure.

3. *Chilled-water temperature.* As the chilled-water outlet temperature decreases, the ratio of steam/refrigeration effect decreases, thus increasing condensing temperatures and/or increasing the condensing-water requirements.

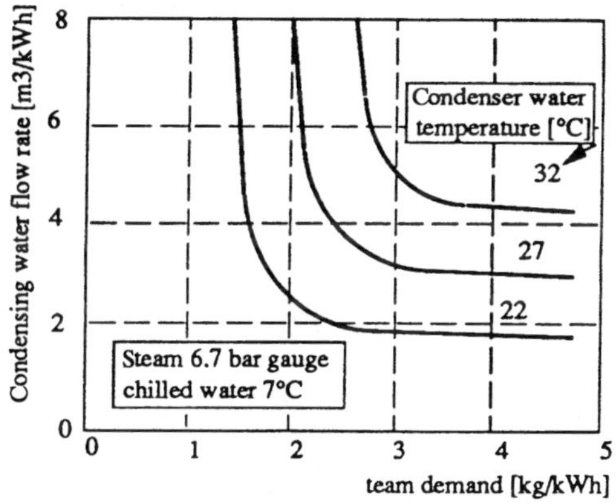


**FIG. 11-107** Steam demand versus condenser-water flow rate.

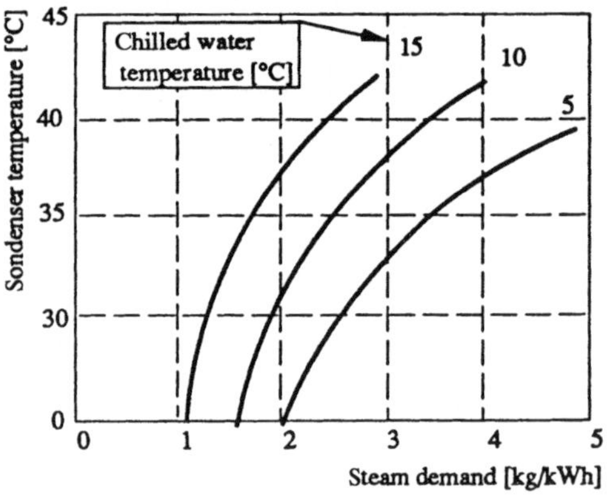


**FIG. 11-108** Steam demand versus chilled-water temperature for typical steam-jet system (ASHRAE 1983 Equipment Handbook).

Unlike other refrigeration systems, the chilled-water flow rate is of no particular importance in steam-jet system design, because there is, due to direct heat exchange, no influence of evaporator tube velocities and related temperature differences on heat-transfer rates. Widely varying return chilled-water temperatures have little effect on steam-jet equipment.

**Multistage Systems** The majority of steam-jet systems being currently installed are multistage. Up to five stage systems are in commercial operation.

**Capacity Control** The simplest way to regulate the capacity of most steam vacuum refrigeration systems is to furnish several primary boosters in parallel and operate only those required to handle the heat load. It is not uncommon to have as many as four main boosters on larger units for capacity variation. A simple automatic on-off type of control may be used for this purpose. By sensing the chilled-water temperature leaving the flash tank, a controller can turn steam on and off to each ejector as required.

Additionally, two other control systems which will regulate steam flow or condenser-water flow to the machine are available. As the condenser-water temperature decreases during various periods of the year, the absolute condenser pressure will decrease. This will permit the ejectors to operate on less steam because of the reduced discharge pressure. Either the steam flow or the condenser water quantities can be reduced in order to lower operating costs at other then design periods. The arrangement selected depends on cost considerations between the two flow quantities. Some systems have been arranged for a combination of the two, automatically reducing steam flow down to a point, followed by a reduction in condenser-water flow. For maximum operating efficiency, automatic control systems are usually justifiable in keeping operating cost to a minimum without excessive operator attention. In general, steam savings of about 10 percent of rated booster flow are realized for each 2.5°C reduction in condensing-water temperature below the design point.

In some cases, with relatively cold inlet condenser water it has been possible to adjust automatically the steam inlet pressure in response to chilled-water outlet temperatures. In general, however, this type of control is not possible because of the differences in temperature between the flash tank and the condenser. Under usual conditions of warm condenser-water temperatures, the main ejectors must compress water vapor over a relatively high ratio, requiring an ejector with entirely different operating characteristics. In most cases, when the ejector steam pressure is throttled, the capacity of the jet remains almost constant until the steam pressure is reduced to a point at which there is a sharp capacity decrease. At this point, the ejectors are unstable, and the capacity is severely curtailed. With a sufficient increase in steam pressure, the ejectors will once again become stable and operate at their deign capacity. In effect, steam jets have a vapor-handling capacity fixed by the pressure at the suction inlet. In order for the ejector to operate along its characteristic pumping curve, it requires a certain minimum steam flow rate which is fixed for any particular pressure in the condenser. (For further information on the design of ejectors, see Sec. 6)

***Further reading and reference:*** *ASHRAE 1983 Equipment Handbook;* Spencer, E., 1961, *New Development in Steam Vacuum Refrigeration,* ASHRAE Transactions Vol. 67, p. 339.

**Refrigerants** A refrigerant is any body or substance which acts as a cooling agent by absorbing heat from another body or substance which has to be cooled. Primary refrigerants are those that are used in the refrigeration systems, where they alternately vaporize and condense as they absorb or and give off heat respectively. Secondary refrigerants are heat transfer fluids or heat carriers. Refrigerant pairs in absorption systems are ammonia-water and lithium bromide-water, while steam (water) is used as a refrigerant in ejector systems. Refrigerants used in the mechanical refrigeration systems are far more numerous.

A list of the most significant refrigerants is presented in the *ASHRAE Handbook* Fundamentals. More data are shown in the Chap. 3 of this handbook—"Physical and Chemical Data." Due to rapid changes in refrigerant, issue readers are advised to consult the most recent data and publications at the time of application.

The first refrigerants were natural: air, ammonia, $CO_2$, $SO_2$, and so on. Fast expansion of refrigeration in the second and third quarters of the 20th century is marked by the new refrigerants, chlorofluorocarbons (CFC) and hydrochlorofluorocarbons (HCFC). They are halocarbons which contain one or more of three halogens chlorine, fluorine, and bromine (Fig. 11-109). These refrigerants introduced many advantageous qualities compared to most of the existing refrigerants: odorless, nonflamable, nonexplosive, compatible with the most engineering materials, reasonably high COP, and nontoxic.

In the last decade, the refrigerant issue is extensively discussed due to the accepted hypothesis that the chlorine and bromine atoms from halocarbons released to the environment were using up ozone in the stratosphere, depleting it specially above the polar regions. Montreal Protocol and later agreements ban use of certain CFCs and halon compounds. It seems that all CFCs and most of the HCFCs will be out of production by the time this text will be published.

Chemical companies are trying to develop safe and efficient refrigerant for the refrigeration industry and application, but uncertainty in CFC and HCFC substitutes is still high. When this text was written HFCs were a promising solution. That is true especially for the R134a which seems to be the best alternative for R12. Substitutes for R22 and R502 are still under debate. Numerous ecologists and chemists are for

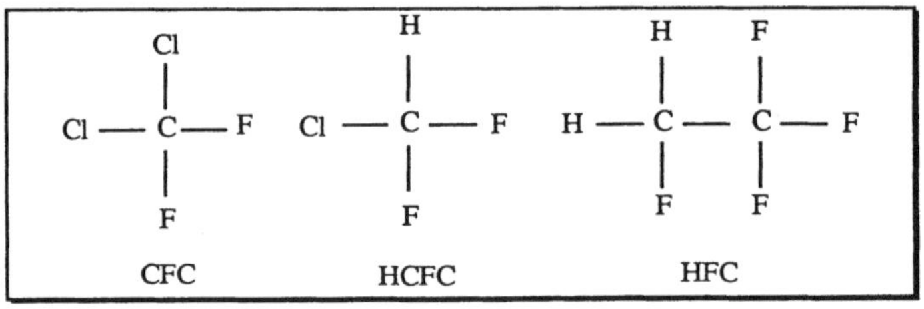


**FIG. 11-109** Halocarbon refrigerants.

an extended ban on HFCs as well, mostly due to significant use of CFCs in production of HFCs. Extensive research is ongoing to find new refrigerants. Many projects are aimed to design and study refrigerant mixtures, both azeotropic (mixture which behaves physically as a single, pure compound) and zeotropic having desirable qualities for the processes with temperature glides in the evaporator and the condenser.

Ammonia (R717) is the single natural refrigerant being used extensively (beside halocarbons). It is significant in industrial applications for its excellent thermodynamic and thermophysical characteristics. Many engineers are considering ammonia as a CFC substitute for various applications. Significant work is being done on reducing the refrigerant inventory and consequently problems related to leaks of this fluid with strong odor. There is growing interest in hydrocarbons in some contries, particularly in Europe. Indirect cooling (secondary refrigeration) is under reconsideration for many applications.

Due to the vibrant refrigerant issue it will be a challenge for every engineer to find the best solution for the particular application, but basic principles are the same. Good refrigerant should be:

- Safe: nontoxic, nonflamable, and nonexplosive
- Environmentally friendly
- Compatible with materials normally used in refrigeration: oils, metals, elastomers, etc.
- Desirable thermodynamic and thermophysical characteristics:
  - High latent heat
  - Low specific volume of vapor
  - Low compression ratio
  - Low viscosity
  - Reasonably low pressures for operating temperatures
  - Low specific heat of liquid
  - High specific heat of vapor
  - High conductivity and other heat transfer related characteristics
  - Reasonably low compressor discharge temperatures
  - Easily detected if leaking
  - High dielectric constant
  - Good stability

**Secondary Refrigerants (Antifreezes or Brines)** These are mostly liquids used for transporting heat energy from the remote heat source (process heat exchanger) to the evaporator of the refrigeration system. Antifreezes or brines do not change state in the process, but there are examples where some secondary refrigerants are either changing state themselves, or just particles which are carried in them.

Indirect refrigeration systems are more prevalent in the chemical industry than in the food industry, commercial refrigeration, or comfort air-conditioning. This is even more evident in the cases where a large amount of heat is to be removed or where a low temperature level is involved. Advantage of an indirect system is centralization of refrigeration equipment, which is specially important for relocation of refrigeration equipment in a nonhazardous area, both for people and equipment.

***Salt Brines*** The typical curve of freezing point is shown in Fig. 11-110. Brine of concentration $x$ (water concentration is 1-$x$) will not solidify at 0°C (freezing temperature for water, point A). When the temperature drops to B, the first crystal of ice is formed. As the temperature decreases to C, ice crystals continue to form and their mixture with the brine solution forms the slush. At the point C there will be part ice in the mixture $l_2/(l_1 + l_2)$, and liquid (brine) $l_1/(l_1 + l_2)$. At point D there is mixture of $m_1$ parts eutectic brine solution $D_1$ [concentration $m_1/(m_1 + m_2)$], and $m_2$ parts of ice [concentration $m_2/(m_1 + m_2)$]. Cooling the mixture below D solidifies the entire solution at the eutectic temperature. Eutectic temperature is the lowest temperature that can be reached with no solidification.

It is obvious that further strengthening of brine has no effect, and can cause a different reaction—salt sometimes freezes out in the installations where concentration is too high.

Sodium chloride, an ordinary salt (NaCl), is the least expensive per volume of any brine available. It can be used in contact with food and in open systems because of its low toxicity. Heat transfer coefficients are relatively high. However, its drawbacks are it has a relatively high freezing point and is highly corrosive (requires inhibitors thus must be checked on a regular schedule).

Calcium chloride ($CaCl_2$) is similar to NaCl. It is the second lowest-cost brine, with a somewhat lower freezing point (used for temperatures as low as −37°C). Highly corrosive and not appropriate for direct contact with food. Heat transfer coefficients are rapidly reduced at

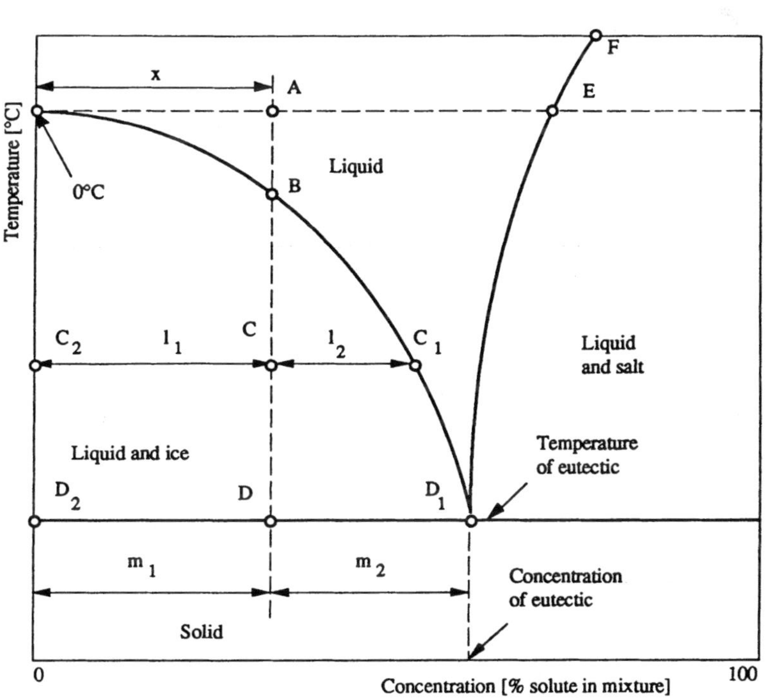


**FIG. 11-110** Phase diagram of the brine.

temperatures below −20°C. The presence of magnesium salts in either sodium or calcium chloride is undesirable because they tend to form sludge. Air and carbon dioxide are contaminants and excessive aeration of the brine should be prevented by use of close systems. Oxygen, required for corrosion, normally comes from the atmosphere and dissolves in the brine solution. Dilute brines dissolve oxygen more readily and are generally more corrosive than concentrated brines. It is believed that even a closed brine system will not prevent the infiltration of oxygen.

To adjust alkaline condition to pH 7.0–8.5 use caustic soda (to correct up to 7.0) or sodium dichromate (to reduce excessive alkalinity below pH 8.5). Such slightly alkaline brines are generally less corrosive than neutral or acid ones, although with high alkalinity the activity may increase.

If the untreated brine has the proper pH value, the acidifying effect of the dichromate may be neutralized by adding commercial flake caustic soda (76 percent pure) in quantity that corresponds to 27 percent of sodium dichromate used. Caustic soda must be thoroughly dissolved in warm water before it is added to the brine.

Recommended inhibitor (sodium dichromate) concentrations are 2 kg/m$^3$ of $CaCl_2$ and 3.2 kg/m$^3$ of NaCl brine. Sodium dichromate when dissolved in water or brine makes the solution acid. Steel, iron, copper, or red brass can be used with brine circulating systems. Calcium chloride systems are generally equipped with all-iron-and-steel pumps and valves to prevent electrolysis in event of acidity. Copper and red brass tubing are used for calcium chloride evaporators. Sodium chloride systems are using all-iron or all-bronze pumps.

**Organic Compounds (Inhibited Glycols)** *Ethylene glycol* is colorless and practically odorless and completely miscible with water. Advantages are low volatility and relatively low corrosivity when properly inhibited. Main drawbacks are relatively low heat-transfer coefficients at lower temperatures due to high viscosities (even higher than for propylene glycol). It is somewhat toxic, but less harmful than methanol water solutions. It is not appropriate for food industry and should not stand in open containers. Preferably waters that are classified as soft and are low in chloride and sulfate ions should be used for preparation of ethylene glycol solution.

Pure ethylene glycol freezes at −12.7°C. Exact composition and temperature for eutectic point are unknown, since solutions in this region turn to viscous, glassy mass that makes it difficult to determine the true freezing point. For the concentrations lower than eutectic, ice forms on freezing, while on the concentrated, solid glycol separates from the solution.

Ethylene glycol normally has pH of 8.8 to 9.2 and should not be used below 7.5. Addition of more inhibitor can not restore the solution to original condition. Once inhibitor has been depleted, it is recommended that the old glycol be removed from the system and the new charge be installed.

*Propylene glycol* is very similar to ethylene glycol, but it is not toxic and is used in direct contact with food. It is more expensive and, having higher viscosity, shows lower heat transfer coefficients.

*Methanol water* is an alcohol-base compound. It is less expensive than other organic compounds and, due to lower viscosity, has better heat transfer and pressure drop characteristics. It is used up to −35°C. Disadvantages are (1) considered more toxic than ethylene glycol and thus more suitable for outdoor applications (2) flammable and could be assumed to be a potential fire hazard.

For ethylene glycol systems copper tubing is often used (up to 3 in), while pumps, cooler tubes, or coils are made of iron, steel, brass, copper, or aluminum. Galvanized tubes should not be used in ethylene glycol systems because of reaction of the inhibitor with the zinc.

Methanol water solutions are compatible with most materials but in sufficient concentration will badly corrode aluminum.

*Ethanol water* is a solution of denatured grain alcohol. Its main advantage is that it is nontoxic and thus is widely used in the food and chemical industry. By using corrosion inhibitors it could be made noncorrosive for brine service. It is more expensive than methanol water and has somewhat lower heat transfer coefficients. As an alcohol derivate it is flammable.

Secondary refrigerants shown below, listed under their generic names, are sold under different trade names. Some other secondary refrigerants appropriate for various refrigeration application will be listed under their trade names. More data could be obtained from the manufacurer.

*Syltherm XLT (Dow Corning Corporation).* A silicone polymer (Dimethyl Polysiloxane); recommended temperature range −70°C to 250°C; odorless; low in acute oral toxicity; noncorrosive toward metals and alloys commonly found in heat transfer systems.

*Syltherm 800 (Dow Corning Corporation).* A silicone polymer (Dimethyl Polysiloxane); recommended temperature range −40°C to 400°C; similar to Syltherm XLT, more appropriate for somewhat higher temperatures; flash point is 160°C.

*D-limonene (Florida Chemicals).* A compound of optically active terpene ($C_{10}H_{16}$) derived as an extract from orange and lemon oils; limited data shows very low viscosity at low temperatures—only one centipoise at −50°C; natural substance having questionable stability.

*Therminol D-12 (Monsanto).* A synthetic hydrocarbon; clear liquid; recommended range −40°C to 250°C; not appropriate for contact with food; precautions against ignitions and fires should be taken with this product; could be found under trade names Santotherm or Gilotherm.

*Therminol LT (Monsanto).* Akylbenzene, synthetic aromatic ($C_{10}H_{14}$); recommended range −70°C to −180°C; not appropriate for contact with food; precautions against ignitions and fire should be taken dealing with this product.

*Dowtherm J (Dow Corning Corporation).* A mixture of isomers of an alkylated aromatic; recommended temperature range −70°C to 300°C; noncorrosive toward steel, common metals and alloys; combustible material; flash point 58°C; low toxic; prolonged and repeated exposure to vapors should be limited 10 ppm for daily exposures of eight hours.

*Dowtherm Q (Dow Corning Corporation).* A mixture of dyphenylehane and alkylated aromatics; recommended temperature range −30°C to 330°C; combustible material; flash point 120°C; considered low toxic, similar to Dowtherm J.

**Safety in Refrigeration Systems** This is of paramount importance and should be considered at every stage of installation.

The design engineer should have safety as the primary concern by choosing suitable system and refrigerant: selecting components, choosing materials and thicknesses of vessels, pipes, and relief valves of pressure vessels, proper venting of machine rooms, and arranging the equipment for convenient access for service and maintenance (piping arrangements, valve location, machine room layout, etc.). He or she should conform to the stipulation of the safety codes, which is also important for the purpose of professional liability.

During construction and installation, the installer's good decisions and judgments are crucial for safety, because design documentation never specifies all details. This is especially important when there is reconstruction or repair while the facility has been charged.

During operation the plant is in the hands of the operating personnel. They should be properly trained and familiar with the installation. Very often, accidents are caused by an improper practice, such as making an attempt to repair when proper preparation is not made. Operators should be trained in first-aid procedures and how to respond to emergencies.

Most frequently needed standards and codes are listed below, and the reader can find comments in: W. F. Stoecker: *Industrial Refrigeration,* Vol. 2. Ch. 12, Business News Publishing Co., Troy, MI. 1995; *ASHRAE Handbook* Refrigeration System and Applications, 1994, Ch. 51. Some important standards and codes on safety that a refrigeration engineer should consult are: ANSI/ASHRAE Standard 15-92—Safety Code for Mechanical Refrigeration, ASHRAE, Atlanta GA, 1992; ANSI/ASHRAE Standard 34-92—Number Designation of Refrigerants, ASHRAE, Atlanta GA, 1992; ANSI/ASME Boiler and Pressure Vessel Code, ASME, New York, 1989; ANSI/ASME Code for Pressure Piping, B31, B31.5–1987, ASME, New York, 1987; ANSI/IIAR 2—1984, Equipment, Design and Installation of Ammonia Mechanical Refrigeration Systems, IIAR, Chicago, 1984; IIAR Minimum Safety Criteria for a Safe Ammonia Refrigeration Systems, Bulletin 109; IIAR, IIAR Start-up, Inspection, and Maintenance of Ammonia Mechanical Refrigeration Systems, Bulletin 110, Chicago, 1988; IIAR Recommended Procedures in Event of Ammonia Spills, Bulletin No. 106, IIAR, Chicago, 1977; A Guide to Good Practices for the Operation of an Ammonia Refrigeration System, IIAR Bulletin R1, 1983.

# CRYOGENIC PROCESSES

## INTRODUCTION

Cryogenics, the production of low temperatures, is a major business in the United States with an annual market in 2003 of almost 14 billion dollars. It is a very diverse supporting technology, a means to an end and not an end in itself. For example, the combined production of oxygen and nitrogen, obtained by the cryogenic separation of air, accounts for 15 percent of the total annual U.S. production of $2.9 \times 10^{11}$ kg (639 billion lb) of organics and inorganics (2003 *C & EN Annual Report*). Liquid hydrogen production, in the last four decades, has risen from laboratory quantities to a level of over 2.2 kg/s, first spurred by nuclear weapons development and later by the United States space program. Similarly, the space age increased the need for liquid helium by more than a factor of ten, requiring the construction of large plants to separate helium from natural gas by cryogenic means. The demands for energy have likewise accelerated the construction of large base-load liquefied natural gas (LNG) plants around the world and have been responsible for the associated domestic LNG industry of today with its use of peak shaving plants.

Freezing as a means of preserving food dates back to 1840. However, today the food industry uses large quantities of liquid nitrogen for this purpose and uses the refrigerant in frozen-food transport systems. In biological applications liquid-nitrogen cooled containers are routinely used to preserve whole blood, tissue, bone marrow, and animal semen for extended periods of time. Cryogenic surgery has become accepted in curing such involuntary disorders as Parkinson's disease. Medical analysis of patients has increased in sophistication with the use of *magnetic resonance imaging* (MRI) which utilizes cryogenically cooled superconducting magnets. Finally, one must recognize the role that cryogenics plays in the chemical-processing industry with the recovery of valuable feedstocks from natural gas streams, upgrading the heat content of fuel gas, purification of various process and waste streams, production of ethylene, as well as other chemical processes.

## PROPERTIES OF CRYOGENIC FLUIDS

There are presently several database programs of thermodynamic properties data developed specifically for fluids commonly associated with low temperature processing including helium, hydrogen, neon, nitrogen, oxygen, argon, and methane. For example, the NIST Standard Reference Database 12, Version 3.0 includes a total of 34 thermophysical properties for seventeen fluids in the database. Cryodata Inc. provides a similar computer version for 28 pure fluids as well as for mixtures incorporating many of these fluids. To fully appreciate the major effort in developing these thermophysical properties, consult the presentation by Jacobsen and coworkers (*Thermodynamics Properties of Cryogenic Fluids*, Plenum Press, New York, 1997). A few peculiarities associated with the fluids of helium, hydrogen, oxygen, and air need to be noted below.

Liquid helium-4 can exist in two different liquid phases: liquid helium I, the normal liquid, and liquid helium II, the superfluid, since under certain conditions the latter fluid acts as if it had no viscosity. The phase transition between the two liquid phases is identified as the lambda line and where this transition intersects the vapor-pressure curve is designated as the lambda point. Thus, there is no triple point for this fluid as for other fluids. In fact, solid helium can only exist under a pressure of 2.5 MPa or more.

A unique property of hydrogen is that it can exist in two different molecular forms: orthohydrogen and parahydrogen. (This is also true for deuterium, an isotope of hydrogen with an atomic mass of 2.) The thermodynamic equilibrium composition of the ortho- and para-varieties is temperature dependent. The equilibrium mixture of 75 percent orthohydrogen and 25 percent parahydrogen at ambient temperatures is recognized as normal hydrogen.

In contrast to other cryogenic fluids, liquid oxygen is slightly magnetic. It is also chemically reactive, particularly with hydrocarbon materials. Oxygen thus presents a safety problem and requires extra precautions in handling.

Since air is a mixture of predominantly nitrogen, oxygen, and a host of lesser impurities, there has been less interest in developing precise thermodynamic properties. A widely used correlation of thermodynamic properties is that published by NIST (NIST Database 72, NIST AIRPROPS, Version 1.0).

## PROPERTIES OF SOLIDS

A knowledge of the properties and behavior of materials used in any cryogenic system is essential for proper design considerations. Often the choice of materials for the construction of cryogenic equipment will be dictated by other considerations besides mechanical properties as, for example, thermal conductivity (heat transfer along a structural member), thermal expansivity (expansion and contraction during cycling between ambient and low temperatures), and density (mass of the system). Since properties at low temperatures are often significantly different from those at ambient temperature, there is no substitute for test data on a truly representative sample specimen when designing for the limit of effectiveness of a cryogenic material or structure. For example, some metals including elements, intermetallic compounds and alloys exhibit the phenomenon of superconductivity at very low temperatures. The properties that are affected when a material becomes superconducting include specific heat, thermal conductivity, electrical resistance, magnetic permeability, and thermoelectric effect. As a result, the use of superconducting metals in the construction of equipment for temperatures lower than 10 K needs to be evaluated carefully. (High temperature superconductors because of their brittle ceramic structure are generally not considered as construction materials.)

**Structural Properties at Low Temperatures** It is most convenient to classify metals by their lattice symmetry for low temperature mechanical properties considerations. The *face-centered-cubic* (fcc) metals and their alloys are most often used in the construction of cryogenic equipment. Al, Cu Ni, their alloys, and the austenitic stainless steels of the 18-8 type are fcc and do not exhibit an impact ductile-to-brittle transition at low temperatures. As a general rule, the mechanical properties of these metals with the exception of 2024-T4 aluminum, improve as the temperature is reduced. Since annealing of these metals and alloys can affect both the ultimate and yield strengths, care must be exercised under these conditions.

The *body-centered-cubic* (bcc) metals and alloys are normally classified as undesirable for low temperature construction. This class includes Fe, the martensitic steels (low carbon and the 400-series stainless steels), Mo, and Nb. If not brittle at room temperature, these materials exhibit a ductile-to-brittle transition at low temperatures. Cold working of some steels, in particular, can induce the austenite-to-martensite transition.

The *hexagonal-close-packed* (hcp) metals generally exhibit mechanical properties intermediate between those of the fcc and bcc metals. For example Zn encounters a ductile-to-brittle transition whereas Zr and pure Ti do not. The latter and their alloys with a hcp structure remain reasonably ductile at low temperatures and have been used for many applications where weight reduction and reduced heat leakage through the material have been important. However, small impurities of O, N, H, and C can have a detrimental effect on the low temperature ductility properties of Ti and its alloys.

Plastics increase in strength as the temperature is decreased, but this is also accompanied by a rapid decrease in elongation in a tensile test and a decrease in impact resistance. Teflon and glass-reinforced plastics retain appreciable impact resistance as the temperature is lowered. The glass-reinforced plastics also have high strength-to-weight and strength-to-thermal conductivity ratios. All elastomers, on the other hand, become brittle at low temperatures. Nevertheless, many of these materials including rubber, Mylar, and nylon can be used for static seal gaskets provided they are highly compressed at room temperature prior to cooling.

The strength of glass under constant loading also increases with decrease in temperature. Since failure occurs at a lower stress when the glass surface contains surface defects, the strength can be improved by tempering the surface.

**Thermal Properties at Low Temperatures** The thermodynamic properties of most gases at low temperature approximate quite clearly those of an ideal gas, and thus the ideal equation of state is adequate for most cryogenic engineering designs. However, the heat capacity of an ideal gas depends on its molecular composition. In most cryogenic gases, electronic and vibrational degrees of freedom are not excited, while rotational contributions to the specific heat at constant-volume conditions have reached their full classical value. Under these conditions the heat capacity depends only on whether the gas is composed of isolated atoms, linear molecules, or nonlinear molecules. The values for $C_v$, the heat capacity at constant volume are $3/2R$, $5/2R$ and $3R$, respectively, where $R$ is the universal gas constant. For $C_p$, the heat capacity at constant pressure, the corresponding values are $5/2R$, $7/2R$, and $4R$, respectively. The $C_v$, data for a number of real gases presented by Barron and White (*Heat Capacity and Thermal Expansion at Low Temperatures,* Kluwer Academic/Plenum Publisher, New York, 1999, p. 132) show that these values provide a good approximation over wide ranges in temperatures, although there are large deviations for hydrogen below about 200 K and for methane at higher temperatures due to vibrational excitation. For solids, the Debye model developed with the aid of statistical mechanics and quantum theory gives a satisfactory representation of the specific heat with temperature. Procedures for calculating values of $\Theta_D$, the Debye characteristic temperature, using either elastic constants, the compressibility, the melting point, or the temperature dependence of the expansion coefficient are outlined by Barron (*Cryogenic Systems, 2d ed.,* Oxford University Press, 1985, pp. 24–29).

Adequate prediction of the thermal conductivity for pure metals can be made by means of the Wiedeman-Franz law which states that the ratio of the thermal conductivity to the product of the electrical conductivity and the absolute temperature is a constant. This ratio for high-conductivity metals extrapolates essentially to the Sommerfeld value of $2.449 \times 10^{-8}$ W $\Omega$/K$^2$ at 0 K, but falls considerably below it at higher temperatures. High-purity aluminum and copper exhibit peaks in thermal conductivity between 20 to 50 K, but these peaks are rapidly suppressed with increased impurity levels and cold work of the metal. Some metals including Monel, Inconel, stainless steel, and structural and aluminum alloys show a steady decrease in thermal conductivity with a decrease in temperature.

All cryogenic liquids except hydrogen and helium have thermal conductivities that increase as the temperature is decreased. For these two exceptions, the thermal conductivity decreases with a decrease in temperature. The kinetic theory of gases correctly predicts the decrease in thermal conductivity of all gases when the temperature is lowered.

The expansion coefficient of a solid can be estimated with the aid of an approximate thermodynamic equation of state for solids which equates the thermal expansion coefficient $\beta$ with the quantity $\gamma C_v \rho / B$ where $\gamma$ is the Grüneisen dimensionless ratio, $C_v$ is the specific heat of the solid, $\rho$ is the density of the material, and $B$ is the bulk modulus. For fcc metals the average value of the Grüneisen constant is near 2.3. However, there is a tendency for this constant to increase with atomic number.

**Electrical Properties at Low Temperatures** The electrical resistivity of most pure metallic elements at ambient and moderately low temperatures is approximately proportional to the absolute temperature. At very low temperatures, however, the resistivity (with the exception of superconductors) approaches a residual value almost independent of temperature. Alloys, on the other hand, have resistivities much higher than those of their constituent elements and resistance-temperature coefficients that are quite low. The electrical resistivity of alloys as a consequence is largely independent of temperature and may often be of the same magnitude as the room temperature value.

**Superconductivity** The physical state in which all resistance to the flow of direct-current electricity disappears is defined as superconductivity. The Bardeen-Cooper-Schriefer (BCS) theory has been reasonably successful in accounting for most of the basic features observed of the superconducting state for *low-temperature superconductors* (LTS) operating below 23 K. The advent of the ceramic *high-temperature superconductors* (HTS) by Bednorz and Miller (*Z. Phys.* B64, 189, 1989) has called for modifications to existing theories which are still actively being debated. The massive interest in the new superconductors that can be cooled with liquid nitrogen is presently underway with many new applications.

**TABLE 11-25 Composition and Critical Temperature $T_c$ for HTS Materials**

| Formula* | Accepted notation | Forms reported | | Critical $T_c$, K |
|---|---|---|---|---|
| Y-Ba-Cu-O† | YBCO | 123 | 124 | 80–92 |
| Bi-Sr-Ca-Cu-O | BSCCO | 2212 | 2223 | 80–110 |
| Tl-Ba-Ca-Cu-O | TBCCO | Several | | to 125 |
| Hg-Ba-Ca-Cu-O | HBCCO | 1201 | 1223 | 95–155‡ |

*Subscripts for compounds are not listed since there are generally several forms that can be produced (see column 3).

†Other rare earths may be substituted for Y (yttrium) providing new compounds with somewhat different properties.

‡Highest $T_c$ obtained while subjecting sample to external pressure.

Three important characteristics of the superconducting state are the critical temperature, the critical magnetic field, and the critical current. These parameters many times can be varied by using different materials or giving them special metallurgical treatments.

The alloy *niobium titanium* (NbTi) and the intermetallic compound of *niobium and tin* ($Nb_3$ Sn) are the most technologically advanced LTS materials presently available. Even though NbTi has a lower critical field and critical current density, it is often selected because its metallurgical properties favor convenient wire fabrication. In contrast, $Nb_3Sn$ is a very brittle material and requires wire fabrication under very well-defined temperature conditions. However, the recent discovery by Akimitsu and coworkers (*Nature* **10,** 63, 2001) of the superconducting property of $MgBr_2$ has spurred new activity in the development of still another LTS conductor for magnetic coils.

There are essentially four families of high-temperature superconductors under investigation for practical magnet applications. Table 11-25 shows that all HTS are copper oxide ceramics even though the oxygen content may vary. However, this variation generally has little effect on the physical properties of importance to superconductivity.

The most widely used development in HTS wire production is the powder-in-tube procedure with BSCCO ceramic materials. In this procedure very fine HTS powder, placed inside of a hollow silver tube, is fused as the tube length is mechanically increased to form a wire. Very high magnetic fields with this wire have been reported at 4 K; however, performance degrades substantially above 20 to 30 K.

HTS materials, because of their ceramic nature, are quite brittle. This has introduced problems relative to the winding of superconducting magnets. One solution is to first wind the magnet with the powder-in-tube wire before the ceramic powder has been bonded and then heat treat the desired configuration to form the final product. Another solution is to form the superconductor into such fine filaments that they remain sufficiently flexible even after the powder has been heat treated.

## REFRIGERATION AND LIQUEFACTION

A process for producing refrigeration or liquefaction at cryogenic temperatures usually involves ambient compression of a process fluid with heat rejection to a coolant. During the compression process, the enthalpy and entropy of the fluid are decreased. At the cryogenic temperature where heat is absorbed, the enthalpy and entropy are increased. The reduction in temperature of the process fluid is usually accomplished by heat exchange with a colder fluid and then followed by an expansion. This expansion may take place using either a throttling device (isenthalpic expansion) with only a reduction in temperature or a work-producing device (isentropic expansion) in which both temperature and enthalpy are decreased. Because of liquid withdrawal, a liquefaction system experiences an unbalanced flow in the heat exchanger while a refrigeration system with no liquid withdrawal system usually operates with a balanced flow in the heat exchanger, except where a portion of the flow is diverted through the work-producing expander.

**Principles** The performance of a real refrigerator is measured by the coefficient of performance, COP, defined as

$$COP = \frac{Q}{W} = \frac{\text{heat removed at low temperature}}{\text{net work input}} \qquad (11\text{-}108)$$

Another means of comparing the performance of a practical refrigerator is by the use of the figure of merit, FOM, defined as

$$FOM = \frac{COP}{COP_i} \qquad (11\text{-}109)$$

where COP is the coefficient of performance of the actual refrigerator system and $COP_i$ is the coefficient of performance for the thermodynamically ideal system. For a liquefier, the FOM is generally specified as

$$FOM = \frac{W_i/\dot{m}_f}{W/\dot{m}_f} \qquad (11\text{-}110)$$

where $W_i$ is the work of compression for the ideal cycle, $W$ is the work of compression for the actual cycle, and $\dot{m}_f$ is the mass rate liquefied in the ideal or actual cycle.

The methods of refrigeration and/or liquefaction generally used include (1) vaporization of a liquid, (2) application of the Joule-Thomson effect in a gas, and (3) expansion of a gas in a work-producing engine. Normal commercial refrigeration generally is accomplished in a vapor-compression process. Temperatures to about 200 K can be obtained by cascading vapor-compression processes in which refrigeration is accomplished by liquid evaporation. Below this temperature, isenthalpic or isentropic expansions are generally used either singly or in combination. With few exceptions, refrigerators using these methods also absorb heat by vaporization of the liquid.

If refrigeration is to be accomplished at a temperature range where no suitable liquid exists to absorb heat by evaporation, then a cold gas must be available to absorb the heat. This is generally accomplished by using a work-producing expansion engine.

**Expansion Types of Refrigerators** A thermodynamic process utilizing isenthalpic expansion to obtain cryogenic temperatures, and commonly referred to as the simple Linde or J-T cycle, is shown schematically with its corresponding temperature-entropy diagram in Fig. 11-111. The gaseous refrigerant is compressed at ambient temperature while essentially rejecting heat isothermally to a coolant. The compressed refrigerant is cooled countercurrently in a heat exchanger by the cold gas stream leaving the liquid reservoir before it enters the throttling valve. Upon expansion, Joule-Thomson cooling further reduces the temperature until, at steady state, a portion of the refrigerant is liquefied. For a refrigerator, the unliquefied fraction and the vapor formed by liquid evaporation from the absorbed heat $Q$ are warmed in the heat exchanger before returning to the intake of the compressor. Assuming no heat inleaks, as well as negligible kinetic and potential energy changes in the fluid, the refrigeration duty $Q$ is equivalent to $\dot{m}(h_1 - h_2)$, where the subscripts refer to the locations shown on Fig. 11-111. Applying Eq. 11-108, the coefficient of performance for the ideal J-T refrigerator is given by

$$COP = \frac{h_1 - h_2}{T_1[s_1 - s_2 - (h_1 - h_2)]} \qquad (11\text{-}111)$$

For a simple J-T liquefier, the liquefied portion is continuously withdrawn from the reservoir and only the unliquefied portion of the fluid is warmed in the countercurrent heat exchanger and returned to the compressor. The fraction $y$ that is liquefied is obtained by applying the first law to the heat exchanger, J-T valve, and liquid reservoir. This results in

$$y = \frac{h_1 - h_2}{h_1 - h_f} \qquad (11\text{-}112)$$

where $h_f$ is the specific enthalpy of the liquid being withdrawn. Maximum liquefaction occurs when the difference between $h_1$ and $h_2$ is maximized. To account for heat inleak, $q_L$, the relation needs to be modified to

$$y = \frac{h_1 - h_2 - q_L}{h_1 - h_f} \qquad (11\text{-}113)$$

with a resultant decrease in the fraction liquefied.

Refrigerants used in this process have a critical temperature well below ambient; consequently liquefaction by direct compression is not possible. In addition, the inversion temperature of the refrigerant must be above ambient temperature to provide initial cooling by the J-T process. Auxiliary refrigeration is required if the simple J-T cycle is to be used to liquefy neon, hydrogen, or helium whose inversion temperatures are below ambient. Liquid nitrogen is the optimum refrigerant for hydrogen and neon liquefaction systems, while liquid hydrogen is the normal refrigerant for helium liquefaction systems.

To reduce the work of compression in this cycle a two-stage or dual-pressure process may be used whereby the pressure is reduced by two successive isenthalpic expansions. Since the isothermal work of compression is approximately proportional to the logarithm of the pressure ratio, and the Joule-Thomson cooling is roughly proportional to the pressure difference, there is a greater reduction in compressor work than in refrigerating performance for this dual-pressure process.

In a work-producing expansion, the temperature of the process fluid is always reduced; hence, cooling does not depend on being below the inversion temperature prior to expansion. Additionally, the work-producing expansion results in a larger amount of cooling than in an isenthalpic expansion over the same pressure difference.

In large systems utilizing expanders, the work produced during expansion is conserved. In small refrigerators, the energy from the expansion is usually expended in a gas or hydraulic pump, or other suitable device. A schematic of a simple cold-gas refrigerator using this expansion principle and the corresponding temperature-entropy diagram is shown in Fig. 11-112. Gas compressed isothermally at ambient temperature is cooled in a heat exchanger by gas being warmed on its return to the compressor intake. Further cooling takes

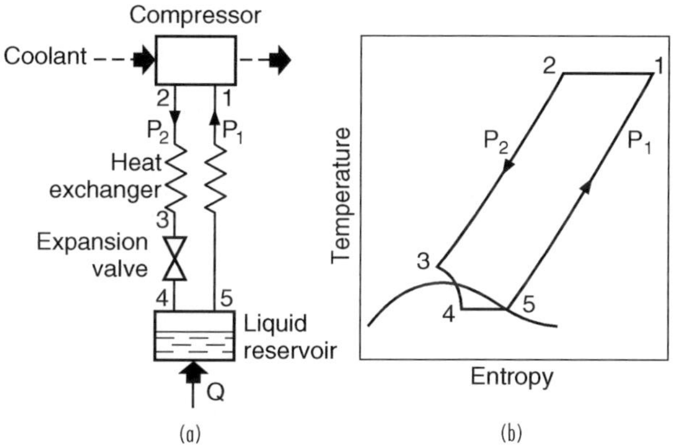


**FIG. 11-111** Refrigerator using simple J-T cycle.

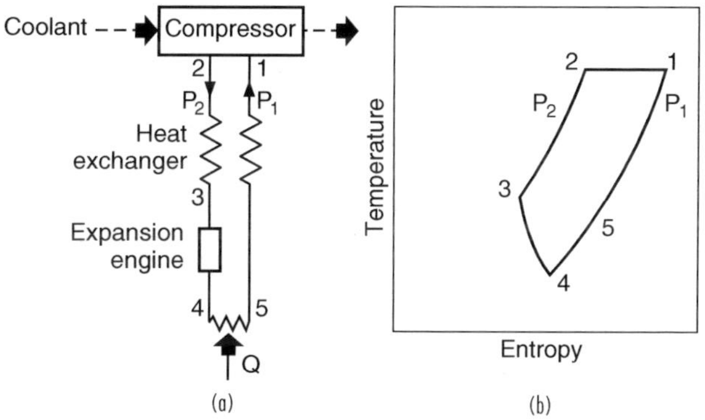


**FIG. 11-112** Cold-gas refrigerator.

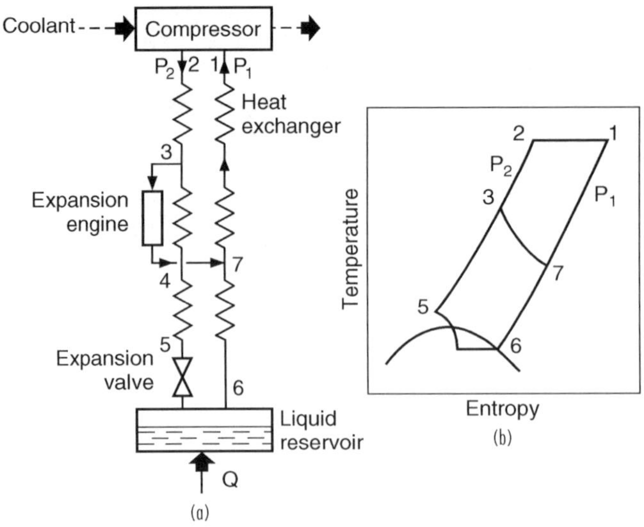


**FIG. 11-113** Claude cycle refrigerator utilizing both expansion processes.

place during the engine expansion. In practice this expansion is never truly isentropic, and is reflected by path 3-4 on the temperature-entropy diagram. This specific refrigerator produces a cold gas which absorbs heat from 4-5 and provides a method of refrigeration that can be used to obtain temperatures between those of the boiling points of the lower-boiling cryogens.

It is not uncommon to utilize both the isentropic and isenthalpic expansions in a cycle. This is done to avoid the technical difficulties associated with the formation of liquid in the expander. The Claude or expansion engine cycle is an example of a combination of these methods and is shown in Fig. 11-113 along with the corresponding temperature-entropy diagram.

The mixed refrigerant cycle was developed to meet the need for liquefying large quantities of natural gas to minimize transportation costs of this fuel. This cycle resembles the classic cascade cycle in principle and may best be understood by referring to that cycle. In the latter, the natural gas stream after purification is cooled successively by vaporization of propane, ethylene, and methane. Each refrigerant may be vaporized at two or three pressure levels to increase the natural gas cooling efficiency, but at a cost of considerable increased process complexity.

Cooling curves for natural gas liquefaction by the cascade process are shown in Fig. 11-114. It is evident that the cascade cycle efficiency can be improved by increasing the number of refrigerants employed.

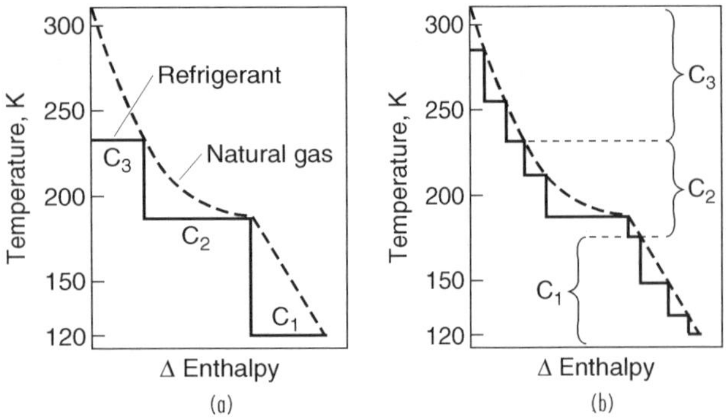


**FIG. 11-114** Three- and nine-level cascade cycle cooling curve for natural gas.

For the same refrigeration capacity, the actual work required for the nine-level cascade cycle depicted is approximately 80 percent of that required by the three-level cascade cycle. This increase in efficiency is achieved by minimizing the temperature difference between the refrigerant and the natural gas stream throughout each increment of the cooling curve.

The mixed refrigerant cycle is a variation of the cascade cycle and involves the circulation of a single multicomponent refrigerant stream. The simplification of the compression and heat exchange services in such a cycle can offer potential for reduced capital expenditure over the conventional cascade cycle.

Figure 11-115 shows the basic concepts for a mixed refrigerant cycle (Gaumer, *Advances in Cryogenic Engineering* Vol. 31, Plenum, New York, 1986, p. 1095). Variations of the cycle are proprietary with those cryogenic engineering firms that have developed the technology. However, all of the mixed refrigerant processes use a carefully prepared refrigerant mix which is repeatedly condensed, vaporized, separated, and expanded. Thus, these processes require

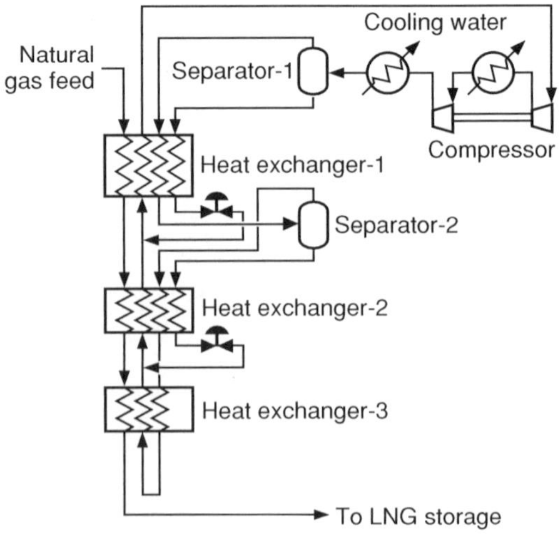


**FIG. 11-115** Mixed-refrigerant cycle.

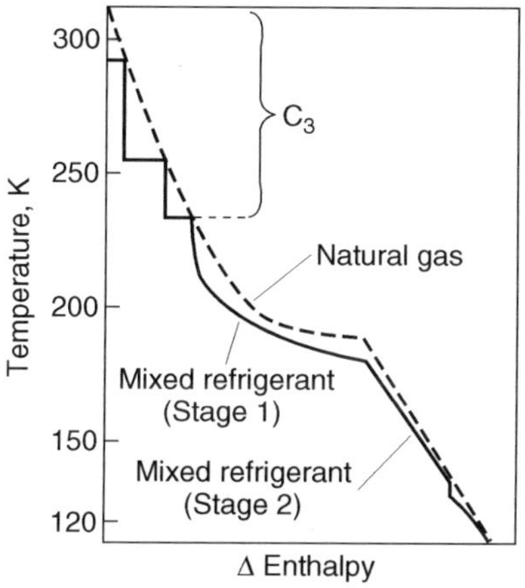


**FIG. 11-116** Propane precooled mixed-refrigerant cycle cooling curve for natural gas.

more complete knowledge of the thermodynamic properties of gaseous mixtures than those required in the expander or classical cascade cycles. This is particularly evident when cooling curves similar to the one shown in Fig. 11-116 are desired. An inspection of the mixed refrigerant cycle also shows that these processes must routinely handle two-phase flows in the heat exchangers.

**Miniature Refrigerators** Expanded space and science projects have provided a need for miniature cryogenic coolers designated as cryocoolers. Such coolers provide useful cooling from a fraction of a watt to several watts at temperature levels from 1 to 90 K. These coolers are used to increase the sensitivity and signal-to-noise ratio of detectors by providing the required cryogenic operating temperatures as well as cooling the optical components to decrease the detector background radiation. The types of coolers developed to meet various specific requirements include solid cryogen coolers, radiative coolers, mechanical coolers, $^3$He adsorption coolers, adiabatic demagnetization refrigerators, and liquid helium storage systems. Mechanical coolers are generally classified as regenerative or recuperative. Regenerative coolers use reciprocating components that move the working fluid back and forth through a regenerator. The recuperative coolers, on the other hand, use countercurrent heat exchangers to accomplish the heat transfer operation. The Stirling, Gifford-McMahon, and pulse tube cycles are typically regenerative coolers while the Joule-Thomson and Brayton cycles are associated with recuperative coolers.

The miniature split single-stage Stirling cooler developed by Philips Laboratories produces 5 W of cooling at 65 K with the aid of linear motors and magnetically moving parts. A smaller split Stirling cycle cooler that uses a stacked diaphragm spring rather than magnetic means to levitate the piston and displacer has been developed at Oxford University (Bradshaw, et al., *Advances in Cryogenic Engineering*, Vol. **31,** Plenum, New York, 1986, p. 801). The promise of higher reliability has spurred interest in the *pulse-tube refrigerator* (PTR). In the latest version of this device, an orifice and reservoir have been added to the warm end of the pulse tube (OPTR) to permit control of the phase shift required for optimum resonance in the system.

The Joule-Thomson cycle has also benefited from creative thinking. For example, Little (*6th International Cryocooler Conference,* Naval Postgraduate School, Monterey, CA, 1989, p. 3) has introduced a new method of fabricating J-T refrigerators using a photolithographic manufacturing process in which gas channels for the heat exchangers, expansion capillary, and liquid reservoir are etched on planar, glass substrates that are fused together to form the sealed refrigerator. These microminiature refrigerators have been made in a wide range of sizes and capacities.

Mixtures of highly polar gases are receiving considerable attention for J-T cycles since the magnitude of the Joule-Thomson coefficient increases with nonideality of the gas. For example, for pure nitrogen with a high pressure of 5 MPa, a low pressure of 0.1 MPa, and an ambient temperature of 300 K, the ideal efficiency at 77 K is only 8.6 percent of Carnot. For a mixture of 40% nitrogen, 17% methane, 15% ethane, and 28% propane, the ideal efficiency for the same operating conditions is 52 percent of Carnot. The higher efficiency possible permits the use of compressors for domestic and commercial refrigeration, thereby reducing the overall costs of the cryogenic unit.

**Thermodynamic Analyses of Cycles** The thermodynamic quality measure of either a piece of equipment or an entire process is its reversibility. The second law, or more precisely the entropy increase, is an effective guide to this degree of irreversibility. However, to obtain a clearer picture of what these entropy increases mean, it has become convenient to relate such an analysis to the additional work that is required to overcome these irreversibilities. The fundamental equation for such an analysis is

$$W = W_{\text{rev}} + T_o \sum \dot{m}\Delta s \tag{11-114}$$

where the total work, $W$, is the sum of the reversible work, $W_{\text{rev}}$, plus a summation of the losses in availability for various unit operations in the analysis. Application of this method has been demonstrated numerically by Timmerhaus and Flynn (*Cryogenic Process Engineering,* Plenum Press, 1989, p. 175).

Numerous analyses and comparisons of refrigeration and liquefaction cycles are available in the literature. Great care must be exercised in accepting these comparisons since it is quite difficult to put all processes on a strictly comparable basis. Many assumptions need to be made in the course of the calculations, and these can have considerable effect on the conclusions. Major factors upon which assumptions generally have to be made include heat leak, temperature differences in the exchangers, efficiencies of compressors and expanders, number of stages of compression, fraction of expander work recovered, state of expander exhaust, purity and condition of inlet gases, pressure drop due to fluid flow, and so on. In view of this fact, differences in power requirements of 10 to 20 percent can readily be due to differences in assumed variables and can negate the advantage of one cycle over another. Barron (*Cryogenic Systems, 2d ed.,* Oxford University Press, New York, 1985, p. 94] has made an analysis of some of the more common liquefaction systems described earlier that emphasize this point rather well.

## PROCESS EQUIPMENT

The equipment normally associated with cryogenic systems includes heat exchangers, compressors, expanders, throttling valves, and storage containers. Since the reciprocal or centrifugal compressors used generally operate at ambient temperatures, their operating principles are not covered here but in Sec. 10. Storage containers, are discussed later in Sec. 11.

**Heat Exchangers** Since most cryogens, with the exception of helium II behave as classical fluids, well-established principles of mechanics and thermodynamics at ambient temperature also apply for cryogens. Thus, similar conventional heat transfer correlations have been formulated for simple low-temperature heat exchangers. These correlations are described in terms of well-known dimensionless quantities such as the Nusselt, Reynolds, Prandtl, and Grashof numbers.

Because of the need to operate more efficiently at low temperatures, the simple heat exchangers have generally been replaced with more sophisticated types. Guidance for the development of such units for low-temperature service include the following factors:

1. Small temperature differences between inlet and exit streams to enhance efficiency.
2. Large surface area-to-volume ratio to minimize heat leak.
3. High heat transfer to reduce surface area.
4. Low mass to minimize start-up time.
5. Multichannel capability to minimize the number of units.
6. High-pressure capability to provide design flexibility.
7. Minimum pressure drop to reduce compression costs.
8. High reliability with minimal maintenance to reduce shutdowns.

Problems sometimes occur in trying to minimize the temperature difference at the cold end of the heat exchanger, particularly if the specific heat of the warm fluid decreases with decreasing temperature as is the case with gaseous hydrogen.

The selection of an exchanger for low-temperature operation is usually determined by the process-design requirements, mechanical-design limitations, and economic considerations. Laboratory needs are generally met by concentric tube and extended surface exchangers, while industrial needs are most often met by the coiled-tube, plate-fin, reversing, and regenerator types of exchangers.

The coiled-tube heat exchanger offers unique advantages, especially when dealing with low-temperature design conditions where (1) simultaneous heat transfer between more than two streams is desired, (2) a large number of heat transfer units is required, and (3) high operating pressures are involved. Heat transfer for single-phase flow of either gas or liquid on the tubeside is generally well represented by either the Colburn correlation or modified forms of the Dittus-Boelter relationship.

The shape of the cooling and warming curves in coiled-tube heat exchangers is affected by the pressure drop in both the tube and shellsides of the heat exchanger. This is particularly important for two-phase flows of multicomponent systems. For example, an increase in pressure drop on the shellside causes boiling to occur at a higher temperature, while an increase in pressure drop on the tubeside will cause condensation to occur at a lower temperature. The net result is both a decrease in the effective temperature difference between the two streams and a requirement for additional heat transfer area to compensate for these losses.

Plate-fin heat exchangers are about nine times as compact as conventional shell-and-tube heat exchangers with the same amount of surface area, weigh less than conventional heat exchangers, and withstand design pressures up to 6 MPa for temperatures between 4 and 340 K. Flow instability frequently becomes a limiting design parameter for plate-fin heat exchangers handling either boiling or condensing two-phase flows. This results in lower optimum economic mass flow velocities for plate-fin heat exchangers when compared with coiled-tube heat exchangers. The use of fins or extended surfaces in plate-fin or similar exchangers greatly increases the heat transfer area. Calculations using finned surfaces are outlined earlier in Sec. 11.

There are two basic approaches to heat-exchanger design for low temperatures: (1) the effectiveness-NTU approach and (2) the *log-mean-temperature-difference* (LMTD) approach. The LMTD approach is used most frequently when all the required mass flows are known and the area of the exchanger is to be determined. The effectiveness-NTU approach is used more often when the inlet temperatures and the flow rates are specified for an exchanger with fixed area and the outlet temperatures are to be determined. Both methods are described earlier in Sec. 11.

System performance in cryogenic liquefiers and refrigerators is directly related to the effectiveness of the heat exchangers used in the system. For example, the liquid yield for a simple J-T cycle as given by Eq. 11-112 needs to be modified to

$$y = \frac{(h_1 - h_2) - (1 - \varepsilon)(h_1 - h_g)}{(h_1 - h_f) - (1 - \varepsilon)(h_1 - h_g)} \qquad (11\text{-}115)$$

if the heat exchanger is less than 100 percent effective. Likewise, the heat exchanger ineffectiveness increases the work required for the system by an amount of

$$\Delta \dot{W} = \dot{m}(h_1 - h_g)(1 - \varepsilon) \qquad (11\text{-}116)$$

Uninterrupted operation of heat exchangers at low temperatures requires removal of essentially all impurities present in the streams that are to be cooled. Equipment is readily available for the satisfactory removal of these impurities by both chemical and physical methods, but at increased operating expense. Another effective method for also accomplishing this impurity removal utilizes reversing heat exchangers. Proper functioning of the reversing heat exchanger is dependent upon the relationship between the pressures and temperatures of the two streams. Since the pressures are generally fixed by other factors, the purification function of the heat exchanger is normally controlled by selecting the right temperature difference throughout the heat exchanger. To assure that reevaporation takes place, these differences must be such that the vapor pressure of the impurity is greater than the partial pressure of the impurity in the purging stream.

Another type of reversing heat exchanger is the regenerator as detailed by Ackerman (*Cryogenic Regenerative Heat Exchangers*, Plenum Press, New York, 1997). As with all reversing heat exchangers, regenerators provide the simultaneous cooling and purification of gases in low-temperature processes. As noted earlier, reversing heat exchangers usually operate continuously. Regenerators do not operate continuously; instead, they operate by periodically storing heat in a packing during the first half of the cycle and then giving up this stored heat to the fluid during the second half of the cycle. Typically, a regenerator consists of two identical columns, which are packed with a porous solid material with a high heat capacity such as metal ribbon, through which the gases flow.

The low cost of the packing material, its large surface-area-per-unit volume, and the low-pressure drops encountered provide compelling arguments for utilizing regenerators. However, the intercontamination of fluid streams, caused by the periodic flow reversals, and the problems associated with designing regenerators to handle three or more fluids, has restricted their use to simple fluids, and favored adoption of plate-fin reversing heat exchangers.

**Expanders** The primary function of cryogenic expansion equipment is the reduction of the temperature of the gas being expanded to provide needed refrigeration. The expansion of a fluid to produce refrigeration may be carried out in two distinct ways: (1) in an expander where mechanical work is produced, and (2) in a Joule-Thomson valve where no work is produced.

***Mechanical Expanders*** Reciprocating expanders are very similar in concept and design to reciprocating compressors. Generally these units are used with inlet pressures of 4 to 20 MPa. These machines operate at speeds up to 500 rpm. The thermal efficiencies (actual enthalpy difference/maximum possible enthalpy difference) range from about 75 percent for small units to 85 percent for large machines.

Turboexpanders have replaced reciprocating expanders in high-power installations as well as in small helium liquefiers. Sizes range from 0.75 to 7500 kW with flow rates up to 28 million $m^3$/day. Today's large-tonnage air-separation plants are a reality due to the development of highly reliable and efficient turboexpanders. These expanders are being selected over other cryogenic equipment because of their ability to handle condensed ethane and heavier hydrocarbons. This type of expander usually weighs and costs less and requires less space and operating personnel.

Turboexpanders can be classified as either axial or radial. Axial flow expanders have either impulse or reaction type blades and are suitable for multistage expanders because they permit a much easier flow path from one stage to the next. However, radial turboexpanders have lower stresses at a given tip speed, which permits them to run at higher speeds. This results in higher efficiencies with correspondingly lower energy requirements. As a consequence, most turboexpanders built today are of the radial type.

***Joule-Thomson Valves*** The principal function of a J-T valve is to obtain isenthalpic cooling of the gas flowing through the valve. These valves generally are needle-type valves modified for cryogenic operation. They are an important component in most refrigeration systems, particularly in the last stage of the liquefaction process. Joule-Thomson valves also offer an attractive alternative to turboexpanders for small-scale gas-recovery applications.

## SEPARATION AND PURIFICATION SYSTEMS

The energy required to reversibly separate gas mixtures is the same as that necessary to isothermally compress each component in the mixture from the partial pressure of the gas in the mixture to the final pressure of the mixture. This reversible isothermal work is given by the familiar relation

$$\frac{-W_i}{\dot{m}} = T(s_1 - s_2) - (h_1 - h_2) \qquad (11\text{-}117)$$

where $s_1$ and $h_1$ refer to conditions before the separation and $s_2$ and $h_2$ refer to conditions after the separation. For a binary system, and assuming a perfect gas for both components, Eq. (11-117) simplifies to

$$\frac{-W_i}{\dot{m}} = RT\left(n_A \ln \frac{P_T}{p_A} + n_B \ln \frac{P_T}{p_B}\right) \qquad (11\text{-}118)$$

where $n_A$ and $n_B$ are the moles of $A$ and $B$ in the mixture, $p_A$ and $p_B$ are the partial pressures of these two components in the mixture, and $P_T$ is the total pressure of the mixture. The figure of merit for a separation system is defined similar to that for a liquefaction system; see Eq. (11-110).

If the mixture to be separated is essentially a binary, both the McCabe-Thiele and the Ponchon-Savarit methods outlined in Sec. 13, with the appropriate cryogenic properties, can be used to obtain the ideal number of stages required. It should be noted, however, that it is not satisfactory in the separation of air to treat it as a binary mixture of oxygen and nitrogen if high purity (99 percent or better) oxygen is desired. The separation of oxygen from argon is a more difficult separation than oxygen from nitrogen and would require correspondingly many more plates. In fact, if the argon is not extracted from air, only 95 percent oxygen would be produced. The other rare gas constituents of air (helium, neon, krypton, and xenon) are present in such small quantities and have boiling points so far removed from those of oxygen and nitrogen that they introduce no important complications.

**Air-Separation Systems** Of the various separation schemes available today, the simplest is known as the Linde single-column system shown in Fig. 11-117 and first introduced in 1902. In it, purified compressed air passes through a precooling heat exchanger (if oxygen gas is the desired product, a three-channel exchanger for air, waste nitrogen, and oxygen gas is used; if liquid oxygen is to be recovered from the bottom of the column, a two-channel exchanger for air and waste nitrogen is employed), then through a coil in the boiler of the rectifying column where it is further cooled (acting at the same time as the boiler heat source); following this, it expands essentially to atmospheric pressure through a J-T valve and reenters the top of the column with the liquid providing the required reflux. Rectification occurs as liquid and vapor on each plate establish equilibrium. If oxygen gas is to be the product, purified air need only be compressed to a pressure of 3 to 6 MPa; if the product is to be liquid oxygen, a compressor outlet pressure of 20 MPa is necessary. Note that the Linde single-column separation system is simply a J-T liquefaction system with a substitution of a rectification column for the liquid reservoir. However, any of the other liquefaction systems discussed earlier could have been used to furnish the liquid for the column.

In a simple single-column process, although the oxygen purity is high, the nitrogen effluent stream is impure. The equilibrium vapor concentration of the overhead nitrogen effluent for an initial liquid mixture of 21 percent oxygen/79 percent nitrogen at 100 kPa is about 6 to 7 percent oxygen. Thus, the nitrogen waste gas stream with such an impurity may only be usable as a purge gas for certain conditions.

The impurity problem noted in the previous paragraph was solved by the introduction of the Linde double-column system shown in Fig. 11-118. Two rectification columns are placed one on top of the other (hence the name double-column system).

In this system the liquid air is introduced at an intermediate point B into the lower column, and a condenser-evaporator at the top of the lower column makes the arrangement a complete reflux distillation column which delivers almost pure nitrogen at E. In order for the column simultaneously to deliver pure oxygen, the oxygen-rich liquid (about 45 percent $O_2$) from the bottom boiler is introduced at an intermediate level C in the upper column. The reflux and rectification in the upper column produce pure oxygen at the bottom and pure nitrogen at the top provided all major impurities are first removed from the column. More than enough liquid nitrogen is produced in the lower column for the needed reflux in both columns. Since the condenser must condense nitrogen vapor by evaporating liquid oxygen, it is necessary to operate the lower column at a higher pressure, about 500 kPa, while the upper column is operated at approximately 110 kPa. This requires a reduction in the pressure of the fluids from the lower column as they are admitted to the upper column.

In the cycle shown, gaseous oxygen and nitrogen are withdrawn at room temperature. Liquid oxygen could be withdrawn from point A and liquid nitrogen from point E, but in this case more refrigeration would be needed.

Even the best modern low-temperature air separation plant has an efficiency only a small fraction of the theoretical optimum, that is, about 15 to 20 percent. The principal sources of inefficiency are threefold: (1) the nonideality of the refrigerating process, (2) the imperfection of the heat exchangers, and (3) losses of refrigeration through heat leak.

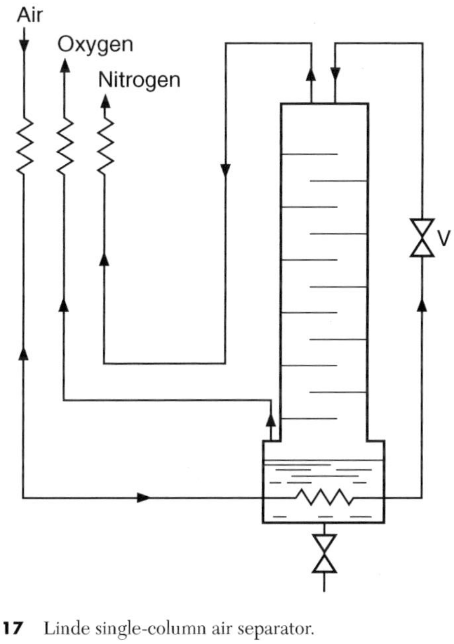


**FIG. 11-117** Linde single-column air separator.

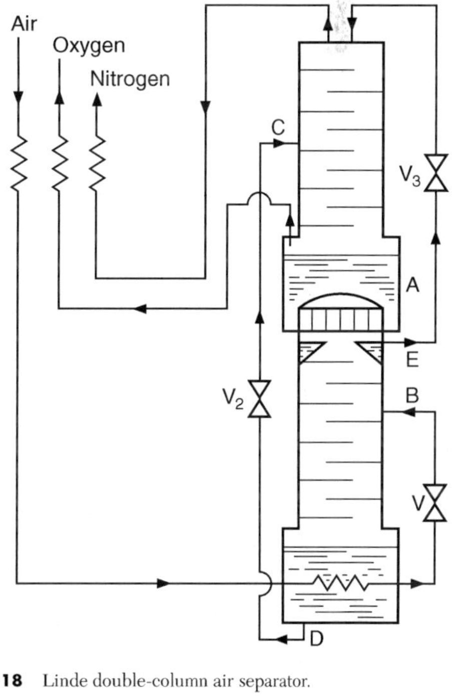


**FIG. 11-118** Linde double-column air separator.

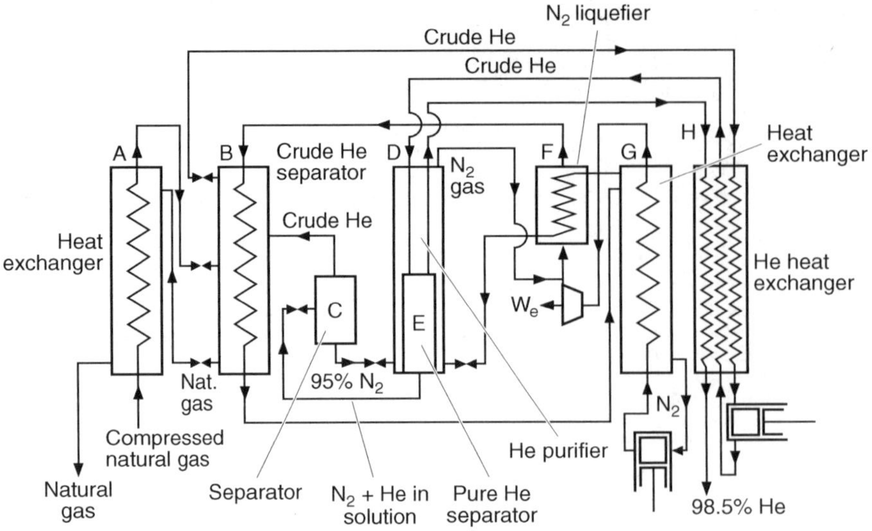


**FIG. 11-119** Typical helium-separation plant as operated by the U.S. Bureau of Mines.

**Helium and Natural-Gas Systems Separation** Helium is produced primarily by separation of helium-rich natural gas. The helium content of the natural gas from plants operated by the U.S. Bureau of Mines normally has varied from 1 to 2 percent while the nitrogen content of the natural gas has varied from 12 to 80 percent. The remainder of the natural gas is methane, ethane, and heavier hydrocarbons.

A Bureau of Mines system for the separation of helium from natural gas is shown in Fig. 11-119. Since the major constituents of natural gas have boiling points very much different from that of helium, a distillation column is not necessary and the separation can be accomplished with condenser-evaporators.

The need to obtain greater recoveries of the $C_2$, $C_3$, and $C_4$'s in natural gas has resulted in the expanded use of low-temperature processing of these streams. The majority of the natural gas processing at low temperatures to recover light hydrocarbons is now accomplished using the turboexpander cycle. Feed gas is normally available from 1 to 10 MPa. The gas is first dehydrated to a dew point of 200 K and lower. After dehydration the feed is cooled with cold residue gas. Liquid produced at this point is separated before entering the expander and sent to the condensate stabilizer. The gas from the separator is expanded in a turboexpander where the exit stream can contain as much as 20 wt % liquid. This two-phase mixture is sent to the top section of the stabilizer which separates the two phases. The liquid is used as reflux in this unit while the cold gas exchanges heat with the fresh feed and is recompressed by the expander-driven compressor. Many variations to this cycle are possible and have been used in actual plants.

**Gas Purification** The nature and concentration of impurities to be removed depends on the type of process involved. For example, in the production of large quantities of oxygen, various impurities must be removed to avoid plugging of the cold process lines or to avoid buildup of hazardous contaminants. The impurities in air that would contribute most to plugging would be water and carbon dioxide. Helium, hydrogen, and neon, on the other hand, will accumulate on the condensing side of the condenser-reboiler located between the two separation columns and will reduce the rate of heat transfer unless removed by intermittent purging. The buildup of acetylene, however, can prove to be dangerous even though the feed concentration in the air is no greater than 0.04 ppm.

Refrigeration purification is a relatively simple method for removing water, carbon dioxide, and certain other contaminants from a process stream by condensation or freezing. (Either regenerators or reversing heat exchangers may be used for this purpose since a flow reversal is periodically necessary to reevaporate and remove the solid deposits.) The effectiveness of this method depends upon the vapor pressure of the impurities relative to that of the major components of the process stream at the refrigeration temperature. Thus, assuming ideal gas behavior, the maximum impurity content in a gas stream after refrigeration would be inversely proportional to its vapor pressure. However, due to the departure from ideality at higher pressures, the impurity content will be considerably higher than predicted for the ideal situation. For example, the actual water vapor content in air will be over four times that predicted by ideal gas behavior at a temperature of 228 K and a pressure of 20 MPa.

Purification by a solid adsorbent is one of the most common low-temperature methods for removing impurities. Materials such as silica gel, carbon, and synthetic zeolites (molecular sieves) are widely used as adsorbents because of their extremely large effective surface areas. Most of the gels and carbon have pores of varying sizes in a given sample, but the synthetic zeolites are manufactured with closely controlled pore-size openings ranging from about 0.4 to 1.3 nm. This makes them even more selective than other adsorbents since it permits separation of gases on the basis of molecular size.

Information needed in the design of low-temperature adsorbers includes the equilibrium between the solid and gas and the rate of adsorption. Equilibrium data for the common systems generally are available from the suppliers of such material. The rate of adsorption is usually very rapid and the adsorption is essentially complete in a relatively narrow zone of the adsorber. If the concentration of the adsorbed gas is more than a trace, then heat of adsorption may also be a factor of importance in the design. (The heat of adsorption is usually of the same order or larger than the normal heat associated with the phase change.) Under such situations it is generally advisable to design the purification in two steps, that is, first removing a significant portion of the impurity either by condensation or chemical reaction and then completing the purification with a low-temperature adsorption system. A scheme combining the condensation and adsorption is shown in Fig. 11-120.

In normal plant operation at least two adsorption purifiers are employed—one in service while the other is desorbed of its impurities. Often there is an advantage in using an additional purifier by placing this unit in series with the adsorption unit to provide a backup if impurities are not trapped by the first unit. Cooling of the purifier must utilize some of the purified gas to avoid adsorption during this period.

Experience in air separation plant operations and other cryogenic processing plants has shown that local freeze-out of impurities such as

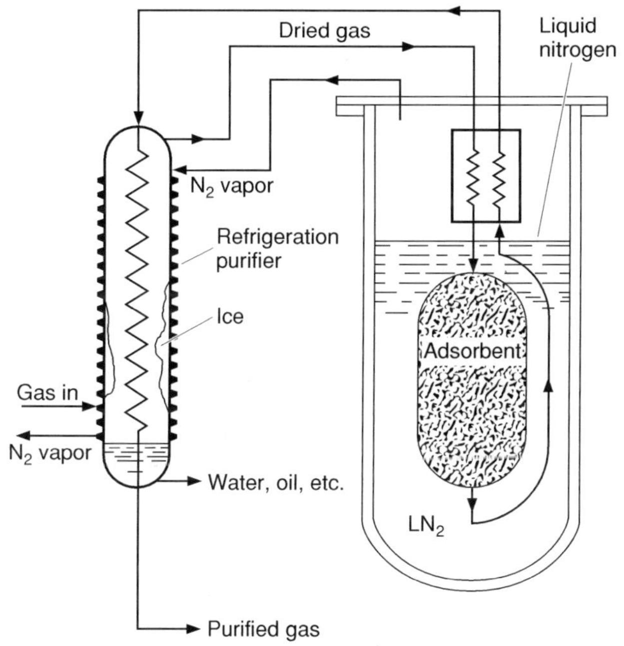


**FIG. 11-120** Purifier using refrigeration and adsorption schemes in series.

carbon dioxide can occur at concentrations well below the solubility limit. For this reason, the carbon dioxide content of the feed gas subject to the minimum operating temperature is usually kept below 50 ppm. The amine process and the molecular sieve adsorption process are the most widely used methods for carbon dioxide removal. The amine process involves adsorption of the impurity by a lean aqueous organic amine solution. With sufficient amine recirculation rate, the carbon dioxide in the treated gas can be reduced to less than 25 ppm. Oxygen is removed by a catalytic reaction with hydrogen to form water.

## STORAGE AND TRANSFER SYSTEMS

Storage vessels range in type from low-performance containers, insulated by rigid foam or fibrous insulation where the liquid in the container evaporates in a few hours, up to high-performance containers, insulated with multilayer insulations where less than 0.1 percent of the fluid contents is evaporated per day. In the more effective units, the storage container consists of an inner vessel which encloses the cryogenic fluid to be stored and an outer vessel or vacuum jacket. The latter maintains the vacuum necessary to make the insulation effective and at the same time serves as a vapor barrier to the migration of water and other condensibles to the cold surface of the inner vessel. Improvements have been made in the insulation used in these containers, but the vacuum-insulated double-walled Dewar is still the basic idea for high-performance cryogenic-fluid container design.

**Insulation Principles** The effectiveness of a liquefier or refrigerator is highly dependent upon the heat leak entering such a system. Since heat removal becomes more costly with a lowering in temperature as demonstrated by the Carnot limitation, most cryogenic systems employ some form of insulation to minimize the effect. The insulation strategy is to minimize radiative heat transfer, minimize convective heat transfer, and use only a minimum of solid conductance media. Factors considered in the selection of the most suitable insulation include its ruggedness, convenience, volume, weight, ease of fabrication and handling, and thermal effectiveness and cost. It is common practice to use an experimentally obtained apparent thermal conductivity to characterize the thermal effectiveness of various insulations. Typical $k_a$ values for insulations used in cryogenic service are listed in Table 11-26.

**TABLE 11-26 Representative Apparent Thermal Conductivity Values**

| Type of insulation | $k_a$, mW/m·K (77–300K) |
|---|---|
| Pure vacuum, $1.3 \times 10^{-4}$ Pa | 5 |
| Foam insulation | 26–35 |
| Nonevacuated powders (perlite, silica aerogel) | 19–44 |
| Evacuated powders and fibers ($1.3 \times 10^{-1}$ Pa) | 1–2 |
| Opacified powdered insulations ($1.3 \times 10^{-1}$ Pa) | $3.5 \times 10^{-1}$ |
| Multilayer insulations ($1.3 \times 10^{-4}$ Pa) | $1.7–4 \times 10^{-2}$ |

**Types of Insulation** Cryogenic insulations have generally been divided into five general categories: high vacuum, multilayer insulation, powder, foam, and special insulations. Each is discussed in turn in the following sections.

***Vacuum Insulation*** Heat transport across an evacuated space ($1.3 \times 10^{-4}$ Pa or lower), is by radiation and by conduction through the residual gas. The heat transfer by radiation generally is predominant and can be approximated by

$$\frac{\dot{Q}_r}{A_1} = \sigma\,(T_2^4 - T_1^4)\left[\frac{1}{e_1} + \frac{A_1}{A_2}\left(\frac{1}{e_2} - 1\right)\right]^{-1} \qquad (11\text{-}119)$$

where $\dot{Q}_r/A_1$ is the heat transfer by radiation-per-unit area, $\sigma$ is the Stefan-Boltzmann constant, and $e$ is the emissivity of the surfaces. Subscript 2 refers to the hot surface and subscript 1 refers to the cold surface. The bracketed term on the right-hand side of this relation is designated as the overall emissivity factor, $F_e$.

The insertion of low-emissivity floating shields within the evacuated space can effectively reduce the heat transport by radiation. The effect of the shields is to greatly reduce the emissivity factor. For example, for $N$ shields or $(N + 2)$ surfaces, an emissivity of the outer and inner surface of $e_o$, and an emissivity of the shields of $e_s$, the emissivity factor reduces to

$$\left[2\left(\frac{1}{e_o} + \frac{1}{e_s} - 1\right) + \frac{(N-1)(2-e_s)}{e_s}\right]^{-1} \qquad (11\text{-}120)$$

In essence, one properly located low-emissivity shield can reduce the radiant heat transfer to around one-half of the rate without the shield, two shields can reduce this to around one-fourth of the rate without the shield, and so on.

***Multilayer Insulation*** Multilayer insulation consists of alternating layers of highly reflecting material, such as aluminum foil or aluminized Mylar, and a low-conductivity spacer material or insulator, such as fiberglass mat or paper, glass fabric, or nylon net, all under high vacuum. When properly applied at the optimum density, this type of insulation can have an apparent thermal conductivity as low as 10 to 50 μW/m-K between 20 and 300 K.

For a highly evacuated (on the order of $1.3 \times 10^{-4}$ Pa) multilayer insulation, heat is transferred primarily by radiation and solid conduction through the spacer material. The apparent thermal conductivity of the insulation material under these conditions may be determined from

$$k_a = \frac{1}{N/\Delta x}\left\{h_s + \frac{\sigma e T_2^3}{2-e}\left[1 + \left(\frac{T_1}{T_2}\right)^2\right]\left(1 + \frac{T_1}{T_2}\right)\right\} \qquad (11\text{-}121)$$

where $N/\Delta x$ is the number of complete layers (reflecting shield plus spacer) of the insulation-per-unit thickness, $h_s$ is the solid conductance of the spacer material, $\sigma$ is the Stefan-Boltzmann constant, $e$ is the effective emissivity of the reflecting shield, and $T_2$ and $T_1$ are the temperatures of the warm and cold sides of the insulation, respectively. It is evident that the apparent thermal conductivity can be reduced by increasing the layer density up to a certain point. It is not obvious from the above relation that a compressive load affects the apparent thermal conductivity and thus the performance of a multilayer insulation. However, under a compressive load the solid conductance increases much more rapidly than $N/\Delta x$ resulting in an overall increase in $k_a$. Plots of heat flux versus compressive load on a logarithmic scale result in straight lines with slopes between 0.5 and 0.67.

The effective thermal conductivity values generally obtained in practice are at least a factor of two greater than the one-dimensional thermal conductivity values measured in the laboratory with carefully controlled techniques. This degradation in insulation thermal performance is caused by the combined presence of edge exposure to isothermal boundaries, gaps, joints, or penetrations in the insulation blanket required for structural supports, fill and vent lines, and high lateral thermal conductivity of these insulation systems.

***Powder Insulation*** A method of realizing some of the benefits of multiple floating shields without incurring the difficulties of awkward structural complexities is to use evacuated powder insulation. The penalty incurred in the use of this type of insulation, however, is a tenfold reduction in the overall thermal effectiveness of the insulation system over that obtained for multilayer insulation. In applications where this is not a serious factor, such as LNG storage facilities, and investment cost is of major concern, even unevacuated powder-insulation systems have found useful applications. The variation in apparent mean thermal conductivity of several powders as a function of interstitial gas pressure is shown in the familiar S-shaped curves of Fig. 11-121.

The apparent thermal conductivity of powder insulation at cryogenic temperatures is generally obtained from

$$k_a = \frac{k_g}{1 - V_r(1 - k_g/k_s)} \qquad (11\text{-}122)$$

where $k_g$ is the thermal conductivity of the gas within the insulation, $k_s$ is the thermal conductivity of the powder, and $V_r$ is the ratio of solid volume to the total volume. The amount of heat transport due to radiation through the powders can be reduced by the addition of metallic powders. A mixture containing approximately 40 to 50 wt % of a metallic powder gives the optimum performance.

***Foam Insulation*** Since foams are not homogeneous materials, their apparent thermal conductivity is dependent upon the bulk density of the insulation, the gas used to foam the insulation, and the mean temperature of the insulation. Heat conduction through a foam is determined by convection and radiation within the cells and by conduction in the solid structure. Evacuation of a foam is effective in reducing its thermal conductivity, indicating a partially open cellular structure, but the resulting values are still considerably higher than either multilayer or evacuated powder insulations.

Data on the thermal conductivity for a variety of foams used at cryogenic temperatures have been presented by Kropschot (*Cryogenic Technology,* R. W. Vance, ed., Wiley, New York, 1963, p. 239). Of all the foams, polyurethane and polystryene have received the widest use at

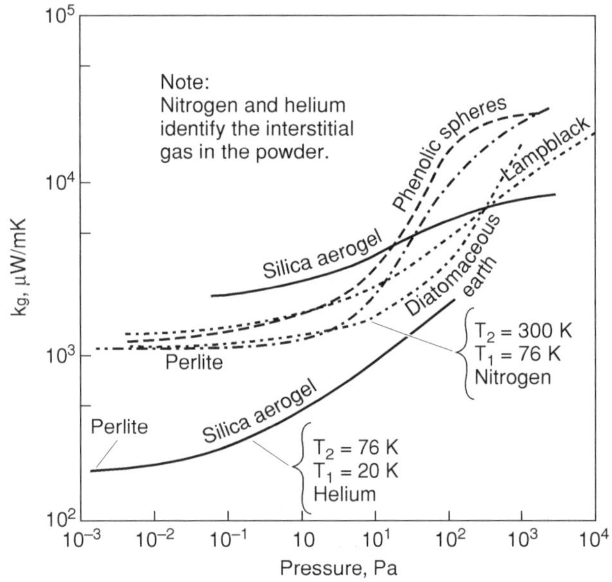


**FIG. 11-121** Apparent mean thermal conductivities of several powder insulations as a function of interstitial gas pressure.

low temperatures. The major disadvantage of foams is that they tend to crack upon repeated thermal cycling and lose their insulation value.

**Storage and Transfer Systems** In general, heat leak into a storage or transfer system for a cryogen is by (1) radiation and conduction through the insulation, and (2) conduction through any inner shell or transfer-line supports, piping leads, and access ports. Conduction losses are reduced by introducing long heat-leak paths, by making the cross sections for heat flow small, and by using materials with low thermal conductivity. Radiation losses, a major factor in the heat leak through insulations, are reduced with the use of radiation shields, such as multilayer insulation, boil-off vapor-cooled shields, and opacifiers in powder insulation.

Several considerations must be met when designing the inner vessel. The material of construction selected must be compatible with the stored cryogen. Nine percent nickel steels are acceptable for the higher-boiling cryogens ($T > 75$ K) while many aluminum alloys and austenitic steels are usually structurally acceptable throughout the entire temperature range. Because of its high thermal conductivity, aluminum is not a recommended material for piping and supports that must cross the insulation space. A change to a material of lower thermal conductivity for this purpose introduces a transition joint of a dissimilar material. Since such transition joints are generally mechanical in nature, leaks into the vacuum space develop upon repeated temperature cycling. In addition, the larger thermal coefficient of expansion of aluminum can pose still further support and cooldown problems.

Economic and cooldown considerations dictate that the shell of the storage container be as thin as possible. As a consequence, the inner container is designed to withstand only the internal pressure and bending forces while stiffening rings are used to support the weight of the fluid. The minimum thickness of the inner shell for a cylindrical vessel under such a design arrangement is given by Sec. VIII of the *ASME Boiler and Pressure Vessel Code.*

Since the outer shell of the storage container is subjected to atmospheric pressure on one side and evacuated conditions going down to $1.3 \times 10^{-4}$ Pa on the other, consideration must be given to provide ample thickness of the material to withstand collapsing or buckling. Failure by elastic instability is covered by the ASME Code, in which design charts are available for the design of cylinders and spheres subjected to external pressure. Stiffening rings are also used on the outer shell to support the weight of the inner container and its contents as well as maintaining the sphericity of the shell.

The outer shell is normally constructed of carbon steel for economic reasons, unless aluminum is required to reduce the weight. Stainless-steel standoffs must be provided on the carbon steel outer shell for all piping penetrations to avoid direct contact with these penetrations when they are cold.

There are a variety of methods for supporting the inner shell within the outer shell and the cold transfer line within the outer line. Materials that have a high strength to thermal conductivity ratio are selected for these supports. Design of these supports for the inner shell must allow for shipping loads which may be several orders higher than in-service loads. Compression supports such as legs or pads may be used, but tension supports are more common. These may take the form of cables, welded straps, threaded bars, or a combination of these to provide restraint of the inner shell in several directions.

Most storage containers for cryogens are designed for a 10 percent ullage volume. The latter permits reasonable vaporization of the contents due to heat leak without incurring too rapid a buildup of the pressure in the container. This, in turn, permits closure of the container for short periods of time to either avoid partial loss of the contents or to transport flammable or hazardous cryogens safely from one location to another.

## CRYOGENIC INSTRUMENTATION

Even though the combined production of cryogenic nitrogen and oxygen exceeds the production of any other chemical in the United States, the cryogenic industry does not appear to warrant a separate product line of instruments for diagnostic and control purposes. Low-temperature thermometry is the one exception. The general approach generally is that instruments developed for the usual CPI needs must be modified or accepted as is for cryogenic use.

Quite often problems arise when instruments for normal service are subjected to low temperature use. Since some metals become brittle at low temperatures, the instrument literally falls apart. Elastomeric gaskets and seals contract faster with decreasing temperatures than the surrounding metal parts, and the seal often is lost. Even hermetically sealed instruments can develop pin holes or small cracks to permit cryogenic liquids to enter these cases with time. Warming the instrument causes the trapped liquid to vaporize, sometimes generating excessive gas pressure and failure of the case.

Therefore, the first task in adapting normal instruments to cryogenic service is simply to give them a severe thermal shock by immersing them in liquid nitrogen repeatedly, and checking for mechanical integrity. This is the general issue; specific issues according to each type of measurement are discussed below.

**Pressure** This parameter is usually measured by the flush-mounted pressure transducer which consists of a force-summing device (bellow, diaphragm, bourdon tube, etc.) that translates the pressure into a displacement. The latter is then measured by an analog device (strain gage, piezoelectric crystal, variable distance between capacitor plates, and the like). Since these elements are likely to be made of different materials (bronze diaphragm, stainless-steel case, semiconductor strain gage), each will react to the temperature change in a different way. This is especially serious during cooldown, when the transient nature of material and construction prohibits all of the pressure-gage elements from being at the same temperature at the same time. Under steady-state conditions it is often possible to provide some temperature compensation through the well-known instrument technique of common-mode-rejection. Such compensation is generally not successful during transient temperature fluctuations. Only two courses of action are open: (1) hand-check each type of pressure transducer for thermal noise by thermally shocking it with immersion in liquid nitrogen; and (2) simplify the pressure-transducer construction to eliminate differences between materials. Some success has been observed in the latter area by manufacturers who make very small pressure sensing elements from a single semiconductor chip. The miniature size of these devices helps to reduce or eliminate temperature gradients across the device. The single-element nature of the pressure-gage assembly reduces differences in materials of construction.

**Liquid Level** The measurements for dense fluids such as liquid oxygen and liquid nitrogen are made in the conventional CPI approach using floats. Sight glasses cannot be used since radiation and thermal conduction would cause the cryogenic fluid within the sight glass to boil. The very light cryogens, liquid helium and liquid hydrogen, cannot sustain a float. Liquid hydrogen has the density of Styrofoam,™ about 70 g/l, making floating devices impractical. Some electrical analog is used for hydrogen and helium, most frequently a linear concentric-tube electrical capacitor. The dielectric constant of cryogens is related to their density by the Clausius-Mosotti relation. As the liquid level rises, the greater dielectric constant of the liquid between the tubes causes the overall capacitance to vary in a linear fashion. For best accuracy, these capacitance liquid-level measuring devices should be calibrated in place.

**Flow** The measurement of cryogenic fluids is most troublesome. Flow rate is not a natural physical parameter, like temperature, but is a derived quantity. A measurement of mass (or volume) must be made over a time interval to derive the flow rate. Because of this, any flow meter is only as good as its calibration. At this time, there is no national capability for calibrating cryogenic flowmeters. From data developed early in the nation's space program, considerable confidence has been developed in turbine-type flowmeters and in pressure-drop-type flowmeters. If all the usual ASTM guidelines are followed for meter installation, and if adequate temperature corrections are applied to changes in dimensions, then such meters can have an accuracy of $\pm 1$ percent of their water calibrations. For very small flow applications, the Coriolis meters are promising. Vortex shedding flow meters appear useful for very large flow rates. Nonetheless, an actual calibration on the cryogen of interest is the only proof of accuracy.

**Temperature** The level of the temperature measurement (4 K, 20 K, 77 K, or higher) is the first issue to be considered. The second issue is the range needed (e.g., a few degrees around 90 K or 1 to 400 K). If the temperature level is that of air separation or liquefacting of natural gas (LNG), then the favorite choice is the platinum resistance thermometer (PRT). Platinum, as with all pure metals, has an electrical resistance that goes to zero as the absolute temperature decreases to zero. Accordingly, the lower useful limit of platinum is about 20 K, or liquid hydrogen temperatures. Below 20 K, semiconductor thermometers (germanium-, carbon-, or silicon-based) are preferred. Semiconductors have just the opposite resistance-temperature dependence of metals—their resistance increases as the temperature is lowered, as fewer valence electrons can be promoted into the conduction band at lower temperatures. Thus, semiconductors are usually chosen for temperatures from about 1 to 20 K.

If the temperature range of interest is large, say 1 to 400 K, then diode thermometers are recommended. Diodes have other advantages compared to resistance thermometers. By contrast, diode thermometers are very much smaller and faster. By selection of diodes all from the same melt, they may be made interchangeable. That is, one diode has the same calibration curve as another, which is not always the case with either semiconductor or metallic-resistance thermometers. It is well known, however, that diode thermometers may rectify an ac field, and thus may impose a dc noise on the diode output. Adequate shielding is required.

Special applications, such as in high-magnetic fields, require special thermometers. The carbon-glass and strontium-titinate resistance thermometers have the least magnetoresistance effects.

Thermocouples are unsurpassed for making temperature-difference measurements. The thermoelectric power of thermocouple materials makes them adequate for use at liquid-air temperatures and above. At 20 K and below, the thermoelectric power drops to a few μV/K, and their use in this range is as much art as science.

A descriptive flowchart has been prepared by Sparks (*Materials at Low Temperatures,* ASM, Metal Park, OH, 1983) to show the temperature range of cryogenic thermometers in general use today. Pavese and Molinar (*Modern Gas-Based Temperature and Pressure Measurements,* Plenum, New York, 1992) provide details on gas- and vapor-pressure thermometry at these temperatures.

## SAFETY

Past experience has shown that cryogenic fluids can be used safely in industrial environments as well as in typical laboratories provided all facilities are properly designed and maintained, and personnel handling these fluids are adequately trained and supervised. There are many hazards associated with cryogenic fluids. However, the principal ones are those associated with the response of the human body and the surroundings to the fluids and their vapors, and those associated with reactions between the fluids and their surroundings. Edeskuty and Stewart (*Safety in Handling Cryogenic Fluids,* Plenum Press, New York, 1996) provide a detailed examination of these various hazards.

**Physiological Hazards** Severe cold "burns" may be inflicted if the human body comes in contact with cryogenic fluids or with surfaces cooled by cryogenic fluids. Damage to the skin or tissue is similar to an ordinary burn. Because the body is composed mainly of water, the low temperature effectively freezes the tissue—damaging or destroying it. The severity of the burn depends upon the contact area and the contact time with prolonged contact resulting in deeper burns. Cold burns are accompanied by stinging sensations and pain similar to those of ordinary burns. The ordinary reaction is to withdraw that portion of the body that is in contact with the cold surface. Severe burns are seldom sustained if withdrawal is possible. Cold gases may not be damaging if the turbulence in the gas is low, particularly since the body can normally adjust for a heat loss of 95 $J/m^2s$ for an area of limited exposure. If the heat loss becomes much greater than this, the skin temperature drops and freezing of the affected area may ensue. Freezing of facial tissue will occur in about 100 s if the heat loss is 2,300 $J/m^2s$.

**Materials and Construction Hazards** Construction materials for noncryogenic service usually are chosen on the basis of tensile strength, fatigue life, weight, cost, ease of fabrication, corrosion resistance, and so on. When working with low temperatures the designer must consider the ductility of the material since low temperatures, as noted earlier, have the effect of making some construction materials brittle or less ductile. Some materials become brittle at low temperatures but still can absorb considerable impact, while others become brittle and lose their impact strength.

**Flammability and Explosion Hazards** In order to have a fire or an explosion requires the combination of an oxidant, a fuel, and an ignition source. Generally the oxidizer will be oxygen. The latter may be available from a variety of sources including leakage or spillage, condensation of air on cryogenically cooled surfaces below 90 K, and buildup, as a solid impurity in liquid hydrogen. The fuel may be almost any noncompatible material or flammable gas; compatible materials can also act as fuels in the presence of extreme heat (strong ignition sources). The ignition source may be a mechanical or electrostatic spark, flame, impact, heat by kinetic effects, friction, chemical reaction, and so on. Certain combinations of oxygen, fuel, and ignition sources will always result in fire or explosion. The order of magnitude of flammability and detonability limits for fuel-oxidant gaseous mixtures of two widely used cryogens is shown in Table 11-27.

**High-Pressure Gas Hazards** Potential hazards also exist in highly compressed gases because of their stored energy. In cryogenic systems such high pressures are obtained by gas compression during liquefaction or refrigeration, by pumping of liquids to high pressure followed by evaporation, and by confinement of cryogenic liquids with subsequent evaporation. If this confined gas is suddenly released through a rupture or break in a line, a significant thrust may be experienced. For example, the force generated by the rupture of a 2.5-cm diameter valve located on a 13.9-MPa pressurized gas cylinder would be over 6670 N.

**TABLE 11-27 Flammability and Detonability Limits of Hydrogen and Methane Gas**

| Mixture | Flammability Limits (mol %) | Detonability Limits (mol %) |
|---|---|---|
| $H_2$-air | 4–75 | 20–65 |
| $H_2$-$O_2$ | 4–95 | 15–90 |
| $CH_4$-air | 5–15 | 6–14 |
| $CH_4$-$O_2$ | 5–61 | 10–50 |

## SUMMARY

It is obvious that the best designed facility is no better than the attention that is paid to safety. The latter is not considered once and forgotten. Rather, it is an ongoing activity that requires constant attention to every conceivable hazard that might be encountered. Because of its importance, safety, particularly at low temperatures, has received a large focus in the literature with its own safety manual prepared by NIST as well as by the British Cryogenics Council.

# EVAPORATORS

GENERAL REFERENCES: Badger and Banchero, *Introduction to Chemical Engineering,* McGraw-Hill, New York, 1955. Standiford, *Chem. Eng.,* **70,** 158–176 (Dec. 9, 1963). *Testing Procedure for Evaporators,* American Institute of Chemical Engineers, 1979. *Upgrading Evaporators to Reduce Energy Consumption,* ERDA Technical Information Center, Oak Ridge, Tenn., 1977.

## PRIMARY DESIGN PROBLEMS

**Heat Transfer** This is the most important single factor in evaporator design, since the heating surface represents the largest part of evaporator cost. Other things being equal, the type of evaporator selected is the one having the highest heat-transfer cost coefficient under desired operating conditions in terms of J/s·K (British thermal units per hour per degree Fahrenheit) per dollar of installed cost. When power is required to induce circulation past the heating surface, the coefficient must be even higher to offset the cost of power for circulation.

**Vapor-Liquid Separation** This design problem may be important for a number of reasons. The most important is usually prevention of entrainment because of value of product lost, pollution, contamination of the condensed vapor, or fouling or corrosion of the surfaces on which the vapor is condensed. Vapor-liquid separation in the vapor head may also be important when spray forms deposits on the walls, when vortices increase head requirements of circulating pumps, and when short circuiting allows vapor or unflashed liquid to be carried back to the circulating pump and heating element.

Evaporator performance is rated on the basis of **steam economy**—kilograms of solvent evaporated per kilogram of steam used. Heat is required (1) to raise the feed from its initial temperature to the boiling temperature, (2) to provide the minimum thermodynamic energy to separate liquid solvent from the feed, and (3) to vaporize the solvent. The first of these can be changed appreciably by reducing the boiling temperature or by heat interchange between the feed and the residual product and/or condensate. The greatest increase in steam economy is achieved by reusing the vaporized solvent. This is done in a **multiple-effect evaporator** by using the vapor from one effect as the heating medium for another effect in which boiling takes place at a lower temperature and pressure. Another method of increasing the utilization of energy is to employ a **thermocompression** evaporator, in which the vapor is compressed so that it will condense at a temperature high enough to permit its use as the heating medium in the same evaporator.

**Selection Problems** Aside from heat-transfer considerations, the selection of type of evaporator best suited for a particular service is governed by the characteristics of the feed and product. Points that must be considered are crystallization, salting and scaling, product quality, corrosion, and foaming. In the case of a **crystallizing evaporator,** the desirability of producing crystals of a definite uniform size usually limits the choice to evaporators having a positive means of circulation. **Salting,** which is the growth on body and heating-surface walls of a material having a solubility that increases with increase in temperature, is frequently encountered in crystallizing evaporators. It can be reduced or eliminated by keeping the evaporating liquid in close or frequent contact with a large surface area of crystallized solid. **Scaling** is the deposition and growth on body walls, and especially on heating surfaces, of a material undergoing an irreversible chemical reaction in the evaporator or having a solubility that decreases with an increase in temperature. Scaling can be reduced or eliminated in the same general manner as salting. Both salting and scaling liquids are usually best handled in evaporators that do not depend on boiling to induce circulation. **Fouling** is the formation of deposits other than salt or scale and may be due to corrosion, solid matter entering with the feed, or deposits formed by the condensing vapor.

**Product Quality** Considerations of product quality may require low holdup time and low-temperature operation to avoid thermal degradation. The low holdup time eliminates some types of evaporators, and some types are also eliminated because of poor heat-transfer characteristics at low temperature. Product quality may also dictate special materials of construction to avoid metallic contamination or a catalytic effect on decomposition of the product. **Corrosion** may also influence evaporator selection, since the advantages of evaporators having high heat-transfer coefficients are more apparent when expensive materials of construction are indicated. Corrosion and erosion are frequently more severe in evaporators than in other types of equipment because of the high liquid and vapor velocities used, the frequent presence of solids in suspension, and the necessary concentration differences.

## EVAPORATOR TYPES AND APPLICATIONS

Evaporators may be classified as follows:

1. Heating medium separated from evaporating liquid by tubular heating surfaces.
2. Heating medium confined by coils, jackets, double walls, flat plates, etc.
3. Heating medium brought into direct contact with evaporating liquid.
4. Heating by solar radiation.

By far the largest number of industrial evaporators employ tubular heating surfaces. Circulation of liquid past the heating surface may be induced by boiling or by mechanical means. In the latter case, boiling may or may not occur at the heating surface.

**Forced-Circulation Evaporators (Fig. 11-122 *a,b,c*)** Although it may not be the most economical for many uses, the forced-circulation (FC) evaporator is suitable for the widest variety of evaporator applications. The use of a pump to ensure circulation past the heating surface makes possible separating the functions of heat transfer, vapor-liquid separation, and crystallization. The pump withdraws liquor from the flash chamber and forces it through the heating element back to the flash chamber. Circulation is maintained regardless of the evaporation rate; so this type of evaporator is well suited to **crystallizing operation,** in which solids must be maintained in suspension at all times. The liquid velocity past the heating surface is limited only by the pumping power needed or available and by accelerated corrosion and erosion at the higher velocities. **Tube velocities** normally range from a minimum of about 1.2 m/s (4 ft/s) in salt evaporators with copper or brass tubes and liquid containing 5 percent or more solids up to about 3 m/s (10 ft/s) in caustic evaporators having nickel tubes and liquid containing only a small amount of solids. Even higher velocities can be used when corrosion is not accelerated by erosion.

Highest heat-transfer coefficients are obtained in FC evaporators when the liquid is allowed to boil in the tubes, as in the type shown in Fig. 11-122*a*. The heating element projects into the vapor head, and the liquid level is maintained near and usually slightly below the top tube sheet. This type of FC evaporator is not well suited to salting solutions because boiling in the tubes increases the chances of salt deposit on the walls and the sudden flashing at the tube exits promotes excessive nucleation and production of fine crystals. Consequently, this type of evaporator is seldom used except when there are headroom limitations or when the liquid forms neither salt nor scale.

**Swirl Flow Evaporators** One of the most significant problems in the thermal design of once-through, tube-side evaporators is the poor predictability of the loss of $\Delta T$ upon reaching the critical heat flux condition. This situation may occur through flashing due to a high wall temperature or due to process needs to evaporate most of, if not all, the liquid entering the evaporator. It is the result of sensible heating of the vapor phase which accumulates at the heat-transfer surface, dries out the tube wall, and blocks the transfer of heat to the remaining liquid.

In some cases, even with correctly predicted heat-transfer coefficients, the unexpected $\Delta T$ loss can reduce the actual performance of the evaporator by as much as 200 percent below the predicted performance. The best approach is to maintain a high level of mixing of the phases through the heat exchanger near the heat-transfer surface.

The use of swirl flow, whereby a rotational vortex is imparted to the boiling fluid to centrifuge the liquid droplets out to the tube wall, has proved to be the most reliable means to correct for and eliminate this loss of $\Delta T$. The use of this technique almost always corrects the design to operate as well as or better than predicted. Also, the use of swirl flow eliminates the need to determine between horizontal or vertical orientation for most two-phase velocities. Both orientations work about the same in swirl flow.

Many commercially viable methods of inducing swirl flow inside of tubes are available in the form of either swirl flow tube inserts (twisted tapes, helical cores, spiral wire inserts) or special tube configurations (Twisted Tube®, internal spiral fins). All are designed to impart a natural swirl component to the flow inside the tubes. Each has been proved to solve the problem of tube-side vaporization at high vapor qualities up to and including complete tube-side vaporization.

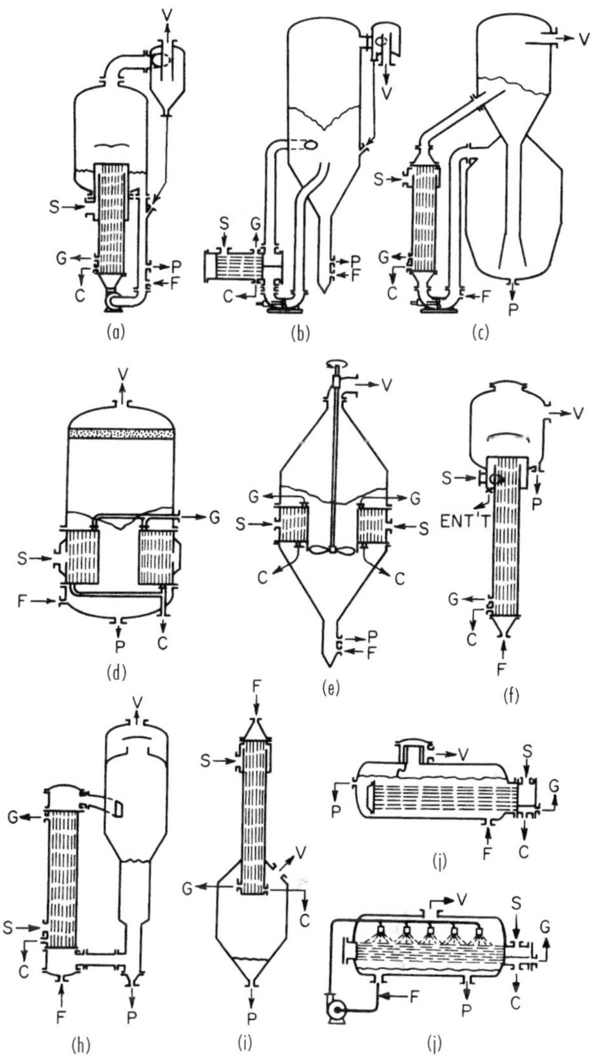


**FIG. 11-122** Evaporator types. (*a*) Forced circulation. (*b*) Submerged-tube forced circulation. (*c*) Oslo-type crystallizer. (*d*) Short-tube vertical. (*e*) Propeller calandria. (*f*) Long-tube vertical. (*g*) Recirculating long-tube vertical. (*h*) Falling film. (*i,j*) Horizontal-tube evaporators. C = condensate; F = feed; G = vent; P = product; S = steam; V = vapor; ENT'T = separated entrainment outlet.

By far the largest number of forced-circulation evaporators are of the submerged-tube type, as shown in Fig. 11-122*b*. The heating element is placed far enough below the liquid level or return line to the flash chamber to prevent boiling in the tubes. Preferably, the hydrostatic head should be sufficient to prevent boiling even in a tube that is plugged (and hence at steam temperature), since this prevents salting of the entire tube. Evaporators of this type sometimes have horizontal heating elements (usually two-pass), but the vertical single-pass heating element is used whenever sufficient headroom is available. The vertical element usually has a lower friction loss and is easier to clean or retube than a horizontal heater. The submerged-tube forced-circulation evaporator is relatively immune to salting in the tubes, since no supersaturation is generated by evaporation in the tubes. The tendency toward scale formation is also reduced, since supersaturation in the heating element is generated only by a controlled amount of heating and not by both heating and evaporation.

The type of **vapor** head used with the FC evaporator is chosen to suit the product characteristics and may range from a simple centrifugal separator to the crystallizing chambers shown in Fig. 11-122*b* and *c*. Figure 11-122*b* shows a type frequently used for common salt. It is designed to circulate a slurry of crystals throughout the system. Figure 11-122*c* shows a submerged-tube FC evaporator in which heating, flashing, and crystallization are completely separated. The crystallizing solids are maintained as a fluidized bed in the chamber below the vapor head and little or no solids circulate through the heater and flash chamber. This type is well adapted to growing coarse crystals, but the crystals usually approach a spherical shape, and careful design is required to avoid production of tines in the flash chamber.

In a submerged-tube FC evaporator, all heat is imparted as sensible heat, resulting in a temperature rise of the circulating liquor that reduces the overall temperature difference available for heat transfer. Temperature rise, tube proportions, tube velocity, and head requirements on the circulating pump all influence the selection of circulation rate. Head requirements are frequently difficult to estimate since they consist not only of the usual friction, entrance and contraction, and elevation losses when the return to the flash chamber is above the liquid level but also of increased friction losses due to flashing in the return line and vortex losses in the flash chamber. Circulation is sometimes limited by vapor in the pump suction line. This may be drawn in as a result of inadequate vapor-liquid separation or may come from vortices near the pump suction connection to the body or may be formed in the line itself by short circuiting from heater outlet to pump inlet of liquor that has not flashed completely to equilibrium at the pressure in the vapor head.

---

Advantages of forced-circulation evaporators:
1. High heat-transfer coefficients
2. Positive circulation
3. Relative freedom from salting, scaling, and fouling

Disadvantages of forced-circulation evaporators:
1. High cost
2. Power required for circulating pump
3. Relatively high holdup or residence time

Best applications of forced-circulation evaporators:
1. Crystalline product
2. Corrosive solutions
3. Viscous solutions

Frequent difficulties with forced-circulation evaporators:
1. Plugging of tube inlets by salt deposits detached from walls of equipment
2. Poor circulation due to higher than expected head losses
3. Salting due to boiling in tubes
4. Corrosion-erosion

---

**Short-Tube Vertical Evaporators (Fig. 11-122*d*)** This is one of the earliest types still in widespread commercial use. Its principal use at present is in the evaporation of cane-sugar juice. Circulation past the heating surface is induced by boiling in the tubes, which are usually 50.8 to 76.2 mm (2 to 3 in) in diameter by 1.2 to 1.8 m (4 to 6 ft) long. The body is a vertical cylinder, usually of cast iron, and the tubes are expanded into horizontal tube sheets that span the body diameter. The circulation rate through the tubes is many times the feed rate; so there must be a return passage from above the top tube sheet to below the bottom tube sheet. Most commonly used is a central well or **downtake** as shown in Fig. 11-122*d*. So that friction losses through the downtake do not appreciably impede circulation up through the tubes, the area of the downtake should be of the same order of magnitude as the combined cross-sectional area of the tubes. This results in a downtake almost half of the diameter of the tube sheet.

Circulation and heat transfer in this type of evaporator are strongly affected by the liquid "level." Highest heat-transfer coefficients are achieved when the level, as indicated by an external gauge glass, is only about halfway up the tubes. Slight reductions in level below the optimum result in incomplete wetting of the tube walls with a consequent increased tendency to foul and a rapid reduction in capacity. When this type of evaporator is used with a liquid that can deposit salt or scale, it is customary to operate with the liquid level appreciably higher than the optimum and usually appreciably above the top tube sheet.

Circulation in the standard short-tube vertical evaporator is dependent entirely on boiling, and when boiling stops, any solids present settle out of suspension. Consequently, this type is seldom used as a crystallizing evaporator. By installing a propeller in the downtake, this objection can be overcome. Such an evaporator, usually called a **propeller calandria,** is illustrated in Fig. 11-122*e*. The propeller is usually placed as low as possible to reduce cavitation and is shrouded by an extension of the downtake well. The use of the propeller can sometimes double the capacity of a short-tube vertical evaporator. The evaporator shown in Fig. 11-122*e* includes an elutriation leg for salt manufacture similar to that used on the FC evaporator of Fig. 11-122*b*. The shape of the bottom will, of course, depend on the particular application and on whether the propeller is driven from above or below. To avoid salting when the evaporator is used for crystallizing solutions, the liquid level must be kept appreciably above the top tube sheet.

---

Advantages of short-tube vertical evaporators:
1. High heat-transfer coefficients at high temperature differences
2. Low headroom
3. Easy mechanical descaling
4. Relatively inexpensive

Disadvantages of short-tube vertical evaporators:
1. Poor heat transfer at low temperature differences and low temperature
2. High floor space and weight
3. Relatively high holdup
4. Poor heat transfer with viscous liquids

Best applications of short-tube vertical evaporators:
1. Clear liquids
2. Crystalline product if propeller is used
3. Relatively noncorrosive liquids, since body is large and expensive if built of materials other than mild steel or cast iron
4. Mild scaling solutions requiring mechanical cleaning, since tubes are short and large in diameter

---

**Long-Tube Vertical Evaporators (Fig. 11-122*f*, *g*, *h*)** More total evaporation is accomplished in this type than in all others combined because it is normally the **cheapest per unit of capacity.** The long-tube vertical (LTV) evaporator consists of a simple one-pass vertical shell-and-tube heat exchanger discharging into a relatively small vapor head. Normally, no liquid level is maintained in the vapor head, and the residence time of liquor is only a few seconds. The tubes are usually about 50.8 mm (2 in) in diameter but may be smaller than 25.4 mm (1 in). Tube length may vary from less than 6 to 10.7 m (20 to 35 ft) in the rising film version and to as great as 20 m (65 ft) in the falling film version. The evaporator is usually operated single-pass, concentrating from the feed to discharge density in just the time that it takes the liquid and evolved vapor to pass through a tube. An extreme case is the caustic high concentrator, producing a substantially anhydrous product at 370°C (700°F) from an inlet feed of 50 percent NaOH at 149°C (300°F) in one pass up 22-mm- (8/8-in-) outside-diameter nickel tubes 6 m (20 ft) long. The largest use of LTV evaporators is for concentrating black liquor in the pulp and paper industry. Because of the long tubes and relatively high heat-transfer coefficients, it is possible to achieve higher single-unit capacities in this type of evaporator than in any other.

The LTV evaporator shown in Fig. 11-122*f* is typical of those commonly used, especially for black liquor. Feed enters at the bottom of the tube and starts to boil partway up the tube, and the mixture of liquid and vapor leaving at the top at high velocity impinges against a deflector placed above the tube sheet. This deflector is effective both as a primary separator and as a foam breaker.

In many cases, as when the ratio of feed to evaporation or the ratio of feed to heating surface is low, it is desirable to provide for **recirculation of product** through the evaporator. This can be done in the type shown in Fig. 11-122*f* by adding a pipe connection between the product line and the feed line. Higher recirculation rates can be achieved in the type shown in Fig. 11-122*g*, which is used widely for condensed milk. By extending the enlarged portion of the vapor head still lower to provide storage space for liquor, this type can be used as a batch evaporator.

Liquid temperatures in the tubes of an LTV evaporator are far from uniform and are difficult to predict. At the lower end, the liquid is usually not boiling, and the liquor picks up heat as sensible heat. Since entering liquid velocities are usually very low, true heat-transfer coefficients are low in this nonboiling zone. At some point up the tube, the liquid starts to boil, and from that point on the liquid temperature decreases because of the reduction in static, friction, and acceleration heads until the vapor-liquid mixture reaches the top of the tubes at substantially vapor-head temperature. Thus the true temperature difference in the boiling zone is always less than the total temperature difference as measured from steam and vapor-head temperatures.

Although the true heat-transfer coefficients in the boiling zone are quite high, they are partially offset by the reduced temperature difference. The point in the tubes at which boiling starts and at which the maximum temperature is reached is sensitive to operating conditions, such as feed properties, feed temperature, feed rate, and heat flux. Figure 11-123 shows typical variations in liquid temperature in tubes of an LTV evaporator operating at a constant terminal temperature difference. Curve 1 shows the normal case in which the feed is not boiling at the tube inlet. Curve 2 gives an indication of the temperature difference lost when the feed enters at the boiling point. Curve 3 is for exactly the same conditions as curve 2 except that the feed contained 0.01 percent Teepol to reduce surface tension [Coulson and Mehta, *Trans. Inst. Chem. Eng.*, **31,** 208 (1953)]. The surface-active agent yields a more intimate mixture of vapor and liquid, with the result that liquid is accelerated to a velocity more nearly approaching the vapor velocity, thereby increasing the pressure drop in the tube. Although the surface-active agent caused an increase of more than 100 percent in the true heat-transfer coefficient, this was more than offset by the reduced temperature difference so that the net result was a reduction in evaporator capacity. This sensitivity of the LTV evaporator to changes in operating conditions is less pronounced at high than at low temperature differences and temperature levels.

The **falling-film** version of the LTV evaporator (Fig. 11-122*h*) eliminates these problems of hydrostatic head. Liquid is fed to the tops of the tubes and flows down the walls as a film. Vapor-liquid separation usually takes place at the bottom, although some evaporators of this type are arranged for vapor to rise through the tube countercurrently to the liquid. The pressure drop through the tubes is usually very small, and the boiling-liquid temperature is substantially the same as the vapor-head temperature. The falling-film evaporator is widely used for concentrating **heat-sensitive materials,** such as fruit juices, because the holdup time is very small, the liquid is not overheated during passage through the evaporator, and heat-transfer coefficients are high even at low boiling temperatures.

The principal problem with the falling-film LTV evaporator is that of **feed distribution** to the tubes. It is essential that all tube surfaces be wetted continually. This usually requires recirculation of the liquid unless the ratio of feed to evaporation is quite high. An alternative to the simple recirculation system of Fig. 11-122*h* is sometimes used when the feed undergoes an appreciable concentration change and the product is viscous and/or has a high boiling point rise. The feed chamber and vapor head are divided into a number of liquor compartments, and separate pumps are used to pass the liquor through the various banks of tubes in series, all in parallel as to steam and vapor pressures. The actual distribution of feed to the individual tubes of a falling-film evaporator may be accomplished by orifices at the inlet to each tube, by a perforated plate above the tube sheet, or by one or more spray nozzles.

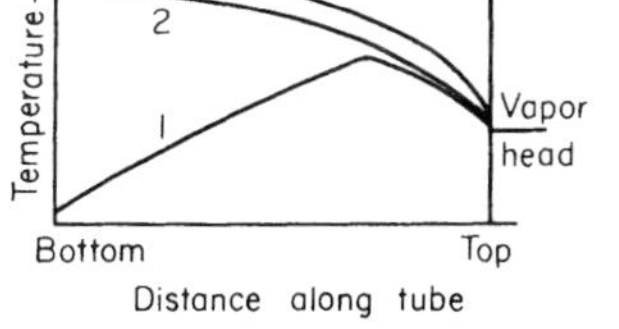


**FIG. 11-123** Temperature variations in a long-tube vertical evaporator.

Both rising- and falling-film LTV evaporators are generally unsuited to salting or severely scaling liquids. However, both are widely used for black liquor, which presents a mild scaling problem, and also are used to carry solutions beyond saturation with respect to a crystallizing salt. In the latter case, deposits can usually be removed quickly by increasing the feed rate or reducing the steam rate in order to make the product unsaturated for a short time. The falling-film evaporator is not generally suited to liquids containing solids because of difficulty in plugging the feed distributors. However, it has been applied to the evaporation of saline waters saturated with $CaSO_4$ and containing 5 to 10 percent $CaSO_4$ seeds in suspension for scale prevention (Anderson, ASME Pap. 76-WA/Pwr-5, 1976).

Because of their simplicity of construction, compactness, and generally high heat-transfer coefficients, LTV evaporators are well suited to service with corrosive liquids. An example is the reconcentration of rayon spin-bath liquor, which is highly acid. These evaporators employ impervious graphite tubes, lead, rubber-covered or impervious graphite tube sheets, and rubber-lined vapor heads. Polished stainless-steel LTV evaporators are widely used for food products. The latter evaporators are usually similar to that shown in Fig. 11-122*g*, in which the heating element is at one side of the vapor head to permit easy access to the tubes for cleaning.

Advantages of long-tube vertical evaporators:
1. Low cost
2. Large heating surface in one body
3. Low holdup
4. Small floor space
5. Good heat-transfer coefficients at reasonable temperature differences (rising film)
6. Good heat-transfer coefficients at all temperature differences (falling film)

Disadvantages of long-tube vertical evaporators:
1. High headroom
2. Generally unsuitable for salting and severely scaling liquids
3. Poor heat-transfer coefficients of rising-film version at low temperature differences
4. Recirculation usually required for falling-film version

Best applications of long-tube vertical evaporators:
1. Clear liquids
2. Foaming liquids
3. Corrosive solutions
4. Large evaporation loads
5. High temperature differences—rising film, low temperature differences—falling film
6. Low-temperature operation—falling film
7. Vapor compression operation—falling film

Frequent difficulties with long-tube vertical evaporators:
1. Sensitivity of rising-film units to changes in operating conditions
2. Poor feed distribution to falling-film units

**Horizontal-Tube Evaporators (Fig. 11-122*i*)** In these types the steam is inside and the liquor outside the tubes. The submerged-tube version of Fig. 11-122*i* is seldom used except for the preparation of boiler feedwater. Low entrainment loss is the primary aim: the horizontal cylindrical shell yields a large disengagement area per unit of vessel volume. Special versions use deformed tubes between restrained tube sheets that crack off much of a scale deposit when sprayed with cold water. By showering liquor over the tubes in the version of Fig. 11-122*f* hydrostatic head losses are eliminated and heat-transfer performance is improved to that of the falling-film tubular type of Fig. 11-122*h*. Originally called the Lillie, this evaporator is now also called the spray-film or simply the horizontal-tube evaporator. Liquid distribution over the tubes is accomplished by sprays or perforated plates above the topmost tubes. Maintaining this distribution through the bundle to avoid overconcentrating the liquor is a problem unique to this type of evaporator. It is now used primarily for seawater evaporation.

---

Advantages of horizontal-tube evaporators:
1. Very low headroom
2. Large vapor-liquid disengaging area—submerged-tube type
3. Relatively low cost in small-capacity straight-tube type
4. Good heat-transfer coefficients
5. Easy semiautomatic descaling—bent-tube type

Disadvantages of horizontal-tube evaporators:
1. Unsuitable for salting liquids
2. Unsuitable for scaling liquids—straight-tube type
3. High cost—bent-tube type
4. Maintaining liquid distribution—film type

Best applications of horizontal-tube evaporators:
1. Limited headroom
2. Small capacity
3. Nonscaling nonsalting liquids—straight-tube type
4. Severely scaling liquids—bent-tube type

---

**Miscellaneous Forms of Heating Surface** Special evaporator designs are sometimes indicated when heat loads are small, special product characteristics are desired, or the product is especially difficult to handle. **Jacketed kettles,** frequently with agitators, are used when the product is very viscous, batches are small, intimate mixing is required, and/or ease of cleaning is an important factor. Evaporators with steam in coiled **tubes** may be used for small capacities with scaling liquids in designs that permit "cold shocking," or complete withdrawal of the coil from the shell for manual scale removal. Other designs for scaling liquids employ flat-plate heat exchangers, since in general a scale deposit can be removed more easily from a flat plate than from a curved surface. One such design, the **channel-switching evaporator,** alternates the duty of either side of the heating surface periodically from boiling liquid to condensing vapor so that scale formed when the surface is in contact with boiling liquid is dissolved when the surface is next in contact with condensing vapor.

**Agitated thin-film evaporators** employ a heating surface consisting of one large-diameter tube that may be either straight or tapered, horizontal or vertical. Liquid is spread on the tube wall by a rotating assembly of blades that either maintain a close clearance from the wall or actually ride on the film of liquid on the wall. The expensive construction limits application to the most difficult materials. High agitation [on the order of 12 m/s (40 ft/s) rotor-tip speed] and power intensities of 2 to 20 kW m$^2$ (0.25 to 2.5 hp/ft$^2$) permit handling extremely viscous materials. Residence times of only a few seconds permit concentration of heat-sensitive materials at temperatures and temperature differences higher than in other types [Mutzenberg, Parker, and Fischer. *Chem. Eng.*, **72,** 175–190 (Sept. 13, 1965)]. High feed-to-product ratios can be handled without recirculation.

Economic and process considerations usually dictate that agitated thin-film evaporators be operated in single-effect mode. Very high temperature differences can then be used: many are heated with Dowtherm or other high-temperature media. This permits achieving reasonable capacities in spite of the relatively low heat-transfer coefficients and the small surface that can be provided in a single tube [to about 20 m$^2$ (200 ft$^2$)]. The structural need for wall thicknesses of 6 to 13 mm (¼ to ½ in) is a major reason for the relatively low heat-transfer coefficients when evaporating water-like materials.

**Evaporators without Heating Surfaces** The **submerged-combustion** evaporator makes use of combustion gases bubbling through the liquid as the means of heat transfer. It consists simply of a tank to hold the liquid, a burner and gas distributor that can be lowered into the liquid, and a combustion-control system. Since there are no heating surfaces on which scale can deposit, this evaporator is well suited to use with severely scaling liquids. The ease of constructing the tank and burner of special alloys or nonmetallic materials makes practical the handling of highly corrosive solutions. However, since the vapor is mixed with large quantities of noncondensable gases, it is impossible to reuse the heat in this vapor, and installations are usually limited to areas of low fuel cost. One difficulty frequently encountered in the use of submerged-combustion evaporators is a high entrainment loss. Also, these evaporators cannot be used when control of crystal size is important.

**Disk** or **cascade evaporators** are used in the pulp and paper industry to recover heat and entrained chemicals from boiler stack gases and to effect a final concentration of the black liquor before it is burned in the boiler. These evaporators consist of a horizontal shaft on which are mounted disks perpendicular to the shaft or bars parallel to the shaft. The assembly is partially immersed in the thick black liquor so that films of liquor are carried into the hot-gas stream as the shaft rotates.

Some forms of **flash evaporators** require no heating surface. An example is a recrystallizing process for separating salts having normal solubility curves from salts having inverse solubility curves, as in separating sodium chloride from calcium sulfate [Richards, *Chem. Eng.*, **59**(3), 140 (1952)]. A suspension of raw solid feed in a recirculating brine stream is heated by direct steam injection. The increased temperature and dilution by the steam dissolve the salt having the normal solubility curve. The other salt remains undissolved and is separated from the hot solution before it is flashed to a lower temperature. The cooling and loss of water on flashing cause recrystallization of the salt having the normal solubility curve, which is separated from the brine before the brine is mixed with more solid feed for recycling to the heater. This system can be operated as a multiple effect by flashing down to the lower temperature in stages and using flash vapor from all but the last stage to heat the recycle brine by direct injection. In this process no net evaporation occurs from the total system, and the process cannot be used to concentrate solutions unless heating surfaces are added.

## UTILIZATION OF TEMPERATURE DIFFERENCE

Temperature difference is the driving force for evaporator operation and usually is limited, as by compression ratio in vapor-compression evaporators and by available steam-pressure and heat-sink temperature in single- and multiple-effect evaporators. A fundamental objective of evaporator design is to make as much of this total temperature difference available for heat transfer as is economically justifiable. Some losses in temperature difference, such as those due to *boiling point rise* (BPR), are unavoidable. However, even these can be minimized, as by passing the liquor through effects or through different sections of a single effect in series so that only a portion of the heating surface is in contact with the strongest liquor.

Figure 11-124 shows approximate BPR losses for a number of process liquids. A correlation for concentrated solutions of many inorganic salts at the atmospheric pressure boiling point [Meranda and Furter, *J. Ch. and E. Data* **22,** 315-7 (1977)] is

$$\text{BPR} = 104.9N_2^{1.14} \tag{11-123}$$

where $N_2$ is the mole fraction of salts in solution. Correction to other pressures, when heats of solution are small, can be based on a constant ratio of vapor pressure of the solution to that of water at the same temperature.

The principal reducible loss in $\Delta T$ is that due to friction and to entrance and exit losses in vapor piping and entrainment separators. Pressure-drop losses here correspond to a reduction in condensing temperature of the vapor and hence a loss in available $\Delta T$. These losses become most critical at the low-temperature end of the evaporator, both because of the increasing specific volume of the vapor and because of the reduced slope of the vapor-pressure curve. Sizing of vapor lines is part of the economic optimization of the evaporator, extra costs of larger vapor lines being balanced against savings in $\Delta T$, which correspond to savings in heating-surface requirements. It should be noted that entrance and exit losses in vapor lines usually exceed by severalfold the straight-pipe friction losses, so they cannot be ignored.

## VAPOR-LIQUID SEPARATION

Product losses in evaporator vapor may result from foaming, splashing, or entrainment. Primary separation of liquid from vapor is accomplished in the vapor head by making the horizontal plan area large

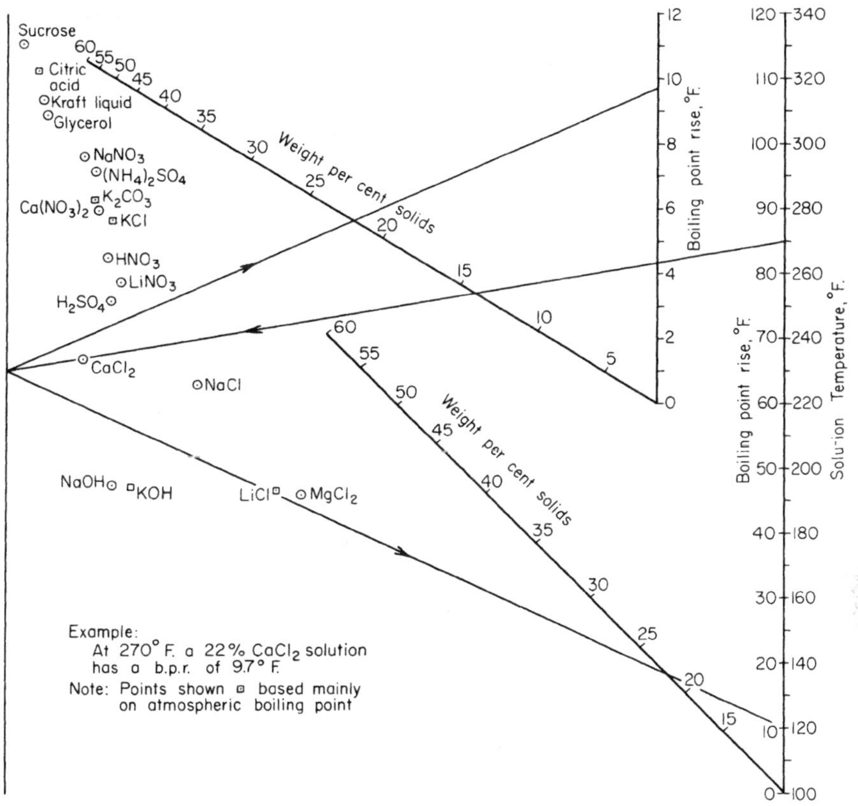


**FIG. 11-124** Boiling-point rise of aqueous solutions. °C = 5/9 (°F − 32).

enough so that most of the entrained droplets can settle out against the rising flow of vapor. Allowable velocities are governed by the Souders-Brown equation: $V = k\sqrt{(\rho_l - \rho_v)/\rho_v}$, in which $k$ depends on the size distribution of droplets and the decontamination factor $F$ desired. For most evaporators and for $F$ between 100 and 10,000, $k \cong 0.245/(F - 50)^{0.4}$ (Standiford, *Chemical Engineers' Handbook,* 4th ed., McGraw-Hill, New York, 1963, p. 11–35). Higher values of $k$ (to about 0.15) can be tolerated in the falling-film evaporator, where most of the entrainment separation occurs in the tubes, the vapor is scrubbed by liquor leaving the tubes, and the vapor must reverse direction to reach the outlet.

**Foaming losses** usually result from the presence in the evaporating liquid of colloids or of surface-tension depressants and finely divided solids. Antifoam agents are often effective. Other means of combating foam include the use of steam jets impinging on the foam surface, the removal of product at the surface layer, where the foaming agents seem to concentrate, and operation at a very low liquid level so that hot surfaces can break the foam. Impingement at high velocity against a baffle tends to break the foam mechanically, and this is the reason that the long-tube vertical, forced-circulation, and agitated-film evaporators are particularly effective with foaming liquids. Operating at lower temperatures and/or higher-dissolved solids concentrations may also reduce foaming tendencies.

**Splashing losses** are usually insignificant if a reasonable height has been provided between the liquid level and the top of the vapor head. The height required depends on the violence of boiling. Heights of 2.4 to 3.6 m (8 to 12 ft) or more are provided in short-tube vertical evaporators, in which the liquid and vapor leaving the tubes are projected upward. Less height is required in forced-circulation evaporators, in which the liquid is given a centrifugal motion or is projected downward as by a baffle. The same is true of long-tube vertical evaporators, in which the rising vapor-liquid mixture is projected against a baffle.

**Entrainment losses by flashing** are frequently encountered in an evaporator. If the feed is above the boiling point and is introduced above or only a short distance below the liquid level, entrainment losses may be excessive. This can occur in a short-tube-type evaporator if the feed is introduced at only one point below the lower tube sheet (Kerr, Louisiana Agric. Expt. Stn. Bull. 149, 1915). The same difficulty may be encountered in forced-circulation evaporators having too high a temperature rise through the heating element and thus too wide a flashing range as the circulating liquid enters the body. Poor vacuum control, especially during startup, can cause the generation of far more vapor than the evaporator was designed to handle, with a consequent increase in entrainment.

**Entrainment separators** are frequently used to reduce product losses. There are a number of specialized designs available, practically all of which rely on a change in direction of the vapor flow when the vapor is traveling at high velocity. Typical separators are shown in Fig. 11-122, although not necessarily with the type of evaporator with which they may be used. The most common separator is the cyclone, which may have either a top or a bottom outlet as shown in Fig. 11-122*a* and *b* or may even be wrapped around the heating element of the next effect as shown in Fig. 11-122*f*. The separation efficiency of a cyclone increases with an increase in inlet velocity, although at the cost of some pressure drop, which means a loss in available temperature difference. Pressure drop in a cyclone is from 10 to 16 velocity heads [Lawrence, *Chem. Eng. Prog.,* **48,** 241 (1952)], based on the velocity in the inlet pipe. Such cyclones can be sized in the same manner as a cyclone dust collector (using velocities of about 30 m/s (100 ft/s) at atmospheric pressure) although sizes may be increased somewhat in order to reduce losses in available temperature difference.

**Knitted wire mesh** serves as an effective entrainment separator when it cannot easily be fouled by solids in the liquor. The mesh is available in woven metal wire of most alloys and is installed as a blanket across the top of the evaporator (Fig. 11-122*d*) or in a monitor of reduced diameter atop the vapor head. These separators have low-pressure drops, usually on the order of 13 mm (½ in) of water, and collection efficiency is above 99.8 percent in the range of vapor velocities from 2.5 to 6 m/s (8 to 20 ft/s) [Carpenter and Othmer, *Am. Inst. Chem. Eng. J.*, **1**, 549 (1955)]. Chevron (hook-and-vane) type separators are also used because of their higher-allowable velocities or because of their reduced tendency to foul with solids suspended in the entrained liquid.

## EVAPORATOR ARRANGEMENT

**Single-Effect Evaporators** Single-effect evaporators are used when the required capacity is small, steam is cheap, the material is so corrosive that very expensive materials of construction are required, or the vapor is so contaminated that it cannot be reused. Single-effect evaporators may be operated in batch, semibatch, or continuous-batch modes or continuously. Strictly speaking, **batch evaporators** are ones in which filling, evaporating, and emptying are consecutive steps. This method of operation is rarely used since it requires that the body be large enough to hold the entire charge of feed and the heating element be placed low enough so as not to be uncovered when the volume is reduced to that of the product. The more usual method of operation is **semibatch,** in which feed is continually added to maintain a constant level until the entire charge reaches final density. **Continuous-batch** evaporators usually have a continuous feed and, over at least part of the cycle, a continuous discharge. One method of operation is to circulate from a storage tank to the evaporator and back until the entire tank is up to desired concentration and then finish in batches. **Continuous evaporators** have essentially continuous feed and discharge, and concentrations of both feed and product remain substantially constant.

**Thermocompression** The simplest means of reducing the energy requirements of evaporation is to compress the vapor from a single-effect evaporator so that the vapor can be used as the heating medium in the same evaporator. The compression may be accomplished by mechanical means or by a steam jet. In order to keep the compressor cost and power requirements within reason, the evaporator must work with a fairly narrow temperature difference, usually from about 5.5 to 11°C (10° to 20°F). This means that a large evaporator heating surface is needed, which usually makes the vapor-compression evaporator more expensive in first cost than a multiple-effect evaporator. However, total installation costs may be reduced when purchased power is the energy source, since the need for boiler and heat sink is eliminated. Substantial savings in operating cost are realized when electrical or mechanical power is available at a low cost relative to low-pressure steam, when only high-pressure steam is available to operate the evaporator, or when the cost of providing cooling water or other heat sink for a multiple-effect evaporator is high.

**Mechanical thermocompression** may employ reciprocating, rotary positive-displacement, centrifugal, or axial-flow compressors. Positive-displacement compressors are impractical for all but the smallest capacities, such as portable seawater evaporators. Axial-flow compressors can be built for capacities of more than 472 $m^3/s$ ($1 \times 10^6$ $ft^3/min$). Centrifugal compressors are usually cheapest for the intermediate-capacity ranges that are normally encountered. In all cases, great care must be taken to keep entrainment at a minimum, since the vapor becomes superheated on compression and any liquid present will evaporate, leaving the dissolved solids behind. In some cases a vapor-scrubbing tower may be installed to protect the compressor. A mechanical recompression evaporator usually requires more heat than is available from the compressed vapor. Some of this extra heat can be obtained by preheating the feed with the condensate and, if possible, with the product. Rather extensive heat-exchange systems with close approach temperatures are usually justified, especially if the evaporator is operated at high temperature to reduce the volume of vapor to be compressed. When the product is a solid, an elutriation leg such as that shown in Fig. 11-122*b* is advantageous, since it cools the product almost to feed temperature. The remaining heat needed to maintain the evaporator in operation must be obtained from outside sources.

While theoretical compressor power requirements are reduced slightly by going to lower evaporating temperatures, the volume of vapor to be compressed and hence compressor size and cost increase so rapidly that low-temperature operation is more expensive than high-temperature operation. The requirement of low temperature for fruit-juice concentration has led to the development of an evaporator employing a **secondary fluid,** usually Freon or ammonia. In this evaporator, the vapor is condensed in an exchanger cooled by boiling Freon. The Freon, at a much higher vapor density than the water vapor, is then compressed to serve as the heating medium for the evaporator. This system requires that the latent heat be transferred through two surfaces instead of one, but the savings in compressor size and cost are enough to justify the extra cost of heating surface or the cost of compressing through a wider temperature range.

**Steam-jet thermocompression** is advantageous when steam is available at a pressure appreciably higher than can be used in the evaporator. The steam jet then serves as a reducing valve while doing some useful work. The efficiency of a steam jet is quite low and falls off rapidly when the jet is not used at the vapor-flow rate and terminal pressure conditions for which it was designed. Consequently multiple jets are used when wide variations in evaporation rate are expected. Because of the low first cost and the ability to handle large volumes of vapor, steam-jet thermocompressors are used to increase the economy of evaporators that must operate at low temperatures and hence cannot be operated in multiple effect. The steam-jet thermocompression evaporator has a heat input larger than that needed to balance the system, and some heat must be rejected. This is usually done by venting some of the vapor at the suction of the compressor.

**Multiple-Effect Evaporation** Multiple-effect evaporation is the principal means in use for economizing on energy consumption. Most such evaporators operate on a continuous basis, although for a few difficult materials a continuous-batch cycle may be employed. In a multiple-effect evaporator, steam from an outside source is condensed in the heating element of the first effect. If the feed to the effect is at a temperature near the boiling point in the first effect, 1 kg of steam will evaporate almost 1 kg of water. The first effect operates at (but is not controlled at) a boiling temperature high enough so that the evaporated water can serve as the heating medium of the second effect. Here almost another kilogram of water is evaporated, and this may go to a condenser if the evaporator is a double-effect or may be used as the heating medium of the third effect. This method may be repeated for any number of effects. Large evaporators having six and seven effects are common in the pulp and paper industry, and evaporators having as many as 17 effects have been built. As a first approximation, the **steam economy** of a multiple-effect evaporator will increase in proportion to the number of effects and usually will be somewhat less numerically than the number of effects.

The increased steam economy of a multiple-effect evaporator is gained at the expense of evaporator first cost. The total heat-transfer surface will increase substantially in proportion to the number of effects in the evaporator. This is only an approximation since going from one to two effects means that about half of the heat transfer is at a higher temperature level, where heat-transfer coefficients are generally higher. On the other hand, operating at lower temperature differences reduces the heat-transfer coefficient for many types of evaporator. If the material has an appreciable boiling-point elevation, this will also lower the available temperature difference. The only accurate means of predicting the changes in steam economy and surface requirements with changes in the number of effects is by detailed heat and material balances together with an analysis of the effect of changes in operating conditions on heat-transfer performance.

The approximate **temperature distribution** in a multiple-effect evaporator is under the control of the designer, but once built, the evaporator establishes its own equilibrium. Basically, the effects are a number of series resistances to heat transfer, each resistance being approximately proportional to $1/U_nA_n$. The total available temperature drop is divided between the effects in proportion to their resistances. If one effect starts to scale, its temperature drop will increase at the expense of the temperature drops across the other effects. This provides a convenient means of detecting a drop in heat-transfer coefficient in an effect of an operating evaporator. If the steam pressure and final vacuum do not change, the temperature in the effect that is scaling will decrease and the temperature in the preceding effect will increase.

The feed to a multiple-effect evaporator is usually transferred from one effect to another in series so that the ultimate product concentration is reached only in one effect of the evaporator. In **backward-feed** operation, the raw feed enters the last (coldest) effect, the discharge from this effect becomes the feed to the next-to-the-last effect, and so on until product is discharged from the first effect. This method of operation is advantageous when the feed is cold, since much less liquid must be heated to the higher temperature existing in the early effects. It is also used when the product is so viscous that high temperatures are needed to keep the viscosity low enough to give reasonable heat-transfer coefficients. When product viscosity is high but a hot product is not needed, the liquid from the first effect is sometimes flashed to a lower temperature in one or more stages and the flash vapor added to the vapor from one or more later effects of the evaporator.

In **forward-feed** operation, raw feed is introduced in the first effect and passed from effect to effect parallel to the steam flow. Product is withdrawn from the last effect. This method of operation is advantageous when the feed is hot or when the concentrated product would be damaged or would deposit scale at high temperature. Forward feed simplifies operation when liquor can be transferred by pressure difference alone, thus eliminating all intermediate liquor pumps. When the feed is cold, forward feed gives a low steam economy since an appreciable part of the prime steam is needed to heat the feed to the boiling point and thus accomplishes no evaporation. If forward feed is necessary and feed is cold, steam economy can be improved markedly by preheating the feed in stages with vapor bled from intermediate effects of the evaporator. This usually represents little increase in total heating surface or cost since the feed must be heated in any event and shell-and-tube heat exchangers are generally less expensive per unit of surface area than evaporator heating surface.

**Mixed-feed** operation is used only for special applications, as when liquor at an intermediate concentration and a certain temperature is desired for additional processing.

**Parallel feed** involves the introduction of raw feed and the withdrawal of product at each effect of the evaporator. It is used primarily when the feed is substantially saturated and the product is a solid. An example is the evaporation of brine to make common salt. Evaporators of the types shown in Fig. 11-122*b* or *e* are used, and the product is withdrawn as a slurry. In this case, parallel feed is desirable because the feed washes impurities from the salt leaving the body.

**Heat-recovery systems** are frequently incorporated in an evaporator to increase the steam economy. Ideally, product and evaporator condensate should leave the system at a temperature as low as possible. Also, heat should be recovered from these streams by exchange with feed or evaporating liquid at the highest possible temperature. This would normally require separate liquid-liquid heat exchangers, which add greatly to the complexity of the evaporator and are justifiable only in large plants. Normally, the loss in thermodynamic availability due to flashing is tolerated since the flash vapor can then be used directly in the evaporator effects. The most commonly used is a **condensate flash** system in which the condensate from each effect but the first (which normally must be returned to the boiler) is flashed in successive stages to the pressure in the heating element of each succeeding effect of the evaporator. Product flash tanks may also be used in a backward- or mixed-feed evaporator. In a forward-feed evaporator, the principal means of heat recovery may be by use of **feed preheaters** heated by vapor bled from each effect of the evaporator. In this case, condensate may be either flashed as before or used in a separate set of exchangers to accomplish some of the feed preheating. A feed preheated by last-effect vapor may also materially reduce condenser water requirements.

**Seawater Evaporators** The production of potable water from saline waters represents a large and growing field of application for evaporators. Extensive work done in this field to 1972 was summarized in the annual *Saline Water Conversion Reports* of the Office of Saline Water, U.S. Department of the Interior. **Steam economies** on the order of 10 kg evaporation/kg steam are usually justified because (1) unit production capacities are high, (2) fixed charges are low on capital used for public works (i.e., they use long amortization periods and have low interest rates, with no other return on investment considered), (3) heat-transfer performance is comparable with that of pure water, and (4) properly treated seawater causes little deterioration due to scaling or fouling.

Figure 11-125*a* shows a **multiple-effect** (falling-film) flow sheet as used for seawater. Twelve effects are needed for a steam economy of 10. Seawater is used to condense last-effect vapor, and a portion is then treated to prevent scaling and corrosion. Treatment usually consists of acidification to break down bicarbonates, followed by deaeration, which also removes the carbon dioxide generated. The treated seawater is then heated to successively higher temperatures by a portion of the vapor from each effect and finally is fed to the evaporating surface of the first effect. The vapor generated therein and the partially concentrated liquid are passed to the second effect, and so on until the last effect. The feed rate is adjusted relative to the steam rate so that the residual liquid from the last effect can carry away all the salts in solution, in a volume about one-third of that of the feed. Condensate formed in each effect but the first is flashed down to the following effects in sequence and constitutes the product of the evaporator.

As the feed-to-steam ratio is increased in the flow sheet of Fig. 11-125*a*, a point is reached where all the vapor is needed to preheat the feed and none is available for the evaporator tubes. This limiting case is the **multistage flash evaporator,** shown in its simplest form in Fig. 11-125*b*. Seawater is treated as before and then pumped through a number of feed heaters in series. It is given a final boost in temperature with prime steam in a **brine heater** before it is flashed down in series to provide the vapor needed by the feed heaters. The amount of steam required depends on the approach-temperature difference in the feed heaters and the flash range per stage. Condensate from the feed heaters is flashed down in the same manner as the brine.

Since the flow being heated is identical to the total flow being flashed, the temperature rise in each heater is equal to the flash range in each flasher. This temperature difference represents a loss from the temperature difference available for heat transfer. There are thus two ways of increasing the steam economy of such plants: increasing the heating surface and increasing the number of stages. Whereas the number of effects in a multiple-effect plant will be about 20 percent greater than the steam economy, the number of stages in a flash plant will be 3 to 4 times the steam economy. However, a large number of stages can be provided in a single vessel by means of internal bulkheads. The heat-exchanger tubing is placed in the same vessel, and the tubes usually are continuous through a number of stages. This requires ferrules or special close tube-hole clearances where the tubes pass through the internal bulkheads. In a plant for a steam economy of 10, the ratio of flow rate to heating surface is usually such that the seawater must pass through about 152 m of 19-mm (500 ft of ¾-in) tubing before it reaches the brine heater. This places a limitation on the physical arrangement of the vessels.

Inasmuch as it requires a flash range of about 61°C (110°F) to produce 1 kg of flash vapor for every 10 kg of seawater, the multistage flash evaporator requires handling a large volume of seawater relative to the product. In the flow sheet of Fig. 11-125*b* all this seawater must be deaerated and treated for scale prevention. In addition, the last-stage vacuum varies with the ambient seawater temperature, and ejector equipment must be sized for the worst condition. These difficulties can be eliminated by using the **recirculating multistage flash** flow sheet of Fig. 11-125*c*. The last few stages, called the **reject stages,** are cooled by a flow of seawater that can be varied to maintain a reasonable last-stage vacuum. A small portion of the last-stage brine is blown down to carry away the dissolved salts, and the balance is recirculated to the **heat-recovery stages.** This arrangement requires a much smaller makeup of fresh seawater and hence a lower treatment cost.

The multistage flash evaporator is similar to a multiple-effect forced-circulation evaporator, but with all the forced-circulation heaters in series. This has the advantage of requiring only one large-volume forced-circulation pump, but the sensible heating and short-circuiting losses in available temperature differences remain. A disadvantage of the flash evaporator is that the liquid throughout the system is at almost the discharge concentration. This has limited its industrial use to solutions in which no great concentration differences are required between feed and product and to where the liquid can be heated through wide temperature ranges without scaling. A partial remedy is

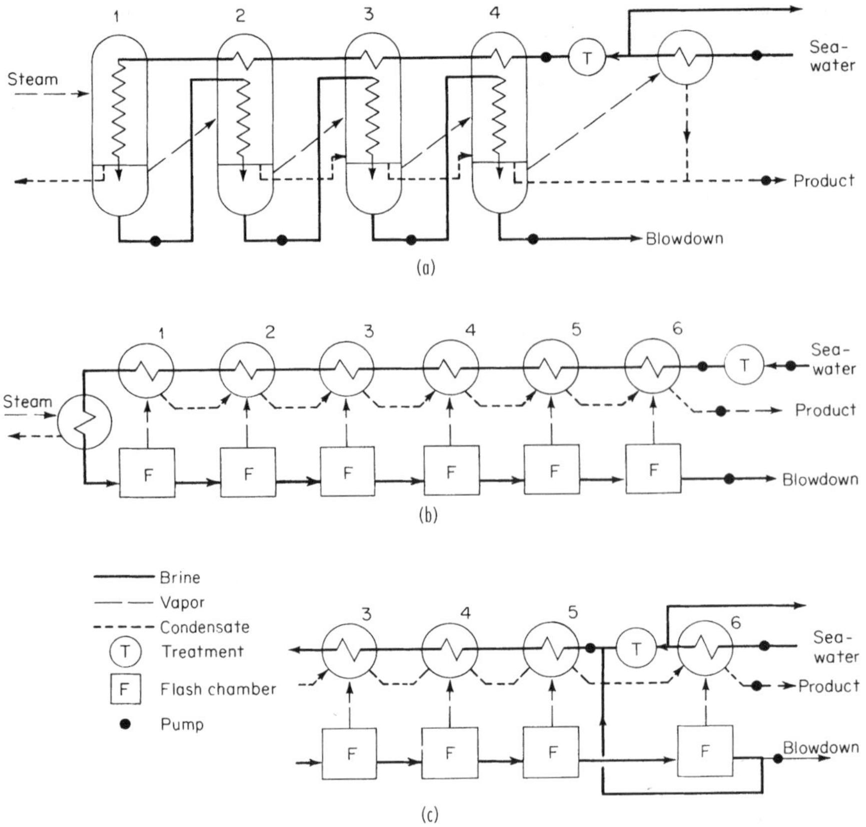


**FIG. 11-125** Flow sheets for seawater evaporators. (*a*) Multiple effect (falling film). (*b*) Multistage flash (once-through). (*c*) Multistage flash (recirculating).

to arrange several multistage flash evaporators in series, the heat-rejection section of one being the brine heater of the next. This permits independent control of concentration but eliminates the principal advantage of the flash evaporator, which is the small number of pumps and vessels required. An unusual feature of the flash evaporator is that fouling of the heating surfaces reduces primarily the steam economy rather than the capacity of the evaporator. Capacity is not affected until the heat-rejection stages can no longer handle the increased flashing resulting from the increased heat input.

## EVAPORATOR CALCULATIONS

**Single-Effect Evaporators** The heat requirements of a single-effect continuous evaporator can be calculated by the usual methods of stoichiometry. If enthalpy data or specific heat and heat-of-solution data are not available, the heat requirement can be estimated as the sum of the heat needed to raise the feed from feed to product temperature and the heat required to evaporate the water. The latent heat of water is taken at the vapor-head pressure instead of at the product temperature in order to compensate partially for any heat of solution. If sufficient vapor-pressure data are available for the solution, methods are available to calculate the true latent heat from the slope of the Dühring line [Othmer, *Ind. Eng. Chem.*, **32,** 841 (1940)].

The heat requirements in batch evaporation are the same as those in continuous evaporation except that the temperature (and sometimes pressure) of the vapor changes during the course of the cycle. Since the enthalpy of water vapor changes but little relative to temperature, the difference between continuous and batch heat requirements is almost always negligible. More important usually is the effect of variation of fluid properties, such as viscosity and boiling-point rise, on heat transfer. These can only be estimated by a step-by-step calculation.

In selecting the **boiling temperature,** consideration must be given to the effect of temperature on heat-transfer characteristics of the type of evaporator to be used. Some evaporators show a marked drop in coefficient at low temperature—more than enough to offset any gain in available temperature difference. The condenser cooling-water temperature and cost must also be considered.

**Thermocompression Evaporators** Thermocompression-evaporator calculations [Pridgeon, *Chem. Metall. Eng.*, **28,** 1109 (1923); Peter, *Chimia (Switzerland)*, **3,** 114 (1949); Petzold, *Chem. Ing. Tech.*, **22,** 147 (1950); and Weimer, Dolf, and Austin, *Chem. Eng. Prog.*, **76**(11), 78 (1980)] are much the same as single-effect calculations with the added complication that the heat supplied to the evaporator from compressed vapor and other sources must exactly balance the heat requirements. Some knowledge of compressor efficiency is also required. Large axial-flow machines on the order of 236-$m^3$/s (500,000-$ft^3$/min) capacity may have efficiencies of 80 to 85 percent. Efficiency drops to about 75 percent for a 14-$m^3$/s (30,000-$ft^3$/min) centrifugal compressor. Steam-jet compressors have thermodynamic efficiencies on the order of only 25 to 30 percent.

**Flash Evaporators** The calculation of a heat and material balance on a flash evaporator is relatively easy once it is understood that the temperature rise in each heater and temperature drop in each flasher must all be substantially equal. The steam economy $E$, kg evaporation/kg of 1055-kJ steam (lb/lb of 1000-Btu steam) may be approximated from

$$E = \left(1 - \frac{\Delta T}{1250}\right) \frac{\Delta T}{Y + R + \Delta T/N} \tag{11-124}$$

where $\Delta T$ is the total temperature drop between feed to the first flasher and discharge from the last flasher. °C; $N$ is the number of flash stages; $Y$ is the approach between vapor temperature from the first flasher and liquid leaving the heater in which this vapor is condensed. °C (the approach is usually substantially constant for all stages); and $R$. °C. is the sum of the boiling-point rise and the short-circuiting loss in the first flash stage. The expression for the mean effective temperature difference $\Delta t$ available for heat transfer then becomes

$$\Delta t = \frac{\Delta T}{N \ln \dfrac{1 - \Delta T/1250 - RE/\Delta T}{1 - \Delta T\ 1250 - RE/\Delta T - E/N}} \qquad (11\text{-}125)$$

**Multiple-Effect Evaporators** A number of approximate methods have been published for estimating performance and heating-surface requirements of a multiple-effect evaporator [Coates and Pressburg, *Chem. Eng.*, **67**(6), 157 (1960); Coates, *Chem. Eng. Prog.*, **45,** 25 (1949); and Ray and Carnahan, *Trans. Am. Inst. Chem. Eng.*, **41,** 253 (1945)]. However, because of the wide variety of methods of feeding and the added complication of feed heaters and condensate flash systems, the only certain way of determining performance is by detailed heat and material balances. Algebraic solutions may be used, but if more than a few effects are involved, trial-and-error methods are usually quicker. These frequently involve trial-and-error within trial-and-error solutions. Usually, if condensate flash systems or feed heaters are involved, it is best to start at the first effect. The basic steps in the calculation are then as follows:

1. Estimate temperature distribution in the evaporator, taking into account boiling-point elevations. If all heating surfaces are to be equal, the temperature drop across each effect will be approximately inversely proportional to the heat-transfer coefficient in that effect.
2. Determine total evaporation required, and estimate steam consumption for the number of effects chosen.
3. From assumed feed temperature (forward feed) or feed flow (backward feed) to the first effect and assumed steam flow, calculate evaporation in the first effect. Repeat for each succeeding effect, checking intermediate assumptions as the calculation proceeds. Heat input from condensate flash can be incorporated easily since the condensate flow from the preceding effects will have already been determined.
4. The result of the calculation will be a feed to or a product discharge from the last effect that may not agree with actual requirements. The calculation must then be repeated with a new assumption of steam flow to the first effect.
5. These calculations should yield liquor concentrations in each effect that make possible a revised estimate of boiling-point rises. They also give the quantity of heat that must be transferred in each effect. From the heat loads, assumed temperature differences, and heat-transfer coefficients, heating-surface requirements can be determined. If the distribution of heating surface is not as desired, the entire calculation may need to be repeated with revised estimates of the temperature in each effect.
6. If sufficient data are available, heat-transfer coefficients under the proposed operating conditions can be calculated in greater detail and surface requirements readjusted.

Such calculations require considerable judgment to avoid repetitive trials but are usually well worth the effort. Sample calculations are given in the American Institute of Chemical Engineers *Testing Procedure for Evaporators* and by Badger and Banchero, *Introduction to Chemical Engineering,* McGraw-Hill, New York, 1955. These balances may be done by computer but programming time frequently exceeds the time needed to do them manually, especially when variations in flow sheet are to be investigated. The MASSBAL program of SACDA, London, Ont., provides a considerable degree of flexibility in this regard. Another program, not specific to evaporators, is ASPEN PLUS by Aspen Tech., Cambridge, MA. Many such programs include simplifying assumptions and approximations that are not explicitly stated and can lead to erroneous results.

**Optimization** The primary purpose of evaporator design is to enable production of the necessary amount of satisfactory product at the lowest total cost. This requires economic-balance calculations that may include a great number of variables. Among the possible variables are the following:

1. Initial steam pressure versus cost or availability.
2. Final vacuum versus water temperature, water cost, heat-transfer performance, and product quality.
3. Number of effects versus steam, water, and pump power cost.
4. Distribution of heating surface between effects versus evaporator cost.
5. Type of evaporator versus cost and continuity of operation.
6. Materials of construction versus product quality, tube life, evaporator life, and evaporator cost.
7. Corrosion, erosion, and power consumption versus tube velocity.
8. Downtime for retubing and repairs.
9. Operating-labor and maintenance requirements.
10. Method of feeding and use of heat-recovery systems.
11. Size of recovery heat exchangers.
12. Possible withdrawal of steam from an intermediate effect for use elsewhere.
13. Entrainment separation requirements.

The type of evaporator to be used and the materials of construction are generally selected on the basis of past experience with the material to be concentrated. The method of feeding can usually be decided on the basis of known feed temperature and the properties of feed and product. However, few of the listed variables are completely independent. For instance, if a large number of effects is to be used, with a consequent low temperature drop per effect, it is impractical to use a natural-circulation evaporator. If expensive materials of construction are desirable, it may be found that the forced-circulation evaporator is the cheapest and that only a few effects are justifiable.

The variable having the greatest influence on total cost is the number of effects in the evaporator. An economic balance can establish the optimum number where the number is not limited by such factors as viscosity, corrosiveness, freezing point, boiling-point rise, or thermal sensitivity. Under present United States conditions, savings in steam and water costs justify the extra capital, maintenance, and power costs of about seven effects in large commercial installations when the properties of the fluid are favorable, as in black-liquor evaporation. Under governmental financing conditions, as for plants to supply fresh water from seawater, evaporators containing from 12 to 30 or more effects can be justified.

As a general rule, the optimum number of effects increases with an increase in steam cost or plant size. Larger plants favor more effects, partly because they make it easier to install heat-recovery systems that increase the steam economy attainable with a given number of effects. Such recovery systems usually do not increase the total surface needed but do require that the heating surface be distributed between a greater number of pieces of equipment.

The most common evaporator design is based on the use of the same heating surface in each effect. This is by no means essential since few evaporators are "standard" or involve the use of the same patterns. In fact, there is no reason why all effects in an evaporator must be of the same type. For instance, the cheapest salt evaporator might use propeller calandrias for the early effects and forced-circulation effects at the low-temperature end, where their higher cost per unit area is more than offset by higher heat-transfer coefficients.

Bonilla [*Trans. Am. Inst. Chem. Eng.*, **41,** 529 (1945)] developed a simplified method for distributing the heating surface in a multiple-effect evaporator to achieve minimum cost. If the cost of the evaporator per unit area of heating surface is constant throughout, then minimum cost and area will be achieved if the ratio of area to temperature difference $A/\Delta T$ is the same for all effects. If the cost per unit area $z$ varies, as when different tube materials or evaporator types are used, then $zA/\Delta T$ should be the same for all effects.

## EVAPORATOR ACCESSORIES

**Condensers** The vapor from the last effect of an evaporator is usually removed by a condenser. **Surface condensers** are employed when mixing of condensate with condenser cooling water is not desired. They are for the most part shell-and-tube condensers with vapor on the shell side and a multipass flow of cooling water on the tube side. Heat loads, temperature differences, sizes, and costs are usually of the same order of magnitude as for another effect of the evaporator. Surface condensers use more cooling water and are so much more expensive that they are never used when a direct-contact condenser is suitable.

The most common type of direct-contact condenser is the countercurrent **barometric condenser,** in which vapor is condensed by rising against a rain of cooling water. The condenser is set high enough so that water can discharge by gravity from the vacuum in the condenser. Such condensers are inexpensive and are economical on water consumption. They can usually be relied on to maintain a vacuum corresponding to a saturated-vapor temperature within 2.8°C (5°F) of the water temperature leaving the condenser [How, *Chem. Eng.*, **63**(2), 174 (1956)]. The ratio of water consumption to vapor condensed can be determined from the following equation:

$$\frac{\text{Water flow}}{\text{Vapor flow}} = \frac{H_v - h_2}{h_2 - h_1} \qquad (11\text{-}126)$$

where $H_v$ = vapor enthalpy and $h_1$ and $h_2$ = water enthalpies entering and leaving the condenser. Another type of direct-contact condenser is the **jet** or **wet condenser,** which makes use of high-velocity jets of water both to condense the vapor and to force noncondensable gases out the tailpipe. This type of condenser is frequently placed below barometric height and requires a pump to remove the mixture of water and gases. Jet condensers usually require more water than the more common barometric-type condensers and cannot be throttled easily to conserve water when operating at low evaporation rates.

**Vent Systems** Noncondensable gases may be present in the evaporator vapor as a result of leakage, air dissolved in the feed, or decomposition reactions in the feed. When the vapor is condensed in the succeeding effect, the noncondensables increase in concentration and impede heat transfer. This occurs partially because of the reduced partial pressure of vapor in the mixture but mainly because the vapor flow toward the heating surface creates a film of poorly conducting gas at the interface. (See page 11-14 for means of estimating the effect of noncondensable gases on the steam-film coefficient.) The most important means of reducing the influence of noncondensables on heat transfer is by properly channeling them past the heating surface. A positive vapor-flow path from inlet to vent outlet should be provided, and the path should preferably be tapered to avoid pockets of low velocity where noncondensables can be trapped. Excessive clearances and low-resistance channels that could bypass vapor directly from the inlet to the vent should be avoided [Standiford, *Chem. Eng. Prog.*, **75,** 59–62 (July 1979)].

In any event, noncondensable gases should be vented well before their concentration reaches 10 percent. Since gas concentrations are difficult to measure, the usual practice is to overvent. This means that an appreciable amount of vapor can be lost.

To help conserve steam economy, venting is usually done from the steam chest of one effect to the steam chest of the next. In this way, excess vapor in one vent does useful evaporation at a steam economy only about one less than the overall steam economy. Only when there are large amounts of noncondensable gases present, as in beet-sugar evaporation, is it desirable to pass the vents directly to the condenser to avoid serious losses in heat-transfer rates. In such cases, it can be worthwhile to recover heat from the vents in separate heat exchangers, which preheat the entering feed.

The noncondensable gases eventually reach the condenser (unless vented from an effect above atmospheric pressure to the atmosphere or to auxiliary vent condensers). These gases will be supplemented by air dissolved in the condenser water and by carbon dioxide given off on decomposition of bicarbonates in the water if a barometric condenser is used. These gases may be removed by the use of a water-jet-type condenser but are usually removed by a separate vacuum pump.

The vacuum pump is usually of the steam-jet type if high-pressure steam is available. If high-pressure steam is not available, more expensive mechanical pumps may be used. These may be either a water-ring (Hytor) type or a reciprocating pump.

The primary source of noncondensable gases usually is air dissolved in the condenser water. Figure 11-126 shows the dissolved-gas content

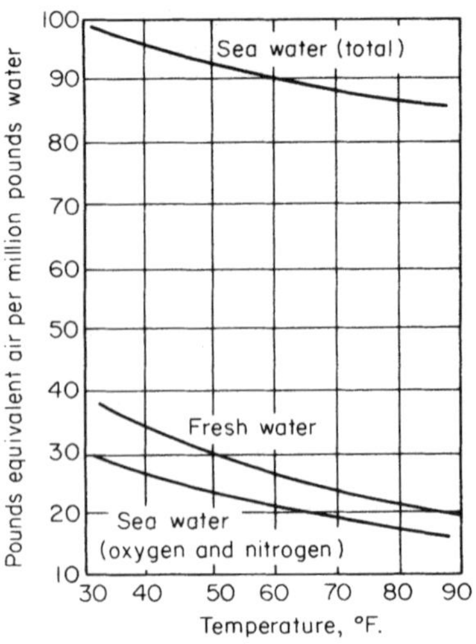


**FIG. 11-126** Gas content of water saturated at atmospheric pressure. °C = 5/9 (°F − 32).

of fresh water and seawater, calculated as equivalent air. The lower curve for seawater includes only dissolved oxygen and nitrogen. The upper curve includes carbon dioxide that can be evolved by complete breakdown of bicarbonate in seawater. Breakdown of bicarbonates is usually not appreciable in a condenser but may go almost to completion in a seawater evaporator. The large increase in gas volume as a result of possible bicarbonate breakdown is illustrative of the uncertainties involved in sizing vacuum systems.

By far the largest load on the vacuum pump is water vapor carried with the noncondensable gases. Standard power-plant practice assumes that the mixture leaving a surface condenser will have been cooled 4.2°C (7.5°F) below the saturation temperature of the vapor. This usually corresponds to about 2.5 kg of water vapor/kg of air. One advantage of the countercurrent barometric condenser is that it can cool the gases almost to the temperature of the incoming water and thus reduce the amount of water vapor carried with the air.

In some cases, as with pulp-mill liquors, the evaporator vapors contain constituents more volatile than water, such as methanol and sulfur compounds. Special precautions may be necessary to minimize the effects of these compounds on heat transfer, corrosion, and condensate quality. They can include removing most of the condensate countercurrent to the vapor entering an evaporator-heating element, channeling vapor and condensate flow to concentrate most of the "foul" constituents into the last fraction of vapor condensed (and keeping this condensate separate from the rest of the condensate), and flashing the warm evaporator feed to a lower pressure to remove much of the foul constituents in only a small amount of flash vapor. In all such cases, special care is needed to properly channel vapor flow past the heating surfaces so there is a positive flow from steam inlet to vent outlet with no pockets, where foul constituents or noncondensibles can accumulate.

**Salt Removal** When an evaporator is used to make a crystalline product, a number of means are available for concentrating and removing the salt from the system. The simplest is to provide settling space in the evaporator itself. This is done in the types shown in Fig. 11-122*b, c,* and *e* by providing a relatively quiescent zone in which the salt can settle. Sufficiently high slurry densities can usually be achieved in this manner to reach the limit of pumpability. The evaporators are usually placed above barometric height so that the slurry can be discharged intermittently on a short time cycle. This permits the use of high velocities in large lines that have little tendency to plug.

If the amount of salts crystallized is on the order of a ton an hour or less, a salt trap may be used. This is simply a receiver that is connected to the bottom of the evaporator and is closed off from the evaporator periodically for emptying. Such traps are useful when insufficient headroom is available for gravity removal of the solids. However, traps require a great deal of labor, give frequent trouble with the shutoff valves, and also can upset evaporator operation completely if a trap is reconnected to the evaporator without first displacing all air with feed liquor.

## EVAPORATOR OPERATION

The two principal elements of evaporator control are *evaporation rate* and *product concentration.* Evaporation rate in single- and multiple-effect evaporators is usually achieved by steam-flow control. Conventional-control instrumentation is used (see Sec. 22), with the added precaution that pressure drop across meter and control valve, which reduces temperature difference available for heat transfer, not be excessive when maximum capacity is desired. Capacity control of thermocompression evaporators depends on the type of compressor; positive-displacement compressors can utilize speed control or variations in operating pressure level. Centrifugal machines normally utilize adjustable inlet-guide vanes. Steam jets may have an adjustable spindle in the high-pressure orifice or be arranged as multiple jets that can individually be cut out of the system.

Product concentration can be controlled by any property of the solution that can be measured with the requisite accuracy and reliability. The preferred method is to impose control on rate of product withdrawal. Feed rates to the evaporator effects are then controlled by their levels. When level control is impossible, as with the rising-film LTV, product concentration is used to control the feed rate—frequently by rationing of feed to steam with the ration reset by product concentration, sometimes also by feed concentration. Other controls that may be needed include vacuum control of the last effect (usually by air bleed to the condenser) and temperature-level control of thermocompression evaporators (usually by adding makeup heat or by venting excess vapor, or both as feed or weather conditions vary). For more control detail, see *Measurement and Control in Water Desalination,* N. Lior, ed., pp. 241–305, Elsevier Science Publ. Co., NY, 1986.

Control of an evaporator requires more than proper instrumentation. Operator logs should reflect changes in basic characteristics, as by use of **pseudo heat-transfer** coefficients, which can detect obstructions to heat flow, hence to capacity. These are merely the ratio of any convenient measure of heat flow to the temperature drop across each effect. **Dilution** by wash and seal water should be monitored since it absorbs evaporative capacity. Detailed tests, routine measurements, and operating problems are covered more fully in *Testing Procedure for Evaporators* (loc. cit.) and by Standiford [*Chem. Eng. Prog.,* **58**(11), 80 (1962)].

By far the best application of computers to evaporators is for working up operators' data into the basic performance parameters such as heat-transfer coefficients, steam economy, and dilution.